Scattering of Waves from Large Spheres

This book describes the scattering of waves, both scalar and electromagnetic, from impenetrable and penetrable spheres. It provides an extensive introduction for those first studying the field, as well as a guide to further analysis and application for researchers in this area.

Although the scattering of plane waves from spheres is an old subject, there is little doubt that it is still maturing as a broad range of new applications demands an understanding of finer details. In this book attention is focused primarily on spherical radii much larger than incident wavelengths, along with the asymptotic techniques required for physical analysis of the scattering mechanisms involved. Applications to atmospheric phenomena such as the rainbow and glory are included, as well as a detailed analysis of optical resonances. Extensions of the theory to inhomogeneous and nonspherical particles, collections of spheres, and bubbles are also discussed.

This book will be of primary interest to graduate students and researchers in physics (particularly in the fields of optics, the atmospheric sciences, and astrophysics), electrical engineering, physical chemistry, and some areas of biology.

WALTER T. GRANDY, JR is Professor of Physics, Emeritus, at the University of Wyoming, where he has been a member of the Department of Physics and Astronomy for over 36 years, serving as Chairman 1971–1979. He has also been a visiting professor at the University of Arizona (1974), the Universidade de São Paulo (1966–1967, 1982), the Universität Tübingen (1979), and the University of Sydney (1988). His major professional interests have been in the fields of statistical mechanics, electrodynamics, classical and quantum scattering theory, and relativistic quantum mechanics. He has numerous publications in these fields and is author of several earlier books, including *Introduction to Electrodynamics and Radiation* (1970), a two-volume work titled *Foundations of Statistical Mechanics* (1987, 1988), and *Relativistic Quantum Mechanics of Leptons and Fields* (1991).

Scattering of Waves from Large Spheres

Walter T. Grandy, Jr

Physics & Astronomy
University of Wyoming

CAMBRIDGE
UNIVERSITY PRESS

CAMBRIDGE UNIVERSITY PRESS
Cambridge, New York, Melbourne, Madrid, Cape Town, Singapore, São Paulo

Cambridge University Press
The Edinburgh Building, Cambridge CB2 2RU, UK

Published in the United States of America by Cambridge University Press, New York

www.cambridge.org
Information on this title: www.cambridge.org/9780521661263

First published 2000
This digitally printed first paperback version 2005

A catalogue record for this publication is available from the British Library

Library of Congress Cataloguing in Publication data
Grandy, Walter T., 1933–
Scattering of waves from large spheres / W. T. Grandy, Jr.
p. cm.
Includes bibliographical references and index.
ISBN 0 521 66126 9 (hardbound)
1. Light–Scattering. 2. Electric waves–Scattering. 3. Sphere.
I. Title
QC427.4.G73 2000
535′.43–dc21   00-020864   CIP

ISBN-13 978-0-521-66126-3 hardback
ISBN-10 0-521-66126-9 hardback

ISBN-13 978-0-521-02124-1 paperback
ISBN-10 0-521-02124-3 paperback

Contents

Preface

Although the scattering of plane waves from spheres is an old subject, there is little doubt that it is still maturing as a broad range of new applications demands an understanding of finer details. The classical theory of electromagnetic scattering from dielectric spheres is due to Lorenz, Mie, and Debye, and has proved to be enormously rich; it is still being developed and continues to yield new insights. Much of this development has been motivated by the availability of small silicon spheres that can be probed precisely with laser light, as well as by new techniques in acoustics, in atmospheric physics, and in the study of biological molecules.

The classic treatise in the subject has long been van de Hulst's *Light Scattering by Small Particles* (1957), supplemented in later years by the application-oriented works of Kerker, *The Scattering of Light and Other Electromagnetic Radiation* (1969), and Bohren and Huffman, *Absorption and Scattering of Light by Small Particles* (1983). These volumes, and others, have contributed greatly to the subject, while concerning themselves primarily (though not exclusively) with scattering from particles whose dimensions are on the order of an incident wavelength or less. Among my reasons for writing the present book, however, is a long-time interest in understanding the detailed physics of the rainbow and glory in terms of modern scattering theory, and these phenomena arise from water droplets whose dimensions are a great deal larger than optical wavelengths. Thus, the time seems ripe for a theoretical exposition extending the earlier works to encompass a broader range of phenomena.

The complete mathematical solution to the problem of light scattering from a sphere was obtained over a century ago as the well-known infinite series of partial waves. This series, which is generically known as the Mie solution, contains in principle all of the physics of the problem for any size of particle and all wavelengths. For particle radii less than a wavelength only a few terms of the series need be retained and a satisfying physical picture is readily constructed. As the particle size increases relative to the wavelength, however, the series becomes very slowly convergent and the job of extracting the important physics from it becomes quite difficult.

For these reasons much of the theoretical work in this field over the past 30 years has been computational in nature, out of necessity. While this is often a useful approach, providing insightful graphical representations of the solutions, it is a difficult (and sometimes impossible) way to uncover the physical origins of many of the interesting phenomena. The computational results represent only a part of the complete picture, and in practice should complement an effective analytic treatment when possible. In the present work the emphasis is therefore on exploring analytically that vast region of relative particle sizes between those for which a pure wave picture of the interaction is effective, and the particle-like regime of geometric optics, where the Mie series tends to obscure its own physical content. In analogy with the quantum-mechanical theory of scattering, this can be referred to as the *semiclassical* domain. Such a task has been aided immensely by the work of Moysés Nussenzveig who, beginning some 30 years ago, developed the relevant asymptotic methods through analytic continuation of the Mie sum into the complex angular momentum plane.

My goal in this monograph has been to provide a self-contained discussion of the scattering of scalar and electromagnetic waves from spherical targets, yet one that calls upon the prospective reader to be somewhat familiar with classical electromagnetic theory and optics, as well as with some elementary quantum mechanics. The latter reflects a desire for mathematical maturity and physical background, rather than for a mastery of content. It is hoped that the level of sophistication asked of the reader is only what is found at the advanced undergraduate and early graduate levels in the physical sciences. The aim is to emphasize strongly the physical mechanisms at work and, in contrast with more application-oriented works, to view the subject more from the approach and language of the theoretical physicist. In this sense Chapter 1 should effectively be a review for the reader, in that the general ideas will be rather familiar. We establish notation and conventions here, as well as provide a perspective for the overall intent of the book, and also set the stage for what follows.

Chapter 2 continues in this vein, but at a higher and more important level. The special case of scalar waves scattering from an impenetrable sphere is treated here by way of an introduction to the mathematical techniques and physical mechanisms that lead to a complete analysis for the transparent sphere in the ensuing chapters. This analysis is carried out in Chapters 3, 4,

5, and 7, with a significant digression in Chapter 6 providing applications to meteorological phenomena. Further applications and extensions of the theory, such as illumination by Gaussian beams rather than plane waves, are reviewed in Chapter 8.

There are six mathematical appendices dedicated to self-containment, which calls for some further comment. In a number of places throughout the text the mathematical development is extensive enough that it is simply not possible to include all the detailed steps in an argument. Much of this detail must be supplied by the reader and involves the asymptotic and other properties of Bessel and Legendre functions. To ease the pain somewhat I have collected together in Appendices A–D what I hope is all the required information on these functions. Appendix E provides a brief introduction to the asymptotic analysis of functions defined by definite integrals, including the saddle-point method of steepest descents. A number of computational issues are discussed briefly in Appendix F, and we have employed the *Mathematica*® system for doing mathematics on a computer for almost all the computations in the book, although not necessarily for all the plotting.

I would like to absolve myself of not forewarning the reader by calling attention here to a number of mathematical notational conventions, some standard, some not, even though they are re-stated later in context. Primes on a function always denote differentiation with respect to the argument, as in $f'(x)$, whereas primes on a variable distinguish x' from x. An asterisk on a variable or function always denotes complex conjugation, and vectors are always denoted by boldface type. In addition, a caret over a vector identifies it as a unit vector, such as $\hat{\mathbf{r}}$. Gaussian units are employed throughout, so that E, D, H, and B all have the same dimensions. Some authors include a factor $(-1)^m$ in the definition of the associated Legendre function $P_\ell^m(\cos\theta)$; we do not, preferring to include that factor in the definition of spherical harmonics. Finally, a *caveat* regarding our notation for Bessel, spherical Bessel, and Ricatti–Bessel functions: other choices are sometimes used by other authors.

Portions of the book were written with the help of some resources provided by the University of Hawaii, which I gratefully acknowledge. I am also indebted to Jennifer Cash for assistance with a number of plots. Professor Lee Schick carefully read the entire manuscript and provided many editorial as well as technical suggestions. We both know what a chore that is, but it is difficult to express my appreciation adequately here, other than to acknowledge that debt.

W. T. Grandy, Jr

1

Classical scattering

Almost all we see and perceive comes to us indirectly by the scattering of light from various objects; that is, by the scattering of electromagnetic radiation over a very restricted interval of the frequency spectrum. Much of this merely illuminates our world and helps us move about, while some exceptional natural scattering phenomena such as rainbows, glories, and halos touch our aesthetic sense. On a technical level, a very large portion of what we have learned about the physical world over the past four millennia has come to us via scattering experiments with both particles and waves, so that a study of scattering theory is an integral part of physics itself.

Classically the most familiar type of scattering is that among particles, such as balls on a pool table – or, more deeply, among gas molecules in the room where we work. Equally evident, however, are the results of scattering of electromagnetic and sound waves, and at first these appear to be entirely different phenomena. Just as modern quantum theory has compelled us to view all matter in terms of a particle–wave dichotomy, however, so have we also learned to view scattering processes as both particle-like and wave-like. That is, at high frequencies and short wavelengths even intrinsically wave-like classical phenomena exhibit particle-like scattering behavior, whereas on the quantum level particle scattering usually must be viewed in terms of waves.

A common experience is to find oneself in a crowded auditorium listening to a speaker who is shielded from view by a pillar, say. While you can hear the speaker perfectly well, you are unable to see him. Although the signals are both transmitted via waves, the wavelength of sound ($\simeq 1\,\mathrm{m}$) is very much greater than that of the scattered light ($\simeq 10^{-7}\,\mathrm{m}$), so that the latter scatters more like a particle, whereas the former wave is able to 'bend' around the pillar while maintaining the correlation of density fluctuations. Almost all physical phenomena exhibit this form of 'complementarity' on one scale or another, and which particular view provides the most useful description

1

in any situation is governed by some combination of the wavelength and the scattering geometry. It will be found useful in what follows to classify scattering behavior in terms of three general domains.

The classical domain: particle and particle-like trajectories; geometric optics.
The wave domain: pure quantum mechanics; pure acoustic and electromagnetic waves; physical optics.
The semiclassical domain: the vast intermediate region between the above two, containing many interesting physical phenomena.

There is a strong analogy here to our understanding of the principal phases of matter, wherein gases are dominated by kinetic energy and the notion of free particles at high temperatures and low densities. The opposite domain of low temperatures and high densities is characterized by the crystal lattice and potential-energy dominance. In the large region between these two extremes lies the liquid state, whose description is much more difficult owing to the equal importance of kinetic energy and potential energy. Similarly, the semiclassical domain is that of intermediate wavelengths in which construction of suitable approximations presents greater difficulties. Nevertheless, it is also a region containing a great deal of interesting physics, and will receive a large measure of attention in that which follows.

Our primary interest in the following chapters will focus on the scattering of classical waves, both scalar and vector, from spherical targets. Specifically, quantum-mechanical scattering will not be discussed at any length, though frequent reference to that theory will be made where it is useful to elucidate various points in the classical theory. Spherical symmetry provides a natural basic model for which the mathematics is entirely tractable, yet corresponds closely to reality over a large range of phenomena.† It introduces considerable simplicity without sacrificing the physics, for only three fundamental parameters are required to characterize the scattering process: the target radius a, the incident wavelength λ, and the index of refraction n of the sphere – geometry, degree of wave-like behavior, and material. Let us begin with the familiar classical domain.

1.1 Particle scattering

Although we shall be interested primarily in the scattering of waves in that which follows, it will be useful to begin by reviewing some elementary aspects of classical particle scattering. This serves not only to affirm common notation, but also to re-introduce geometric arrangements and some fundamental physical ideas that will remain valid throughout the subsequent

---

† Nonspherical targets will be discussed to some extent in Chapter 8.

discussion. In practice the experimenter usually directs a beam of particles at a distribution of targets and arranges an array of detectors at asymptotic distances to record the distribution of scattered particles. Almost always it is safe to assume that neither the particles in the beam nor the scattered particles interact significantly with one another. In addition, we ignore multiple scattering so that we can focus on the truly essential aspects of the process by studying only a single particle scattering from a single target.

When two particles scatter from one another in the rest frame of the observer the process is said to take place in the *laboratory frame*. If both linear momentum and translational kinetic energy are conserved the collision is called *elastic* and we can write

$$\boldsymbol{p}_1 + \boldsymbol{p}_2 = \boldsymbol{p}_1' + \boldsymbol{p}_2', \tag{1.1a}$$

$$T_1 + T_2 = T_1' + T_2', \tag{1.1b}$$

where $T \equiv p^2/2m$, and primed variables here denote scattered quantities. Occasionally it is necessary to conserve explicitly angular momenta as well. These conservation laws provide four equations in 13 unknowns: 12 components of momentum and the ratio of the two masses. In one way or another, then, one must determine or specify nine variables to solve the scattering problem completely.

Consider now the example of particle m_1 incident with momentum $\boldsymbol{p}_1$ upon particle m_2 initially at rest, as in Fig. 1.1. Momentum conservation demands that $\boldsymbol{p}_2'$ be in the plane defined by $\boldsymbol{p}_1$ and $\boldsymbol{p}_1'$, so that the problem is effectively two-dimensional. If m_1, m_2, and $\boldsymbol{p}_1$ are known, and θ_1 is measured, then Eqs. (1.1) yield

$$p_1 = p_1' \cos \theta_1 + p_2' \cos \theta_2, \tag{1.2a}$$

$$p_2'^2 = p_1^2 + p_1'^2 - 2p_1 p_1' \cos \theta_1, \tag{1.2b}$$

$$\frac{p_1'}{p_1} = m_1 \frac{\cos \theta_1}{m_1 + m_2} \pm \left(\frac{m_1^2 \cos^2 \theta_1}{(m_1 + m_2)^2} + \frac{m_2 - m_1}{m_1 + m_2} \right)^{1/2}. \tag{1.2c}$$

These three equations are then solved for p_1', p_2', and θ_2, so that if the masses and initial momenta are known the problem is solved completely by measuring the *scattering angle* θ_1. One could also focus on the recoil angle θ_2, but it is usually θ_1 that is measured.

Note that, for elastic scattering and $m_1 > m_2$, the radical in Eq. (1.2c) implies the existence of a maximum scattering angle

$$\theta_m \equiv |\theta_1|_{\max} = \sin^{-1}(m_2/m_1), \tag{1.3}$$

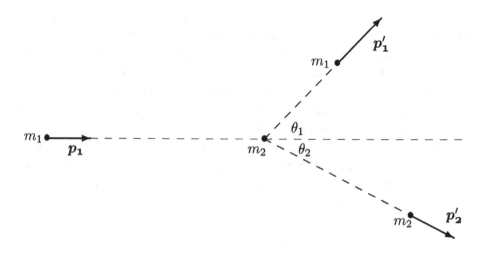

Fig. 1.1. Two-body scattering in the laboratory.

such that $0 \leq \theta_m \leq \pi/2$. However, if $m_1 \leq m_2$, then only the plus sign in Eq. (1.2c) is physical and $0 \leq \theta_1 \leq \pi$ always.

To extend the discussion to relativistic scattering one need only replace $\boldsymbol{p}$ and T by their relativistic counterparts, but this scenario will not be treated further here. If the collision is *inelastic*, then some kinetic energy is either absorbed (an endoergic process) or released (an exoergic process) in the process. Because of their great diversity such processes are difficult to treat in complete generality, but conservation of energy still leads to a definite set of equations to be solved, given sufficient initial data. In later chapters we shall find it relatively simple to account for energy loss in the scattering of waves by absorptive targets.

When Newton's third law is valid, as is usually the case, the two-body system can be reduced to an equivalent one-body problem. In the laboratory frame the equations of motion are

$$m_1 \ddot{\boldsymbol{r}}_1 = \boldsymbol{F}_1, \qquad m_2 \ddot{\boldsymbol{r}}_2 = \boldsymbol{F}_2, \tag{1.4}$$

where dots denote time derivatives, $\boldsymbol{F}_1 = -\boldsymbol{F}_2$, and we shall not consider any external forces to be present. Now introduce relative and center-of-mass coordinates, respectively:

$$\boldsymbol{r} = \boldsymbol{r}_1 - \boldsymbol{r}_2,$$
$$\boldsymbol{R} = \frac{m_1 \boldsymbol{r}_1 + m_2 \boldsymbol{r}_2}{m_1 + m_2}. \tag{1.5}$$

With the definitions of total and reduced masses,

$$M \equiv m_1 + m_2, \qquad \mu \equiv m_1 m_2 / M, \qquad (1.6)$$

respectively, the equations of motion (1.4) become

$$M\ddot{R} = 0, \qquad \mu\ddot{r} = F_1, \qquad (1.7)$$

thereby separating the relative from the center-of-mass motion.

In like manner, one can define center-of-mass and relative velocities, respectively, as

$$V \equiv \dot{R} = \frac{m_1 v_1 + m_2 v_2}{M},$$

$$v \equiv \dot{r} = v_1 - v_2. \qquad (1.8)$$

In terms of these parameters the kinetic energy, angular momentum, and linear momentum of the system can be written, respectively, as

$$T = \tfrac{1}{2} M V^2 + \tfrac{1}{2} \mu v^2, \qquad (1.9a)$$

$$L = M(R \times V) + \mu(r \times v), \qquad (1.9b)$$

$$P = m_1 v_1 + m_2 v_2 = M V. \qquad (1.9c)$$

The absence of any contribution of the form μv to the total linear momentum suggests a very useful coordinate transformation obtained by inverting Eqs. (1.5):

$$r_1 = R + \frac{m_2}{M} r,$$

$$r_2 = R - \frac{m_1}{M} r. \qquad (1.10)$$

In the *center-of-mass* (CM) coordinate system $(r_1 - R, r_2 - R)$ the motion is thus described entirely in terms of the relative coordinate r of a fictitious particle with mass μ – effectively a one-body problem. Furthermore, in this system the total momentum is zero. Alternatively, from Eqs. (1.5) we see that this is simply a transformation to that system in which V is zero.

Figure 1.2 illustrates an actual one-body problem in which a particle of mass m is scattered from a repulsive fixed target or scattering center. The quantity b is called the *impact parameter* and measures the distance of the asymptotic path of the incoming particle from that of a head-on collision, while θ is again the scattering angle. Although we think of this process as taking place in the laboratory system, it is clearly equivalent to 2-body scattering with mass μ in the CM system. This equivalence becomes especially useful when we move from the Newtonian formulation to one in terms of the Lagrangian or Hamiltonian, whereby the particle interaction is described

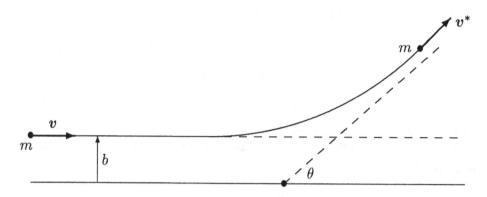

Fig. 1.2. A schematic illustration of either an effective or an actual one-body scattering process, in which a particle is incident with impact parameter b on a fixed scattering center.

by a central potential $V(r) \equiv V(|r_2 - r_1|)$. The potential energy is the same in either scenario, for it depends only on the relative coordinate $|r|$. In this one-body problem the scattering is described completely by conservation of angular momentum L and energy E,

$$L = |L| = mr^2\dot{\varphi}, \qquad E = \tfrac{1}{2}m\dot{r}^2 + U_L(r), \qquad (1.11)$$

where φ is the polar angle in the scattering plane, and U_L is the *effective potential*

$$U_L(r) = V(r) + \frac{L^2}{2mr^2}, \qquad (1.12)$$

a sum of the interaction and centrifugal potentials. In our subsequent investigations it will suffice to focus only on this one-body scenario.

Figure 1.2 indicates that the radial variable ranges from $r = \infty$ to the largest root r_0 of $\dot{r} = 0$, called the *classical distance of closest approach*. Similarly, φ appears to range from π at $r = \infty$ to the scattering angle θ as r again recedes to infinity. However, the latter conclusion is not true in general, as is readily seen by integrating $\dot{\varphi}/\dot{r} = d\varphi/dr$ in Eq. (1.11). The result is called the *deflection angle* Θ,

$$\Theta(L) = \pi - 2L \int_{r_0}^{\infty} \frac{dr}{r^2\sqrt{2m[E - U_L(r)]}}, \qquad (1.13)$$

and ranges over $(0, \pi)$ for a repulsive (positive) interaction. But for an attractive (negative) interaction the particle can go around the scattering center any number of times before emerging in the direction of the scattering angle, so that the two are related by

$$\Theta + 2n\pi = \pm\theta, \tag{1.14}$$

where n is a nonnegative integer such that θ remains within its physical range $(0, \pi)$. Since $|L| = b|p|$, in terms of the asymptotic linear momentum p, we see that Θ is equivalently a function of the impact parameter.

As mentioned earlier, scattering experiments generally involve a beam of particles incident upon a target, so that the particles are distributed more or less uniformly over the azimuthal angle ϕ around the incident z-axis. We define the *effective scattering cross section* $d\sigma$ as the ratio of the number of particles scattered per unit time between angles θ and $\theta + d\theta$ and the number of particles incident per unit time through unit beam cross section. Hence,

$$d\sigma = 2\pi b |db|. \tag{1.15}$$

We are actually interested in the number of outgoing particles in a solid angle $d\Omega = \sin\theta \, d\theta d\phi$, so that the quantity of experimental importance is the *differential scattering cross section*:

$$\frac{d\sigma}{d\Omega} = \frac{b(\theta)}{\sin\theta} \left| \frac{db}{d\Theta} \right|, \tag{1.16}$$

which must necessarily be positive. Note that $d\Theta$ can be replaced by $d\theta$ in this equation, so that the *total scattering cross section* is obtained by integrating over all solid angles:

$$\sigma = \int \left(\frac{d\sigma}{d\Omega} \right) d\Omega. \tag{1.17}$$

If the scattering potential is attractive, or has an attractive portion, then the particle can go around the center many times and $b(\theta)$ is a many-valued function of θ. That is, many different impact parameters can lead to the same scattering angle. Consequently, Eq. (1.16) should be generalized to a sum over all impact parameters leading to a given θ:

$$\frac{d\sigma}{d\Omega} = \sum_j \frac{b_j(\theta)}{\sin\theta} \left| \frac{d\theta}{db_j} \right|^{-1}. \tag{1.18}$$

This expression suggests a number of very interesting phenomena that can arise in processes of this type, to which we shall return presently.

Some simple examples

We shall present four very simple examples of pure classical particle scattering that contain some useful ideas as well as providing fundamental models.[†]
Consider first a particle of mass m with velocity v_1 in a half-space in which its constant potential energy is V_1. It leaves this half-space and enters another in which the potential energy is some other constant V_2, and we wish to determine the final direction of the particle. Let θ_1 and θ_2 be the respective angles v_1 and v_2 make with the normal to the plane separating the two half-spaces before and after penetrating the plane. Because the potential energy is independent of coordinates along this plane, linear momentum must be conserved in those directions. In particular, $v_1 \sin \theta_1 = v_2 \sin \theta_2$, and the relation between v_1 and v_2 is determined by energy conservation. Hence,

$$\frac{\sin \theta_1}{\sin \theta_2} = \sqrt{1 + \frac{2}{mv_1^2}(V_1 - V_2)}. \tag{1.19}$$

We shall see that this is a result reminiscent of, but not equivalent to, Snell's law in geometric optics.

A second example involves the spherical target of Fig. 1.3, an impenetrable sphere of radius a defined such that the interaction with the incident particle is

$$V(r) = \begin{cases} \infty, & r < a \\ 0, & r > a. \end{cases} \tag{1.20}$$

Owing to the impenetrability of the particle, its path is a pair of straight lines symmetric about a radius to the point of impact. Clearly, $b = a \sin \phi$ and the scattering angle is $\theta = \pi - 2\phi$. The differential scattering cross section is then

$$\frac{d\sigma}{d\Omega} = \tfrac{1}{4}a^2, \tag{1.21}$$

and an integration over all solid angles gives the total cross section as $\sigma = \pi a^2$, the geometric cross section of the sphere. This quantity is just the cross section removed from the incident beam and implies the existence of a circular-cylindrical 'shadow' of radius a extending to infinity to the right of the sphere.

Our third example extends the above model to the transparent sphere of Fig. 1.4, which is basically a potential well of radius a and depth V_0:

$$V(r) = \begin{cases} V = -V_0, & r < a \\ 0, & r > a. \end{cases} \tag{1.22}$$

† The first three are taken from Landau and Lifshitz (1960).

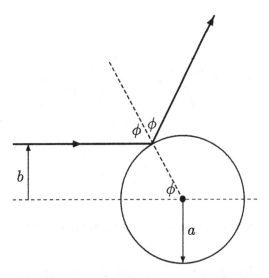

Fig. 1.3. Particle scattering from an impenetrable, or hard, sphere.

From Eq. (1.19) we see that

$$\frac{\sin \alpha}{\sin \beta} = \sqrt{1 + 2V_0/(2mv^2)} \equiv n, \qquad (1.23)$$

where v is the incident velocity. From Fig. 1.4, $\theta = 2(\alpha - \beta)$, and $b = a \sin \alpha$. Elimination of α from these two equations provides the relation between b and θ,

$$b^2 = \frac{a^2 n^2 \sin^2(\theta/2)}{n^2 + 1 - 2n \cos(\theta/2)}, \qquad (1.24)$$

and substitution into Eq. (1.16) yields

$$\frac{d\sigma}{d\Omega} = \frac{a^2 n^2}{4 \cos(\theta/2)} \frac{[n \cos(\theta/2) - 1][n - \cos(\theta/2)]}{[n^2 + 1 - 2n \cos(\theta/2)]^2}. \qquad (1.25)$$

The scattering angle varies from 0 at $b = 0$ to a maximum value θ_m at $b = a$ determined by $\cos(\theta_m/2) = 1/n$. The total cross section is again πa^2.

As a final example, consider the quintessentially long-range potential having the Coulomb or gravitational form: $V(r) = \alpha r^{-1}$. The integral in Eq. (1.13) is evaluated exactly in this case, yielding $\Theta = 2 \tan^{-1}[\alpha/(2Eb)]$. Then, solving this for b, we find for the differential cross section the well-known Rutherford expression

$$\frac{d\sigma}{d\Omega} = \left(\frac{\alpha}{4E \sin^2(\theta/2)} \right)^2. \qquad (1.26)$$

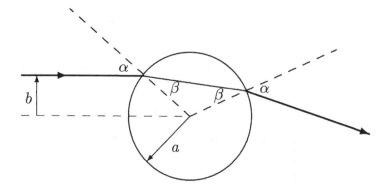

Fig. 1.4. Particle scattering from a transparent, or soft, sphere.

The characteristic divergence in the forward direction is quite evident here.

Singular effects in classical scattering

We now return to Eq. (1.18) and consider the family of effective potentials illustrated in Fig. 1.5, parametrized by the incident angular momentum (or impact parameter). Suppose that $U_L(r)$ achieves a local maximum for $L = L_0$ at $r = r_0$ and energy E. The particle enters an unstable orbit at this point and the integral in Eq. (1.13) diverges logarithmically at the lower limit. The particle will then spiral indefinitely around the scattering center, so that $b(\theta)$ becomes *infinitely* many-valued. This phenomenon of *orbiting*, or *spiral scattering*, appears to have first been discussed extensively by Hirschfelder *et al.* (1954). For $L < L_0$ a potential well develops for $r < r_0$, so that a particle with energy below the top of the barrier finds itself in an oscillating orbit within the well.

The differential cross section of Eq. (1.16) possesses an evident singularity whenever the deflection function passes through a maximum or a minimum:

$$\left(\frac{d\Theta}{db}\right)_{\theta=\theta_R} = 0, \tag{1.27}$$

where θ_R is called the *rainbow angle*. The analogy with the optical rainbow is suggested by the obvious clustering of a large number of particles at a stationary point of Θ, an analogy that will become much clearer in Chapter 4. If the potential is everywhere attractive, meaning that it is negative and monotonically increasing, then Θ vanishes as $b \to 0$. This is also true if the potential is positive, but E is greater than the local maximum. Since Θ must

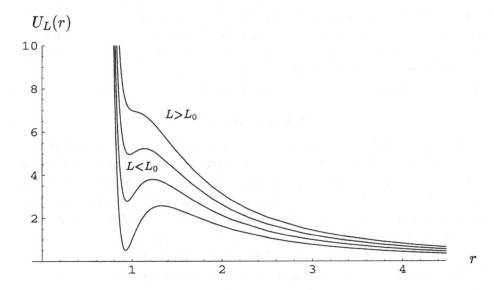

Fig. 1.5. A family of effective potentials $U_L(r)$ for several values of incident angular momentum L, illustrating the formation of a centrifugal barrier when $L < L_0$.

also vanish as $b \to \infty$, it follows that for these potentials and energies either rainbow scattering or orbiting must occur.

Another possible singularity arises in Eq. (1.18) when $b \neq 0$ and θ takes values at nonpositive integral multiples of π. The differential cross section then diverges as $(\sin \theta)^{-1}$ either in the forward or in the backward direction. This cannot happen if the potential is everywhere positive and monotonically decreasing, but can occur whenever the energy is greater than a local maximum. Indeed, whenever orbiting takes place this *glory effect* can as well. Although this forward and backward enhancement of the cross section is referred to as glory scattering, we hasten to point out that the analogous optical effect has an entirely different origin, as will become clear in Chapter 5.

Finally, we note that, if the potential function has an infinite range, as with an exponential or power-law decay, then it deflects *all* impact parameters and there will be a clustering of paths in the forward direction. The differential cross section therefore diverges at $\theta = 0$, and faster than θ^{-1}: as $b \to \infty$ both θ and $d\theta/db$ tend to zero. The Coulomb potential considered above exhibits this behavior very strongly. For cutoff potentials, such as the impenetrable sphere above, this argument is no longer valid – but, as will be seen subsequently, substantial forward scattering is still observed.

As particles and targets become smaller, quantum effects begin to emerge and classical particle scattering becomes a less viable description. Rather, the wave nature of matter asserts itself on this scale as the de Broglie wavelength of a particle becomes relatively larger. Eventually a full wave theory is required to describe the phenomena, albeit in terms of probability-amplitude waves, and much of this will be contained in the classical-wave description developed in the following chapter.

1.2 Geometric optics

A defining characteristic of visible light is the smallness of its wavelength, of order 10^{-5} cm, corresponding to very high frequencies. At such wavelengths light exhibits particle-like behavior much along the lines of Newton's ideas, and the laws of optics can be formulated in a geometric language. Indeed, Hamilton had already expoited the connection in classical mechanics over a century and a half ago. Our object here is to review briefly the salient features of geometric optics, as well as to set the context for later use.

The basic strategy is to mimic time–harmonic plane waves

$$E = E_0 e^{-i\omega t}, \qquad H = H_0 e^{-i\omega t}, \tag{1.28a}$$

by writing

$$E_0 = e(r)e^{ik_0\psi(r)}, \qquad H_0 = h(r)e^{ik_0\psi(r)}, \tag{1.28b}$$

where for an actual plane wave $k_0\psi(r) \to k \cdot r$. The fields e and h are complex, and the real quantity $\psi(r)$ is called the *optical path*. Substitution into Maxwell's equations (e.g., Jackson (1975)) yields a set of equations determining e and h:

$$\nabla\psi \times h + \varepsilon e = \frac{1}{ik_0}\nabla \times h, \tag{1.29a}$$

$$\nabla\psi \times e - \mu h = -\frac{1}{ik_0}\nabla \times e, \tag{1.29b}$$

$$e \cdot \nabla\psi = -\frac{1}{ik_0}(e \cdot \nabla \log \varepsilon + \nabla \cdot e), \tag{1.29c}$$

$$h \cdot \nabla\psi = -\frac{1}{ik_0}(h \cdot \nabla \log \mu + \nabla \cdot h), \tag{1.29d}$$

ε and μ being the permitivity and permeability of the medium, respectively. For short wavelengths, implying very large k_0, these reduce to the geometric-optics approximation:

$$\nabla\psi \times h + \varepsilon e = 0, \tag{1.30a}$$

$$\nabla\psi \times e - \mu h = 0, \tag{1.30b}$$

$$e \cdot \nabla\psi = 0, \tag{1.31a}$$

$$h \cdot \nabla\psi = 0, \tag{1.31b}$$

and the second set can actually be obtained from the first by taking scalar products with $\nabla\psi$.

The independent equations (1.30) provide six equations for the six field components, and nontrivial solutions exist only if the determinant of coefficients vanishes. This provides the condition determining ψ:

$$(\nabla\psi)^2 = \left(\frac{\partial\psi}{\partial x}\right)^2 + \left(\frac{\partial\psi}{\partial y}\right)^2 + \left(\frac{\partial\psi}{\partial z}\right)^2 = n^2(x, y, z), \qquad (1.32)$$

where $n \equiv \sqrt{\varepsilon\mu}$ is the *index of refraction*. This is the *eikonal* equation, and solutions $\psi(\mathbf{r}) = $ constant are the *wavefronts* of the propagating fields.

The time-averaged energy densities are

$$\langle w_e \rangle = \frac{\varepsilon}{16\pi} \mathbf{e} \cdot \mathbf{e}^*, \quad \langle w_m \rangle = \frac{\mu}{16\pi} \mathbf{h} \cdot \mathbf{h}^*, \qquad (1.33)$$

where the * will always denote complex conjugation. Substitution for $\mathbf{e}^*$ and $\mathbf{h}$ from (1.30) shows that the two time averages are identical and equal to $(\mathbf{e} \cdot \mathbf{h} \times \nabla\psi)/(16\pi)$. Thus, from (1.30), the time-averaged Poynting vector is†

$$\langle \mathbf{S} \rangle = v \langle w \rangle \mathbf{s}, \qquad (1.34)$$

with $v = c/n$ and $\langle w \rangle = \langle w_e \rangle + \langle w_m \rangle$, and $\mathbf{s} = \nabla\psi/n$ is a unit vector. Within the geometric-optics approximation, then, the energy propagates in a direction normal to the wavefront, exactly as in the case of plane waves.

One can now define the *geometric light rays* in isotropic media as the orthogonal trajectories to the geometric wavefronts $\psi = $ constant, oriented everywhere in the direction of the average Poynting vector. Let $\mathbf{r} = \mathbf{r}(s)$ describe the position on a ray, such that s is the arc length along the ray. Then $d\mathbf{r}/ds$ is just the unit vector defined above, and

$$n\frac{d\mathbf{r}}{ds} = \nabla\psi, \qquad (1.35)$$

is the equation of the ray. The line integral along a ray is called the *optical length*, $\int n\,ds$, and the length between two points on a ray is just c times the time needed for light to travel between those points.

From Eqs. (1.31) we see that $\mathbf{e}$ and $\mathbf{h}$ are everywhere orthogonal to the ray – indeed, in homogeneous media (n constant) $\mathbf{e}/\sqrt{\mathbf{e} \cdot \mathbf{e}^*}$ and $\mathbf{h}/\sqrt{\mathbf{h} \cdot \mathbf{h}^*}$ remain constant along the ray, implying constant polarization. Because a time-harmonic plane wave in such media also displays this behavior, we conclude that geometric optics exhibits this simplicity primarily because the fields behave *locally* like a plane wave.

It is often useful to consider a narrow tube, or pencil, of rays. The edge

† Time averaging has introduced a factor of $\frac{1}{2}$ into the expressions (1.33).

surface of this tube will begin to turn 'fuzzy' over an interval of a wavelength around the edges, but at very short wavelengths the boundary of the tube should be rather sharp. Define the light intensity as $I \equiv v\langle w \rangle$ and let S be the surface area where a tube of rays intersects the wavefront. Then, in an exercise left to the reader, integration of the conservation law $\nabla \cdot (Is) = 0$ throughout the tube and application of Gauss' theorem demonstrates that $I\,dS$ is constant along a tube of rays. This is the intensity law of geometric optics. In homogeneous media the rays are straight lines, so that, for a cone of rectilinear rays emanating from a point, $I \propto R^{-2}$, where R is the radius of curvature of the surface S.

The laws of geometric optics

The standard boundary conditions of electromagnetic theory require continuity of the normal components of B and D, and the tangential components of H and E (e.g., Jackson (1975)). Thus, a plane wave with wavevector k_1 incident on a plane interface between two media, as in Fig. 1.6, generates both a reflected and a transmitted, or refracted, wave. These two waves must appear if *any* boundary conditions are to be satisfied. Because the spatial and temporal variations must be the same across that plane $z = 0$, meaning all the phases are equal there, we immediately find that $|k_3| = |k_1|$, as well as the *law of reflection*, $\theta_3 = \theta_1$, and *Snell's law of refraction*

$$n_2 \sin \theta_2 = n_1 \sin \theta_1, \tag{1.36}$$

independent of the exact nature of the boundary conditions. Note that this is slightly different from the condition leading to the particle result (1.19), for there the components of momentum are continuous *along* the interface.

To obtain the transmitted and reflected intensities the specific nature of the boundary conditions *is* important. These intensities are defined conventionally in terms of the ratios of plane-wave electric field strengths of unit amplitude, called the *Fresnel coefficients*, and are found separately for components of E parallel ($\parallel$) and perpendicular ($\perp$) to the plane of incidence, the xz-plane. The appropriate boundary matching and utilization of Snell's law yields the following explicit expressions for these coefficients for non-magnetic media (e.g., Born and Wolf (1975)).

Fresnel coefficients ($\parallel$):

$$r_{11}^{\parallel} = \frac{n_2 \cos \theta_1 - n_1 \cos \theta_2}{n_2 \cos \theta_1 + n_1 \cos \theta_2} = \frac{\tan(\theta_1 - \theta_2)}{\tan(\theta_1 + \theta_2)},$$

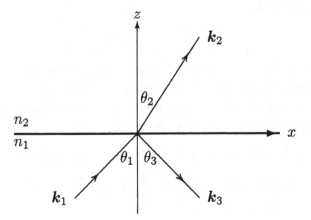

Fig. 1.6. A plane wave with wavevector k_1 incident on the plane interface between two media of refractive indices n_1 and n_2.

$$t_{12}^{\parallel} = \frac{2n_1 \cos\theta_1}{n_2 \cos\theta_1 + n_1 \cos\theta_2} = \frac{2 \cos\theta_1 \sin\theta_2}{\sin(\theta_1 + \theta_2) \cos(\theta_1 - \theta_2)}. \tag{1.37a}$$

The reflectivity and transmissivity coefficients (intensity ratios) are the appropriate components of the time-averaged Poynting vectors, and are given, respectively, by

$$R_{\parallel} = \left(r_{11}^{\parallel}\right)^2, \qquad T_{\parallel} = \frac{n_2 \cos\theta_2}{n_1 \cos\theta_1} \left(t_{12}^{\parallel}\right)^2, \tag{1.37b}$$

and energy conservation is verified by confirming that

$$R_{\parallel} + T_{\parallel} = 1. \tag{1.37c}$$

Fresnel coefficients ($\perp$):

$$r_{11}^{\perp} = \frac{n_1 \cos\theta_1 - n_2 \cos\theta_2}{n_1 \cos\theta_1 + n_2 \cos\theta_2} = \frac{\sin(\theta_2 - \theta_1)}{\sin(\theta_2 + \theta_1)},$$

$$t_{12}^{\perp} = \frac{2n_1 \cos\theta_1}{n_1 \cos\theta_1 + n_2 \cos\theta_2} = \frac{2 \cos\theta_1 \sin\theta_2}{\sin(\theta_2 + \theta_1)}. \tag{1.38a}$$

$$R_{\perp} = \left(r_{11}^{\perp}\right)^2, \qquad T_{\perp} = \frac{n_2 \cos\theta_2}{n_1 \cos\theta_1} \left(t_{12}^{\perp}\right)^2, \tag{1.38b}$$

with

$$R_{\perp} + T_{\perp} = 1. \tag{1.38c}$$

In the important case of normal incidence $\theta_1 = \theta_2 = 0$, both sets of coefficients reduce to

$$r_{11} = \frac{n_1 - n_2}{n_1 + n_2}, \qquad t_{12} = \frac{2n_1}{n_1 + n_2}, \tag{1.39}$$

and we see that, if $n_2 > n_1$, there is a phase reversal for the reflected wave.

One may wish to reverse the directions of the incident and transmitted waves, of course, in which case there will be a reflected wave in medium 2. The following reciprocity relations are readily verified:

$$r_{22}^{\parallel}(\theta_2, \theta_1; n) = r_{11}^{\parallel}(\theta_2, \theta_1; 1/n) = -r_{11}^{\parallel}(\theta_1, \theta_2; n), \tag{1.40a}$$

$$t_{21}^{\parallel}(\theta_2, \theta_1; n) = t_{12}^{\parallel}(\theta_2, \theta_1; 1/n); \tag{1.40b}$$

$$r_{11}^{\perp}(\theta_1, \theta_2; n) = r_{11}^{\parallel}(\theta_1, \theta_2; 1/n), \tag{1.41a}$$

$$t_{12}^{\perp}(\theta_1, \theta_2; n) = \frac{1}{n} t_{12}^{\parallel}(\theta_1, \theta_2; 1/n), \tag{1.41b}$$

where $n \equiv n_2/n_1$ is the relative refractive index.

In a similar manner, when a ray meets an interface the transmitted and incident rays can be visualized as in Fig. 1.7, in which $\boldsymbol{n}$ is a unit normal to the surface and $\boldsymbol{m}$ is a unit vector normal to a plane element of area perpendicular to the surface of discontinuity and bounded by the rectangle C. From Eq. (1.35) $\nabla \times n s = 0$, and Stokes' theorem yields

$$\int \nabla \times ns \cdot m \, dS = \int_C ns \cdot dr. \tag{1.42}$$

As the height δh of the rectangle shrinks to zero this expression implies that

$$n \times (n_2 s_2 - n_1 s_1) = 0, \tag{1.43}$$

so that the tangential component of ns is continuous across the surface. From Fig. 1.7 we see that the refracted ray lies in the plane of incidence formed by the incident ray and the normal to the surface, and that the angles of incidence and refraction are again related by Snell's law.

Just as in the case of plane waves one expects another wave, or ray, to be reflected back into the first medium, and also to lie in the plane of incidence. For this ray we set $n_2 = n_1$, so that $\theta_2 = \pi - \theta_1$. That is, the angle of reflection equals the angle of incidence measured from the normal on the incident side of the surface. Both these laws of geometric optics apply to general waves and surfaces, but the wavelength must be much smaller than the radius of curvature of the surface. As a result, we once again find a localized plane-wave behavior and the laws of reflection and refraction effectively apply to the tangent plane to the surface at the point of impingement of a ray. That is, large spheres are locally flat.

When $n_1 > n_2$ in Fig. 1.6 it is clear that there exists an angle of incidence

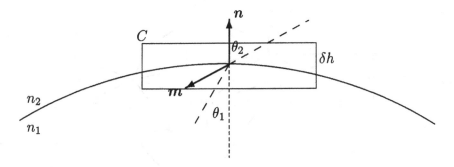

Fig. 1.7. Derivation of the geometric laws of reflection and refraction at the interface of two homogeneous media.

for which the angle of refraction is 90°. This occurs at the *critical angle*, which, from Snell's law, is given by

$$\theta_c \equiv \sin^{-1}\left(\frac{n_2}{n_1}\right). \tag{1.44}$$

Rays with angles of incidence greater than this are totally internally reflected.

A specific extension from plane interfaces to curved is provided by the transparent sphere considered earlier for particles. If a ray of light is incident upon the sphere, as in Fig. 1.8, then many reflections and refractions are possible. If the sphere is a homogeneous drop of water, $n \simeq \frac{4}{3}$, most of the intensity gets transmitted, except for very high angles of incidence. As a consequence, the higher-order internal reflections rapidly become negligible.

It is worth inserting a *caveat* here about the generality of ray optics, because the point becomes central when we examine the actual scattering problem in the following chapter. Geometric optics does not address the question of how the intensity varies transversely to the rays. Thus, when a beam of rays meets a physical boundary the intensity suffers a discontinuity at a light–shadow boundary, for example. Similarly, the intensities of rays reflected and transmitted at an interface between two homogeneous media are not determined by geometric optics, but by the Fresnel formulas, whose origins lie outside the geometric theory. Correcting these shortcomings constitutes much of the mathematical effort to be made in Chapter 5.

Many more details of geometric optics will be discussed in Chapter 4, while we conclude the present survey by noting two further basic results that are described by consideration of rays incident upon a surface of a glass prism of refractive index n, as indicated in Fig. 1.9. The angles θ_1 and θ_2 are those of incidence and emergence of a ray, respectively, while α is fixed and δ is the *angle of deviation*. One is often interested in the angle of incidence

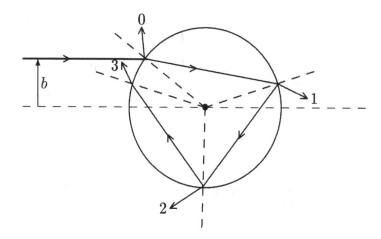

Fig. 1.8. Geometric ray tracing through a homogeneous sphere of refractive index *n*. The incident ray with impact parameter *b* produces an infinite series of relfected and transmitted rays. The initial angle of incidence is also equal to the angle of emergence from the sphere of all subsequent external rays, and the initial angle of refraction gives the angle of all subsequent internally reflected rays. The numbers *m* indicate rays transmitted after *m* internal reflections.

that will produce the minimum angle of deviation, and this obviously follows from the requirement $d\delta/d\theta_1 = 0$. By introducing the internal angles φ_1 and φ_2 and applying Snell's law we obtain the conditions

$$\theta_1 + \theta_2 = \delta + \alpha, \qquad (1.45a)$$

$$\varphi_1 + \varphi_2 = \alpha, \qquad (1.45b)$$

$$\sin\theta_1 = n\sin\varphi_1, \qquad (1.45c)$$

$$\sin\theta_2 = n\sin\varphi_2. \qquad (1.45d)$$

Some algebra then leads to the conclusion that the minimum deviation occurs when the angle of emergence equals the angle of incidence: $\theta_2 = \theta_1$ (and $\varphi_1 = \varphi_2$, as well). One can also show, after further algebra, that $d^2\delta/d\theta_1^2 > 0$ at the extremum angle, so that the deviation is indeed a minimum. If the prism is an isosceles triangle in cross section the ray passes through the prism symmetrically, and the angle of minimum deviation is $\delta_{\min} = 2\theta_1 - \alpha$. In this case we have a simple means for determining the index of refraction of the prism by measuring $\delta_{\min}$:

$$n = \frac{\sin\left[\frac{1}{2}(\alpha + \delta_{\min})\right]}{\sin(\alpha/2)}. \qquad (1.46)$$

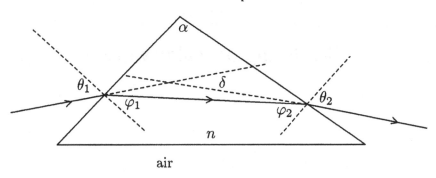

Fig. 1.9. A glass prism of refractive index n in air, illustrating the angle of deviation δ between the incident and emergent rays. These rays make angles θ_1 and θ_2 with the normals to the surfaces, respectively.

If the ray is not monochromatic, but consists of white light, then the prism demonstrates that the refractive index of a medium generally varies a bit with wavelength. Since $n = n(\lambda)$ implies that $\delta = \delta(\lambda)$, we can ask for the change in δ when λ is changed to $\lambda + \Delta\lambda$. Born and Wolf (1975), for example, show that

$$\Delta\delta \propto \frac{dn}{d\lambda}\Delta\lambda, \tag{1.47}$$

implying a rotation of the wavefront. This results in the familiar phenomenon of *chromatic dispersion*, revealing on a detection screen the spectrum of visible light.

As wavelengths become longer we begin to observe deviations from geometric optics, and effects generally collected under the term *diffraction* begin to appear as a consequence of various forms of wave interference. An immediate and common demonstration of these effects is produced by a pencil lying on a well-lit desk top: as the pencil is raised slightly the shadow is seen to be dark and the edges sharp, but when it is raised further the edges begin to blur and the shadow lightens. The light is deflected by the edges of the pencil and begins to fill in behind it. Such diffraction effects were first demonstrated and elucidated in a theory by Thomas Young in 1804, which was further developed by Fresnel, Helmholtz and Kirchhoff. Thus, when λ is much greater than the radius of a sphere, say, then a full wave theory of light scattering is required. We shall see, however, that there is a large region between the two extremes in which much interesting physics is to be found.

2

Scattering of scalar waves

In this chapter we review, as well as extend and elaborate upon, the theory of scalar waves scattering from a spherical target. The context is primarily classical, though almost all the discussion can be applied to quantum-mechanical probability amplitudes, acoustic pressure waves, the Debye potentials of electromagnetic theory (Chapter 3), and basically any kind of purely classical wave phenomenon.

Any function $\psi(r, t)$ describing linear wave motion in a source-free region must satisfy the homogeneous wave equation

$$\nabla^2 \psi - \frac{1}{c^2} \frac{\partial^2 \psi}{\partial t^2} = 0, \tag{2.1}$$

where c is the fundamental speed of propagation in free space. We presume that the wavefunction is susceptible to Fourier decomposition,

$$\psi(r, t) = \int_{-\infty}^{\infty} \Psi(r, \omega) e^{-i\omega t} \, d\omega, \tag{2.2}$$

the integration being over all possible angular frequencies. This linear superposition suggests an implicit harmonic time dependence, making it necessary to study only the scalar Helmholtz equation:

$$\left(\nabla^2 + k^2\right) \psi_k(r) = 0, \qquad k \equiv \omega/c. \tag{2.3}$$

In a homogeneous medium described by an index of refraction† n one must replace k^2 in Eq. (2.3) by $n^2 k^2$, in which n is generally a function of the frequency. (In inhomogeneous media it is a function of position as well). Absorptive media are included by extending the refractive index to complex values, so that $n = m + i\kappa$. The effect of n on wave propagation can be studied simply by examining a plane wave traveling in one dimension – the

† The validity of such a description will be addressed further in Chapter 3.

z-direction, say, for which the phase function is

$$e^{-i\omega t}e^{i\omega(n/c)z} = e^{-i\omega t}e^{-\alpha z/2}e^{im\omega z/c}. \tag{2.4}$$

Absorption is seen to be governed explicitly by the *attenuation coefficient* $\alpha(\omega) = 2\omega\kappa/c$.

Although $n(\omega)$ is usually defined only for $\omega \geq 0$, it can be extended to negative real values by means of the symmetry relation

$$n(-\omega) = n^*(\omega), \qquad \omega \text{ real}. \tag{2.5}$$

This can be derived by analysis of a plane wave propagating through a thin slab, for example (Nussenzveig 1972). Additional symmetries follow from the definition of n:

$$m(-\omega) = m(\omega), \qquad \kappa(-\omega) = -\kappa(\omega), \qquad \alpha(-\omega) = \alpha(\omega). \tag{2.6}$$

As is so often the case, extension into the complex ω-plane leads to further insights, and causality alone ensures that n possesses an analytic continuation into the upper half-plane, where it is regular. We shall explicitly exclude magnetic media here and subsequently, which then allows us to conclude that $n(-\omega^*) = n^*(\omega)$, so that n is real on the imaginary ω-axis. Sometimes it is useful to introduce the complex *dielectric function* $\epsilon(\omega) \equiv n^2(\omega)$, which can also be continued into the upper half-plane.

A major reason for this extension into the complex ω-plane is to examine the possible relationship between the real and imaginary parts of n. Can they be specified independently in a dielectric medium? From a theorem of Titchmarsh (1948) we know that $m + i\kappa$ is the boundary value on the real axis of a function that is regular in the upper half-plane and is a causal transform:

$$n(\omega) - 1 = -\frac{i}{\pi}P\int_{-\infty}^{\infty}\frac{n(\omega') - 1}{\omega' - \omega}\,d\omega', \tag{2.7}$$

a Cauchy principal value. By taking the real part we obtain the *dispersion relation* between the real and imaginary parts of n:

$$m(\omega) - 1 = \frac{1}{\pi}P\int_{-\infty}^{\infty}\frac{\kappa(\omega')}{\omega' - \omega}\,d\omega'. \tag{2.8}$$

By considering also the imaginary part of Eq. (2.7), and employing the symmetry relations (2.6), we find the pair of *Kramers–Kronig relations*

$$m(\omega) - 1 = \frac{2}{\pi}P\int_{0}^{\infty}\omega'\frac{\kappa(\omega')}{\omega'^2 - \omega^2}\,d\omega', \tag{2.9a}$$

$$\kappa(\omega) = -\frac{2\omega}{\pi}P\int_{0}^{\infty}\frac{m(\omega') - 1}{\omega'^2 - \omega^2}\,d\omega', \tag{2.9b}$$

which are valid for any causal model of a dispersive medium.† Thus, the real part of the refractive index for any given frequency depends on the values taken by the attenuation coefficient at all frequencies, and approaches unity as $\omega \to \infty$.

2.1 Spherically symmetric geometry

Unless explicitly noted otherwise, we shall focus primarily on real indices of refraction, and for the moment consider three-dimensional free space ($n = m = 1$). In addition, we assume that a spherical geometry is appropriate in what follows, so that the solutions to Eq. (2.3) in spherical coordinates have the generic form

$$\psi_k(\mathbf{r}) = \sum_{\ell,m} \psi_{k,\ell,m}(\mathbf{r}), \tag{2.10a}$$

with

$$
\begin{aligned}
\psi_{k,\ell,m}(\mathbf{r}) &= R_{k,\ell}(r) Y_\ell^m(\theta, \phi) \\
&= \frac{u_\ell(r,k)}{r} Y_\ell^m(\theta, \phi).
\end{aligned}
\tag{2.10b}
$$

This re-definition of the radial function $R_{k,\ell}$ leads to the following radial equation for u_ℓ:

$$\left(\frac{d^2}{dr^2} + k^2 - \frac{\ell(\ell+1)}{r^2} \right) u_\ell(r,k) = 0. \tag{2.11}$$

In Appendix A we note that the solutions to (2.11) are Ricatti–Bessel functions, so that $r^{-1}u_\ell(r,k)$ is a spherical Bessel function. The $Y_\ell^m(\theta, \phi)$, of course, are the well-known spherical harmonics forming an orthonormal set of basis functions on the surface of the unit sphere. The properties of these and related angular functions are studied in some detail in Appendix D.

The linearly independent solutions to Eq. (2.11) are conventionally taken as the two sets $[z j_\ell(z), z y_\ell(z)]$ and $[z h_\ell^{(1)}(z), z h_\ell^{(2)}(z)]$, where $z = kr$. According to Appendix A the spherical Bessel functions j_ℓ are regular everywhere, the spherical Neumann functions y_ℓ are singular at the origin, and the spherical Hankel functions are linear combinations of the two:

$$h_\ell^{(1,2)}(kr) \equiv j_\ell(kr) \pm i y_\ell(kr) = h_\ell^{(2,1)^*}(kr). \tag{2.12}$$

Because the two Hankel functions of this pair behave asymptotically as

---

† Titchmarsh's theorem is stated explicitly at the end of Chapter 3, where the conditions for the validity of these relations are also discussed at greater length.

outgoing and incoming spherical waves, respectively, it is customary in the scattering problem to take the general solution (2.10a) to be

$$\psi_k(r) = \sum_{\ell,m} \left[A_{\ell m} h_\ell^{(1)}(kr) + B_{\ell m} h_\ell^{(2)}(kr) \right] Y_\ell^m(\theta, \phi), \tag{2.13}$$

where the coefficients are to be determined by the boundary conditions of the specific problem.

A useful example of these solutions is that of a plane wave propagating in free space,

$$\psi_k(r) = e^{ik \cdot r} = e^{ikr \cos \theta}, \tag{2.14}$$

where for convenience we have taken the wavevector k defining the direction of propagation along the positive z-axis. Since this function is independent of the azimuthal angle ϕ (axial symmetry), and cannot have any finite singularities, it must have the following form in spherical coordinates:

$$e^{ikr \cos \theta} = \sum_{\ell=0}^{\infty} c_\ell \, j_\ell(kr) P_\ell(\cos \theta). \tag{2.15}$$

The orthogonality of the Legendre polynomials P_ℓ, Eq. (D.8), and Poisson's integral representation of $j_\ell(kr)$, Eq. (A.15a), allow us to identify c_ℓ directly, so that (2.15) provides what is known as Bauer's formula,

$$e^{ik \cdot r} = \sum_{\ell=0}^{\infty} (2\ell + 1) i^\ell \, j_\ell(kr) P_\ell(\cos \theta). \tag{2.16}$$

This expression is called the *partial-wave expansion* of the plane wave, in which $j_\ell(kr) P_\ell(\cos \theta)$ is the partial wave corresponding to the angular momentum index ℓ. This nomenclature is adopted from quantum mechanics, where ℓ is the quantum number indexing orbital angular momentum. Just as a Fourier–integral synthesis of a wavefunction can be interpreted physically as a linear superposition of plane waves, a partial-wave expansion of the form (2.16) can be thought of as a sum of physical partial waves. Figure 2.1 illustrates how well the plane wave begins to be approximated in this way with only a small number of terms. Both the infinite plane wave and the partial wave are fictions, of course, but each provides a very useful tool and serves as an aid to the intuition.

Consider for a moment a plane wave incident upon the impenetrable sphere of radius a in Fig. 1.3. Spherical symmetry implies that the complete wavefunction outside the sphere, from Eq. (2.13), can be written in the form

$$\psi_k(r) = \frac{1}{2} \sum_{\ell=0}^{\infty} (2\ell + 1) i^\ell \left[A_\ell h_\ell^{(1)}(kr) + B_\ell h_\ell^{(2)}(kr) \right] P_\ell(\cos \theta), \tag{2.17}$$

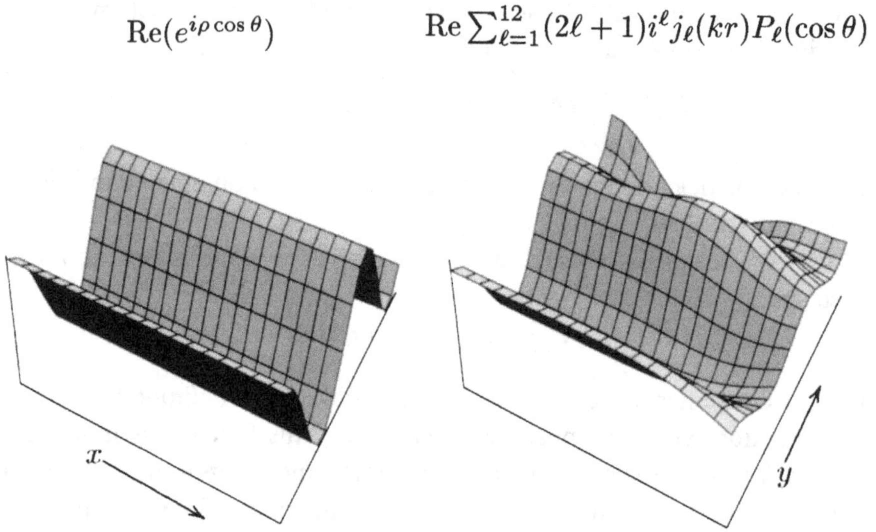

$$\mathrm{Re}(e^{i\rho\cos\theta}) \qquad\qquad \mathrm{Re}\sum_{\ell=1}^{12}(2\ell+1)i^\ell j_\ell(kr)P_\ell(\cos\theta)$$

Fig. 2.1. The partial-wave approximation to an infinite plane wave using the first 12 terms in the sum of Eq. (2.16).

where the factor of $\frac{1}{2}$ is included for later convenience. Impenetrability requires the Dirichlet boundary condition $\psi_k(a,\theta) = 0$, so that

$$S_\ell(\beta) \equiv \frac{A_\ell}{B_\ell} = -\frac{h_\ell^{(2)}(\beta)}{h_\ell^{(1)}(\beta)}, \qquad \beta \equiv ka, \tag{2.18}$$

which is just a phase factor. The quantity β is called the *size parameter*, and is a measure of the range of the interaction in units of incident wavelength.†
The complete unnormalized solution outside the sphere is then

$$\psi_k(r,\theta) = \frac{1}{2}\sum_{\ell=0}^{\infty} B_\ell(2\ell+1)i^\ell\left[h_\ell^{(2)}(kr) + S_\ell(\beta)h_\ell^{(1)}(kr)\right]P_\ell(\cos\theta), \tag{2.19}$$

with B_ℓ to be determined. A similar expression will be found for any other problem with spherical geometry.

We note in passing that this wavefunction can be interpreted as the velocity potential of sound waves scattered from an acoustically soft sphere, or as the Schrödinger wavefunction describing scattering from a hard-sphere potential – the difference lies with the boundary conditions. The focus will remain entirely on the classical interpretation.

Our present interest in scattering phenomena suggests that we are mostly

---

† In the optics literature this parameter is commonly denoted by x, but since x is also the conventional Cartesian coordinate notation we choose to avoid confusion and depart from the optics convention by employing β instead.

concerned with the behavior of $\psi_k(r, \theta)$ in the far, or radiation, zone – that is, asymptotically. (In optics this is known as the Fraunhofer region). One expects the behavior of the wavefunction at large distances to be qualitatively that of the incident plane wave plus an outgoing spherical wave. From the asymptotic limit of j_ℓ in Eq. (A.16a), this suggests the form

$$\psi_k(r, \theta) \xrightarrow[kr \to \infty]{} \sum_{\ell=0}^{\infty} B_\ell (2\ell+1) i^\ell P_\ell(\cos\theta) \left(\frac{\sin(kr - \ell\pi/2)}{kr} + f_\ell(k) \frac{e^{ikr}}{r} \right), \quad (2.20)$$

where $f_\ell(k)$ is determined below. Note that, for large enough r, this will again look like a plane wave, so that we can infer that $B_\ell = 1$.

Again appealing to Appendix A, we can equate Eq. (2.20) with the asymptotic form of Eq. (2.19) to find that

$$\frac{\sin(kr - \ell\pi/2)}{kr} + \frac{e^{ikr}}{r} f_\ell(k) = -\frac{e^{-ikr}}{2kr} \frac{e^{i\ell\pi/2}}{i} + S_\ell \frac{e^{ikr}}{2kr} \frac{e^{-i\ell\pi/2}}{i}. \quad (2.21)$$

This yields the identification of the ℓth *partial scattering amplitude*

$$f_\ell(k) = \frac{e^{-i\ell\pi/2}}{2ik} [S_\ell(\beta) - 1]. \quad (2.22)$$

More generally, we define the total *scattering amplitude* $f(k, \theta)$ by demanding that in the far zone

$$\psi_k(r, \theta) \longrightarrow e^{ik \cdot r} + f(k, \theta) \frac{e^{ikr}}{r}, \quad (2.23)$$

the sum of the incident and outgoing scattered waves. By noting that $i^\ell = e^{i\ell\pi/2}$ we complete the derivation of the partial-wave expansion of the scattering amplitude:

$$f(k, \theta) = \frac{1}{2ik} \sum_{\ell=0}^{\infty} (2\ell + 1) [S_\ell(\beta) - 1] P_\ell(\cos\theta), \quad (2.24)$$

where we recall that $\beta = ka$.

The importance of the scattering amplitude lies with its relation to the measurable cross sections. From the discussion associated with Eqs. (1.15) and (1.16) we know that the differential cross section is the ratio of the scattered intensity to the incident intensity *at the detector*. If the latter lies a distance R from the target this involves the energy flux $|\psi|^2$ in the surface element $R^2 d\Omega$, so that Eq. (2.23) leads to the identification

$$\frac{d\sigma}{d\Omega} = |f(k, \theta)|^2. \quad (2.25)$$

The total cross section for *elastic* scattering is then

$$\sigma_{\text{elas}} \equiv \int |f(k,\theta)|^2 \, d\Omega. \tag{2.26}$$

Clearly, there is a need to understand in some detail the properties of the scattering amplitude.

We noted in Eq. (2.18) that $S_\ell(\beta)$ is just a phase factor, which is a general result for spherically symmetric scattering. We can thus write

$$S_\ell(\beta) \equiv e^{2i\delta_\ell(\beta)}. \tag{2.27}$$

In the quantum-mechanical formalism one introduces an S operator, and in Chapter 3 we shall introduce the S-matrix for the electromagnetic problem. In both cases the idea is that S maps the incident wavefunction into the scattered wavefunction, $\psi_{\text{out}} = S\psi_{\text{in}}$. In the case of spherical scattering the S_ℓ are simply the eigenvalues of the S matrix, and the form (2.27) implies that S must be unitary. Moreover, with this notation the asymptotic form of the expression in brackets in Eq. (2.19) can be written

$$e^{i\delta_\ell} \frac{\sin(kr - \ell\pi/2 + \delta_\ell)}{kr},$$

so that the essential effect of scattering is to shift the phase of each partial wave in the expansion of the incident wave by δ_ℓ. Appropriately enough, the real quantities δ_ℓ are called the *phase shifts* for spherically–symmetric scattering.

Substitution of Eqs. (2.24) and (2.27) into (2.26), along with the orthogonality of the Legendre polynomials, yields a partial-wave expansion for the total elastic cross section:

$$\sigma_{\text{elas}} = \frac{4\pi}{k^2} \sum_{\ell=0}^{\infty} (2\ell + 1) \sin^2 \delta_\ell(\beta). \tag{2.28}$$

A similar calculation with (2.24) demonstrates that this is equivalent to

$$\sigma_{\text{elas}} = \frac{4\pi}{k} \text{Im} f(k,0), \tag{2.29}$$

which is known as the *optical theorem*. This dependence of the total cross section only on the forward scattering of the plane wave is rather general, and has a long history that is nicely summarized by Newton (1976). It can also be derived from the scalar wave equation (2.3), as we now demonstrate.

Consider two different solutions of (2.3) for outgoing wavevectors k and k', with $|k'| = |k|$. The equation and its complex conjugate can then be combined to yield

$$\psi_{k'}^*(r) \nabla^2 \psi_k(r) - \psi_k(r) \nabla^2 \psi_{k'}^*(r) = 0. \tag{2.30}$$

This is valid even if the wave equation has a real inhomogeneous term, as in (2.61) below. It is useful here to introduce the alternative notation $f_k(k')$ for the scattering amplitude, where k' has the magnitude of k but points in the radial direction: $k' = k\hat{r}$, with $\hat{r}$ a unit vector. One advantage of this notation is that it allows for a simple expression of time-reversal invariance in an elastic scattering process: $f_k(k') = f_{k'}(k)$, in which initial and final momenta are interchanged. Now integrate Eq. (2.30) over the volume of a sphere having very large radius r, and employ Green's theorem to obtain a surface integral – we presume the actual wavefunction to vanish as $r \to \infty$, even though we are dealing with only one of its plane-wave components here. Some straightforward algebra yields what is often referred to as a generalized optical theorem:

$$\frac{1}{2i} \left[f_{k'}(k) - f_k^*(k') \right] = \frac{k}{4\pi} \int f_{k_r}^*(k') f_{k_r}(k) \, d\Omega_{k_r}, \tag{2.31}$$

where k_r lies along the radius of the sphere.

For forward scattering, $k' = k$, we regain immediately the optical theorem (2.29), but in its derivation from (2.30) it is now seen as nothing more than an expression of the conservation of incoming flux. (In quantum mechanics it corresponds to zero divergence of the probability current density.) Equation (2.31) also follows from unitarity of the scattering matrix (e.g., Merzbacher (1970), p. 505).

If the scattering target is absorptive, so that it can be described by a complex index of refraction, Eq. (2.4) suggests that the ensuing effects can be accounted for by introducing a complex phase shift, which will yield the expected attenuation. In this event we write

$$S_\ell = \eta_\ell e^{i2\delta_\ell}, \qquad 0 \le \eta_\ell \le 1, \tag{2.32}$$

where the range of the absorption factor η_ℓ is from complete absorption to none. In this situation the total cross section must be written

$$\sigma_{\text{tot}} = \sigma_{\text{elas}} + \sigma_{\text{abs}}, \tag{2.33}$$

where now the elastic and absorption cross sections are identified as

$$\sigma_{\text{elas}} = \frac{\pi}{k^2} \sum_{\ell=0}^{\infty} (2\ell + 1) \left[1 - 2\eta_\ell \cos(2\delta_\ell) + \eta_\ell^2 \right], \tag{2.34a}$$

$$\sigma_{\text{abs}} = \frac{\pi}{k^2} \sum_{\ell=0}^{\infty} (2\ell + 1) \left(1 - \eta_\ell^2 \right). \tag{2.34b}$$

Note that the elastic cross section is regained when $\eta_\ell \to 1$ and that σ_{tot} diverges as $\eta_\ell \to 0$, as it should. The optical theorem is once again expressed by Eq. (2.29), but now for the total cross section. And the S matrix is no longer unitary when absorption is present, so that in general we can conclude only that $S_\ell S_\ell^* \le 1$. All these considerations lead to the conclusion that the total cross section is a measure of the energy removed from the incident wave by the scatterer.

2.2 The impenetrable sphere

Now we can focus explicitly on scattering from an impenetrable sphere and application of the boundary condition has already provided the expression (2.18). This result can also be stated in terms of the phase shifts, following a little algebra:

$$\tan \delta_\ell = \frac{j_\ell(\beta)}{y_\ell(\beta)}, \qquad \beta = ka, \tag{2.35}$$

which is convenient for studying the long wavelength limit. Appendix A provides the limiting expressions for the spherical Bessel functions as $\beta \to 0$, so that

$$\tan \delta_\ell(\beta) \xrightarrow[\beta \to 0]{} - \frac{\beta^{2\ell+1}}{(2\ell + 1)[(2\ell - 1)!!]^2}, \tag{2.36}$$

where $(2\ell - 1)!! = 1 \cdot 3 \cdot 5 \cdots (2\ell - 1)$. We see that $\delta_\ell(\beta)$ falls off rapidly with increasing ℓ, and that all phase shifts vanish as $\beta \to 0$. However, according to Eq. (2.24) the $\ell = 0$ phase shift will actually make a finite contribution to the scattering amplitude, even at $k = 0$. Thus, in quantum-mechanical parlance, s-wave scattering dominates in the long-wavelength, or low-frequency limit, and we readily find that

$$\frac{d\sigma}{d\Omega} \xrightarrow[\beta \to 0]{} a^2. \tag{2.37}$$

It is notable that the scattering is spherically symmetric in this limit, and the total elastic cross section is obviously

$$\sigma_{tot} = 4\pi a^2, \tag{2.38}$$

The factor of 4 difference from the particle result in Chapter 1 is completely a reflection of the wave nature of the process, and its origin will become somewhat clearer subsequently.

At short wavelengths, or high frequencies, the description becomes a bit more complicated, because many more partial waves contribute to the scattering amplitude for large β. To understand this one need only refer to

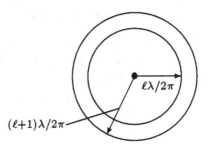

Fig. 2.2. The localization principle associates the ℓth partial wave with an impact parameter located midway between two succcessive angular-momentum rings in the incident beam.

the asymptotic expressions for the spherical Bessel functions in Appendix A and note that they oscillate if $(\ell + \frac{1}{2}) < \beta$, but decay exponentially when $(\ell + \frac{1}{2}) > \beta$. As a consequence, van de Hulst (1957) has noted the existence of a weak *localization principle* when β is large. In quantum wave mechanics the orbital angular momentum appears in terms of the quantum numbers ℓ arising from spherical symmetry as

$$L = \hbar\sqrt{\ell(\ell + 1)} \simeq \hbar(\ell + \tfrac{1}{2}), \qquad \ell \gg 1, \qquad (2.39)$$

where as usual $\hbar$ is Planck's constant divided by 2π. In analogy with the discussion in Chapter 1, we can write the angular momentum for a given ℓ value in terms of a partial impact parameter: $L = pb_\ell = \hbar k b_\ell$, where p is the linear momentum, and thus

$$b_\ell = \frac{\ell + \frac{1}{2}}{k} = \frac{\lambda}{2\pi}(\ell + \tfrac{1}{2}), \qquad (2.40)$$

independent of $\hbar$. Figure 2.2 illustrates how a beam or plane wavefront incident on a target can be viewed in terms of partial waves characterized by annular rings of increasing angular momentum, so that Eq. (2.40) defines the impact parameter at the midpoint between two such rings. Significant interaction with the scatterer, meaning significant phase shifts, occurs for $b_\ell \lesssim a$, or when

$$\ell + \tfrac{1}{2} \lesssim ka = \beta. \qquad (2.41)$$

There is some value in first discussing the limit $ka \gg 1$ in a broad semi-quantitative way that is strictly classical. Then we shall appeal to some methods of quantum mechanics that allow us to understand the approximations in more detail.

The high-frequency limit

For $ka \gg 1$ the scattering is no longer isotropic and we expect the angular distribution to become rather peaked in the forward direction. One way to express this limit, then, is through the inequalities

$$\theta \lesssim \frac{1}{ka} \ll 1, \tag{2.42}$$

followed by introduction of the Green function for the scalar Helmholtz equation:

$$(\nabla^2 + k^2) G_k(r, r') = -4\pi \delta(r - r'). \tag{2.43}$$

We shall also want to employ Green's theorem relating a volume integral to an integral over its bounding surface:

$$\int_V (\phi \nabla^2 \psi - \psi \nabla^2 \phi) \, d^3 r = \oint_S \left(\phi \frac{\partial \psi}{\partial n} - \psi \frac{\partial \phi}{\partial n} \right) dS, \tag{2.44}$$

where $\partial/\partial n$ is the *outward*-directed normal derivative to S – e.g., $\partial \psi/\partial n = n \cdot \nabla \psi$, and n is a unit vector normal to and directed outward from S. Now let $\phi = G$ and take $\psi \to \psi_s$ to be the scattered wavefunction, so that

$$\psi_s(r) = \frac{1}{4\pi} \oint_S \left(\psi_s(r') \frac{\partial G_k(r, r')}{\partial n'} - G_k(r, r') \frac{\partial \psi_s(r')}{\partial n'} \right) dS'. \tag{2.45}$$

A standard exercise in Fourier transformation of Eq. (2.43) confirms that the free-space Green function is e^{ikR}/R, with $R \equiv |r - r'|$. If ψ is not necessarily interpreted as a scattered wavefunction, substitution of G into Eq. (2.43) and imposition of outgoing radiation conditions leads to Kirchhoff's diffraction theory (e.g., Jackson (1975)). In the present context, however, we envision an incident plane wave $\psi_i(r) = A e^{ik \cdot r}$, so that the total wavefunction is $\psi = \psi_i + \psi_s$. In the case of acoustic scattering from an impenetrable sphere of radius a, say, Neumann boundary conditions are adopted:

$$\left(\frac{\partial \psi}{\partial n'} \right)_{r=a} = 0 \longrightarrow \left(\frac{\partial \psi_s}{\partial n} \right)_{r=a} = -\left(\frac{\partial \psi_i}{\partial n'} \right)_{r=a}, \tag{2.46}$$

a substitution to be made in Eq. (2.45).

For very short wavelengths $n \cdot k > 0$ corresponds to the shadow, and $n \cdot k < 0$ to the illuminated part of the surface. That is, in the shadow ψ_s almost cancels ψ_i, including the gradients; but in the lit region, while $\partial \psi_s/\partial n$ cancels $\partial \psi_i/\partial n$, the wavefunctions themselves are approximately equal on the surface. Hence, Eq. (2.45) can be rewritten approximately as

$$\psi_s(r) \simeq \frac{1}{4\pi} \int_{\text{lit}} \left(\psi_i(r') \frac{\partial G_k(r,r')}{\partial n'} + G_k(r,r') \frac{\partial \psi_i(r')}{\partial n'} \right) dS'$$

$$+ \frac{1}{4\pi} \int_{\text{shadow}} \left(G_k(r,r') \frac{\partial \psi_i(r')}{\partial n'} - \psi_i(r') \frac{\partial G_k(r,r')}{\partial n'} \right) dS' \equiv I_1 + I_2, \quad (2.47)$$

representing the reflected and shadow-forming waves, respectively. The integrals are further reduced by noting that in the far zone

$$k|r - r'| \simeq kr - k\hat{r} \cdot r' + \cdots, \qquad \hat{r} \equiv \frac{r}{|r|}, \quad (2.48)$$

and even there $ka^2 \ll r$. Hence, in this asymptotic region the Green function takes the form

$$\frac{e^{ikR}}{R} \longrightarrow \frac{e^{ikr}}{r} e^{-ik' \cdot r'}, \qquad k' \equiv k\hat{r}. \quad (2.49)$$

Although they are quite similar in appearance, owing to the difference in signs, the integrals of Eq. (2.47) must be treated somewhat differently. The integral I_2 can be converted to a line integral by first defining a vector function

$$A \equiv G_k \nabla' \psi_i - \psi_i \nabla' G_k. \quad (2.50a)$$

Then,

$$\nabla' \cdot A = G_k \nabla'^2 \psi_i - \psi_i \nabla'^2 G_k = 4\pi \psi_i(r') \delta(r - r'), \quad (2.50b)$$

which vanishes unless r lies on the surface r'. The vector A must thus be the curl of some other vector B, and from Stokes' theorem

$$\frac{1}{4\pi} \int_{\text{shadow}} \nabla \times B \cdot dS' = \frac{1}{4\pi} \oint_t B \cdot d\ell', \quad (2.51)$$

where t is the terminator between lit and shadow portions of the surface.

The point here is that B does not depend on the shape of the surface, but only on G, ψ, and the shadow line – and this is all we need know about it.†
Hence, we can substitute for the spherical surface an opaque circular disk concentric with the sphere and perpendicular to k. Approximate cancelation can be achieved by choosing the shadow-forming wave to be the negative of a wave radiated by this disk with the amplitude and phase of ψ_i. (This makes contact with Babinet's principle in classical diffraction theory.) With

---

† This observation that the diffraction pattern is largely an edge effect will be of continued significance in the future.

$\psi_i = Ae^{ikz}$, we have

$$I_2 = -\frac{1}{4\pi} \int_0^{2\pi} d\phi' \int_0^a r' \, dr' \left(\frac{e^{ikr}}{r} e^{-ik'\cdot r'} \frac{\partial}{\partial z'} Ae^{ikz'} - Ae^{ikz'} \frac{e^{ikr}}{r} \frac{\partial}{\partial z'} e^{-ik'\cdot r'} \right).$$

(2.52)

Note that

$$\begin{aligned} \mathbf{k'} \cdot \mathbf{r'} &= kr' \cos(\pi/2 - \theta) \cos(\phi - \phi') + kz' \cos\theta \\ &= kr' \sin\theta \cos(\phi - \phi') + kz' \cos\theta, \end{aligned}$$

(2.53)

and that eventually $z' \to 0$ in the integration. The scattering angle is defined by $\mathbf{k} \cdot \mathbf{k'} = kk' \cos\theta$, and now

$$I_2 = -ikA \frac{e^{ikr}}{4\pi r} (1 + \cos\theta) \int_0^{2\pi} d\phi' \int_0^a r' e^{-ikr' \sin\theta \cos(\phi-\phi')} \, dr'.$$

(2.54)

The angular integration in (2.54) is just an integral representation of a cylindrical Bessel function (Abramowitz and Stegun 1964),

$$J_0(z) = \frac{1}{\pi} \int_0^\pi e^{iz \cos\varphi} \, d\varphi,$$

(2.55a)

and from the same source the remaining integral is another Bessel function,

$$\int_0^a x J_0(yx) \, dx = \frac{a}{y} J_1(ay).$$

(2.55b)

Hence,

$$I_2 = -\frac{1}{2} ika^2 A \left(\frac{1 + \cos\theta}{ka \sin\theta} J_1(ka \sin\theta) \right) \frac{e^{ikr}}{r}.$$

(2.56)

The integral I_1 cannot be treated in the same manner, but is susceptible to a saddle-point evaluation by the method of steepest descents (Appendix E). The point on the sphere where the incident wave is specularly reflected in the direction $\mathbf{k'}$ is a saddle point of the integrand, which is just the behavior expected in geometric optics. That is, a point of stationary phase associates a geometric-optics ray with a saddle point, and hence we can evaluate I_1 to first order geometrically. A more careful verification of these interpretations will be given below, but Fig. 2.3 illustrates that the geometric ray reflected from the surface leads to a phase delay, or phase factor, in the scattering amplitude. In this approximation the scattering is isotropic and, since we know that the total cross section in geometric optics is πa^2, this phase factor must be multiplied by a factor of $a/2$. Asymptotically the scattered wavefunction has the behavior

$$\psi_s(r) \longrightarrow Af(k, \theta) \frac{e^{ikr}}{r},$$

(2.57)

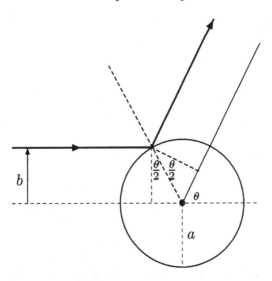

Fig. 2.3. The path difference between a ray reflected from the surface of a sphere and one going through the center is $2a \sin(\theta/2)$.

and thus we identify the short-wavelength scattering amplitude

$$f(k, \theta) \simeq \tfrac{1}{2}ia \left(ie^{-2ika\sin(\theta/2)} - \frac{1 + \cos\theta}{\sin\theta} J_1(ka\sin\theta) \right), \qquad (2.58a)$$

and the differential cross section

$$\frac{d\sigma}{d\Omega} \simeq \frac{a^2}{4} \left[1 + \cot^2(\theta/2) J_1^2(ka\sin\theta) \right]. \qquad (2.58b)$$

These two terms constitute, respectively, the geometric-optics contribution and that of the unavoidable diffraction effects even in the short-wavelength limit. (A cross-term that is negligible in the limit $ka \gg 1$, but which would contribute in the semiclassical domain, has been omitted.)

The total cross section follows from integration over all solid angles, and the first term obviously yields a contribution πa^2. The second integration is readily performed by recalling that, when ka is very large, J_1^2 will be negligible except when $\theta \simeq \epsilon \ll 1$. Hence, for fixed ϵ,

$$I \equiv \lim_{ka\to\infty} \int \cot^2(\theta/2) J_1^2(ka\sin\theta)\, d\Omega \simeq 8\pi \lim_{ka\to\infty} \int_0^{ka\sin\epsilon} \frac{J_1^2(x)}{x}\, dx = 4\pi, \qquad (2.59)$$

and therefore, in the limit $\sigma = 2\pi a^2$, *twice* the geometric cross section. This result is known as the *extinction paradox* in optics, as will be seen again in later chapters, and is clearly a manifestation of the intrinsic wave nature of the scattering, even in this limit.

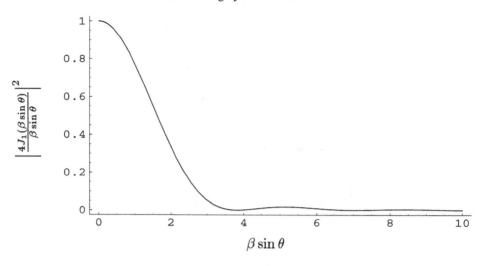

Fig. 2.4. The classical Airy pattern (normalized to unity) showing the forward diffraction peak.

With $\theta \ll 1$, the diffraction contribution to the scattering amplitude (2.58a) can be written

$$f(k, \theta) \simeq ia \frac{J_1(\beta \sin \theta)}{\sin \theta},\qquad(2.60)$$

where again the size parameter is $\beta = ka$. This is the classical *Airy diffraction pattern*, illustrated in Fig. 2.4. About 84% of the intensity goes into the central, or *forward diffraction peak*, which in this approximation has an angular width $\theta \lesssim \beta^{-1}$. In optics this peak is the origin of the *corona* which is often seen around the moon when it is viewed through a significant amount of water vapor. Occasionally it is surrounded by concentric colored rings.

We see from the preceding analysis that, even in the geometric-optics limit there remain residual wave effects, and one never attains true particle-like behavior. Well short of this limit the effects of interference and diffraction are readily apparent, so that a systematic method for describing this mix of ray and wave phenomena is desirable. Just such a technique has long been employed for similar calculations in the quantum theory, as well as in classical electromagnetic theory itself.

Semiclassical approximations

In the sense discussed earlier, we define the realm between pure geometric ray optics and pure wave mechanics as the semiclassical domain. The similarity of the ranges of phenomena in classical and quantum mechanics was first explored in depth by Ford and Wheeler (1959a,b), and the results were extended considerably by Berry and Mount (1972).

In a general medium described by a variable index of refraction $n(r)$ the wave equation (2.3) takes the form

$$\left[\nabla^2 + k^2(r)\right]\psi_k(r) = 0, \tag{2.61}$$

where $k^2(r) = k^2 n^2(r)$. This is identical to the stationary-state Schrödinger equation if we make the following replacements:

$$n(r) \longrightarrow \sqrt{1 - V(r)/E}, \qquad k \longrightarrow \sqrt{2mE/\hbar^2}, \tag{2.62}$$

for particle energy E and the spherically symmetric potential $V(r)$. The so-called *WKB method* (after Wentzel, Kramers, and Brillouin) in quantum mechanics begins with this interpretation of Eq. (2.61), and we shall illustrate the main ideas at first with just a one-dimensional model (e.g., Merzbacher (1970)).

The basic idea follows from the observation that, if V is constant, the solution is a plane wave, so that a tentative guess for a solution if $V(x)$ is slowly varying might be

$$\psi(x) = e^{iu(x)}. \tag{2.63}$$

Substitution into (2.61) provides a nonlinear equation determining $u(x)$,

$$i\frac{d^2u}{dx^2} - \left(\frac{du}{dx}\right)^2 + k^2(x) = 0. \tag{2.64}$$

For a plane wave the second derivative is zero, so one might expect u'' to remain small when V is slowly varying. Indeed, if we set $u'' = 0$ then (2.64) is just the eikonal equation (1.32), and the zeroth-order solution u_0 is simply the integral of $k(x)$. This suggests the use of an iterative technique in Eq. (2.64) for obtaining successive approximations. For example, the first-order solution is, up to a constant,

$$u_1(x) = \pm \int \sqrt{k^2(x) + iu_0''(x)}\,dx = \pm \int \sqrt{k^2(x) \pm ik'(x)}\,dx, \tag{2.65}$$

and the signs are chosen to be consistent with those for the solution u_0. For convergence it is clear that we must have $|k'| \ll k^2$, and with this we can expand the square root in (2.65) to obtain what is called the *WKB approximation*:

$$\psi(x) \simeq \frac{1}{\sqrt{k(x)}} e^{\pm i \int k(x)\,dx}. \tag{2.66}$$

Although we have ignored the constant of integration throughout, leaving it for subsequent normalization, the integral here is still indefinite.

It is important to note that the convergence condition will break down where $k(x)$ vanishes – at the classical turning points $V(x) = E$, say – or where it varies very rapidly, such as in a region where V is very steep. A study of these possible difficulties leads to a set of *connection formulas* at the turning points, which allow the solutions to be continued smoothly across these points where ψ changes qualitatively. A familiar example illustrates the procedure.

Consider the potential barrier of Fig. 2.5 and imagine particles incident from the left with insufficient energy to pass through classically. This is a standard problem in quantum mechanics, but here we presume the WKB approximation to hold in all three of the regions indicated and write the solution to the Schrödinger equation as

$$
\psi(x) = \begin{cases} \dfrac{A}{\sqrt{k(x)}} \exp\left(i \int_a^x k\,dx\right) + \dfrac{B}{\sqrt{k(x)}} \exp\left(-i \int_a^x k\,dx\right), & x < a, \\[3mm] \dfrac{C}{\sqrt{\kappa(x)}} \exp\left(-\int_a^x \kappa\,dx\right) + \dfrac{D}{\sqrt{\kappa(x)}} \exp\left(\int_a^x \kappa\,dx\right), & a < x < b, \\[3mm] \dfrac{F}{\sqrt{k(x)}} \exp\left(i \int_b^x k\,dx\right) + \dfrac{G}{\sqrt{k(x)}} \exp\left(-i \int_b^x k\,dx\right), & b < x, \end{cases}
$$

$$(2.67)$$

where $\kappa(x) \equiv ik(x)$ when $V > E$. The connection formulas here play the role of continuity conditions between different regions and thus lead to relations among the constants. With the definition

$$
\Delta \equiv \exp\left(\int_a^b \kappa(x)\,dx\right), \tag{2.68}
$$

which measures the height and thickness of the barrier as a function of E, we find that

$$
\begin{pmatrix} A \\ B \end{pmatrix} = \frac{1}{2} \begin{pmatrix} 2\Delta + \dfrac{1}{2\Delta} & i\left(2\Delta - \dfrac{1}{2\Delta}\right) \\[3mm] -i\left(2\Delta - \dfrac{1}{2\Delta}\right) & 2\Delta + \dfrac{1}{2\Delta} \end{pmatrix} \begin{pmatrix} F \\ G \end{pmatrix}. \tag{2.69}
$$

From these relations we can calculate the transmission coefficient as the ratio of the transmitted to the incident flux. Under the condition that no wave is incident from the right $G = 0$, and we obtain

$$
T = \frac{|F|^2}{|A|^2} = \frac{4}{\left(2\Delta + \dfrac{1}{2\Delta}\right)^2}. \tag{2.70}
$$

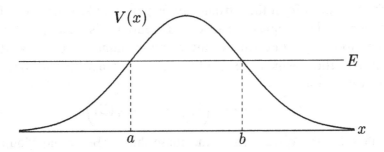

Fig. 2.5. A possible potential barrier for illustrating transmission in the semiclassical approximation.

When the barrier is both wide and high $\Delta \gg 1$; hence a measure of the opacity of the barrier in this case is

$$T \simeq \frac{1}{\Delta^2} = \exp\left(-2\int_a^b \kappa(x)\,dx\right). \tag{2.71}$$

As one might expect, the WKB method becomes a bit more complicated in three dimensions, and is applied primarily to the radial equation appropriate for spherical symmetry (e.g., Eq. (2.11)):

$$\frac{d^2 u_\ell(r)}{dr^2} = -\left(\frac{2m}{\hbar^2}[E - V(r)] - \frac{\ell(\ell+1)}{r^2}\right)u_\ell(r)$$
$$= -k^2(r)u_\ell(r), \tag{2.72}$$

with $u_\ell(0) = 0$ and the radial form of $k(r)$. Owing to the singular nature of the centrifugal potential, a straightforward application of the WKB method to this equation yields very unsatisfactory results – for example, the derived phase shifts exhibit unphysical behavior. The solution to these problems was worked out by Langer (1937) and amounts to making the approximation (2.39) in $k(r)$, which, in the present context, is known as the *Langer modification*. Thus, the WKB approximation for the radial equation is valid for $\ell \gg 1$, which is precisely the domain of semiclassical scattering. We then have

$$u_\ell^{\text{WKB}}(r) = \frac{1}{\sqrt{k_{\text{eff}}(r)}}\exp\left(\pm i\int k_{\text{eff}}(r)\,dr\right), \tag{2.73a}$$

where the effective radial wavenumber is given by

$$k_{\text{eff}}(r) \equiv \left(\frac{2m}{\hbar^2}[E - V(r)] - \frac{(\ell + \frac{1}{2})^2}{r^2}\right)^{1/2}. \tag{2.73b}$$

At this point it is somewhat instructive to outline briefly how one might make the transition from the partial-wave expansion (2.24) (in the WKB approximation) to the completely classical differential cross section of Eq. (1.16). According to (2.27), to do this we must first examine the phase shifts δ_ℓ in this approximation, which are identified from the asymptotic form of the radial wavefunction:

$$u_\ell(r) \xrightarrow[r \gg a]{} \sin\left(k(r) - \frac{\ell\pi}{2} + \delta_\ell(E)\right). \tag{2.74}$$

One verifies that the WKB form of the phase shift is (Berry and Mount 1972)

$$\hbar\delta_\ell^{\text{WKB}} = \int_{r_0}^{\infty} [k_{\text{eff}}(r) - 2mE]\, dr - r_0\sqrt{2mE} + L\pi/2, \tag{2.75}$$

where r_0 is the outermost turning point of the potential. Note that the right-hand side of this equation is entirely classical.

The WKB approximation corresponds to large ℓ, so in the partial-wave sum it is also necessary to employ the asymptotic form of the Legendre polynomial. If neighborhoods of the forward and backward directions are excluded we can write (Szegö 1959)

$$P_\ell(\cos\theta) \simeq \sqrt{\frac{2}{\pi(\ell + \frac{1}{2})\sin\theta}} \cos\left((\ell + \tfrac{1}{2})\theta - \frac{\pi}{4}\right). \tag{2.76}$$

Also, the completeness relation for Legendre polynomials (D.9) allows one to conclude that the term of unity in Eq. (2.24) contributes only in the forward direction, and so can be omitted in this approximation. (See, also, Eq. (2.122) below).

Finally, we must convert the partial-wave sum into an integral, and this cannot be done too cavalierly. In fact, the correct way to do this is through application of Poisson's sum formula (Morse and Feshbach 1953, p. 467):

$$\sum_{\ell=0}^{\infty} g(\ell + \tfrac{1}{2}) = \sum_{m=-\infty}^{\infty} e^{-im\pi} \int_0^{\infty} g(\lambda) e^{2\pi im\lambda}\, d\lambda, \tag{2.77}$$

where the interpolation to continuous values is made through the identification $\lambda = \ell + \frac{1}{2}$. Although this formula will prove very useful subsequently, here we shall make the gross approximation of retaining only the term $m = 0$ to obtain the complete classical limit (rather than a correct semiclassical form). Substitution of the above approximations into Eq. (2.77) for this single term yields, in place of (2.24),

$$f(k,\theta) \simeq \frac{1}{ik\sqrt{2\pi\sin\theta}} \int_0^{\infty} \sqrt{\lambda}\left(e^{i\phi_+} + e^{i\phi_-}\right) d\lambda, \tag{2.78a}$$

where

$$\phi_\pm(\lambda,\theta) \equiv 2\delta_\lambda^{\text{WKB}} \pm (\lambda\theta - \pi/4). \tag{2.78b}$$

Now, much of the point of this exercise lies with the observation that the integrands in (2.78a) oscillate very rapidly in the classical limit, so that the dominant contributions to the integrals arise from those points where the phase is stationary – that is, from points where $d\phi_\pm/d\lambda$ vanish. One readily verifies that these are the points for which

$$2\frac{d\delta_\lambda^{\text{WKB}}}{d\lambda} = \Theta(\lambda), \tag{2.79}$$

where $\Theta(\lambda)$ is the classical deflection angle (1.13) for continuous angular momentum. Thus, by (2.78b), the stationary-phase point $\bar{\lambda}$ is determined by

$$\Theta(\bar{\lambda}) = \pm\theta, \tag{2.80}$$

and from Eq. (1.14) we infer that the signs $\pm$ correspond respectively to repulsive or attractive scattering. The requisite integrals can be evaluated by the well-known method of stationary phase (Appendix E) and we find from (E.17) that, up to overall numerical phase factors,

$$f(k,\theta) \simeq \frac{1}{ik}\left(\frac{\bar{\lambda}}{|\theta'(\bar{\lambda})|\sin\theta}\right)^{1/2} e^{i\left(2\delta_\lambda^{\text{WKB}}(\bar{\lambda})\pm\bar{\lambda}\theta\right)}. \tag{2.81}$$

The argument of the exponential in this equation is just the classical action, verifying that the stationary-phase path satisfies the classical action principle. We therefore affirm that the classical trajectory is associated with paths of stationary phase in the semiclassical approximation.

One now sees that $|f(k,\theta)|^2$ coincides with the classical differential cross section. The work of Ford and Wheeler (1959a) showed that, when several classical paths are present, the scattering amplitude becomes a sum over contributions of the type (2.81), provided that the stationary-phase points are first order and well-separated from one another. Since these paths will lead to interference terms in the cross section, one must average over these oscillations to obtain finally the classical result (1.16).

As an example of this classical transition, one finds for the impenetrable sphere a real saddle point

$$\bar{\lambda} = ka\cos(\theta/2), \tag{2.82}$$

which is also a point of stationary phase. According to the localization principle (2.40), the associated impact parameter is $b = a\cos(\theta/2)$, and from Fig. 2.3 this is exactly the impact parameter of an incident ray reflected in the direction θ. Once again, real stationary-phase (and saddle-) point contributions are equivalent to the WKB approximation. Moreover, stationarity implies that Fermat's variational principle is satisfied, and that is exactly what we expect for a geometric-optics ray.

Because the WKB method of quantum mechanics is so well developed it has been useful to examine the semiclassical approximation to wave scattering in that context – there are many sources available for reference. Having pursued this path, however, let us now observe that the entire formalism can equally well be derived simply by substituting $\psi(x) = e^{iu(x)}$ into the Schrödinger equation and expanding the result in powers of $\hbar$. Exactly the same idea has been pursued in classical electromagnetic theory by Keller *et al.* (1956) without any reference to quantum ideas. The crux of the method is that the scalar Helmholtz equation (2.3) in three dimensions is presumed to have the following asymptotic expansion for very short wavelengths:

$$u(r) \simeq e^{ik\Psi} \sum_{n=0}^{\infty} \frac{v_n(r)}{(ik)^n}. \tag{2.83}$$

Substitution of this expansion into Eq. (2.3) leads to the determining equations

$$(\nabla\Psi)^2 = 1, \tag{2.84a}$$

$$2\nabla v_n \cdot \nabla\Psi + v_n \nabla^2\Psi = -\nabla^2 v_{n-1}, \qquad n = 0, 1, \ldots, \tag{2.84b}$$

where v_{-1} is taken to be zero. No refractive index is included, its possible presence being subsumed in subsequent boundary conditions, so that Eq. (2.84a) is just the eikonal equation of geometric optics that determines the phase function Ψ. Equations (2.84b) form a recursive set for the successive determination of the v_n. Introducing the arc length s along a ray, we see that $\nabla v_n \cdot \nabla\Psi = dv_n/ds$, so that Eqs. (2.84b) are simply ordinary differential equations along the rays. The solutions are readily verified to be

$$v_n(s) = v_n(s_0) \exp\left(-\tfrac{1}{2}\int_{s_0}^{s} \nabla^2\Psi \, ds'\right) - \tfrac{1}{2}\int_{s_0}^{s} \exp\left(-\tfrac{1}{2}\int_{\tau}^{s} \nabla^2\Psi \, ds'\right) \nabla^2 v_{n-1}(\tau)\, d\tau. \tag{2.85}$$

With these results we can now construct asymptotic solutions to various boundary-value problems, and our particular interest here is the impenetrable sphere. Indeed, Keller *et al.* have made this application to obtain the reflected wave resulting from an incident plane wave, and their result through second order is

$$u_{\text{ref}}(r) \simeq -\frac{a}{2r} e^{ik[r-2a\sin(\theta/2)]} \left(1 + \frac{i}{2\beta \sin^3(\theta/2)} + O(\beta^{-2})\right), \tag{2.86}$$

again with $\beta = ka$. The leading term agrees with the geometric-optics contribution to Eq. (2.58a), as it must, and the second provides the first in a series of higher-order corrections. For this expansion in $[2\beta \sin^3(\theta/2)]^{-1}$ to be useful the second term must be much smaller than unity, leading to the following condition for applicability of the WKB approximation in this problem:

$$\theta \gg 2^{1/3}\gamma, \qquad \gamma \equiv (2/\beta)^{1/3}. \tag{2.87}$$

According to Eq. (2.60) classical diffraction theory is valid for $\theta \lesssim \beta^{-1}(\ll \gamma)$, which means that there exists a region from $\theta > \beta^{-1}$ up to the WKB domain where neither approximation applies. This gap has been designated the *penumbra region* and studied in some detail by both Fock (1965) and Nussenzveig (1965). Clearly the gap widens as β decreases. The problem of bridging this gap will be addressed below, but at this point it may be helpful to discuss the origin of the problem briefly in terms of the behavior of the wavefunction around the sphere. This behavior was studied in detail by Nussenzveig (1965), and Fig. 2.6 illustrates various physical domains of interest.

The parallel dashed lines in Fig. 2.6 represent the boundary of the cylindrical shadow predicted by geometric optics when a plane wave is incident from the left. With the center of the sphere as the origin, the coordinates of a point on this shadow boundary are r and $\theta_0 = \sin^{-1}(a/r)$, which provide a reference for delineating other regions. Diffraction into the region behind the sphere, however, illuminates much of this cylinder to some extent, so that the dominant feature of this kind is the deep shadow indicated by the shaded cone and defined by $r \ll a/\gamma$, $(\theta_0 - \theta) \gg \gamma$, where γ is defined in Eq. (2.87). The neighborhood of the deep-shadow boundary within the the same angular sector constitutes the Fresnel region, in which the diffraction pattern is similar to that from a straight edge.

As indicated, the parameter γ defines the extent of the penumbra separating the shadowed and illuminated hemispheres, and therefore extends the geometric boundary between shadow and strong illumination. The latter is the domain $(\theta - \theta_0) \gg \gamma$, where the wavefunction is dominated by the WKB approximation and consists mainly of the incident plus reflected waves, extending all the way to the backward direction.

In the far zone, $r \gg ka^2$, or the Fraunhofer region, we reach the realm of scattering measurements, and centered on the axis are the forward diffraction peak and the Airy pattern. According to the earlier discussion this pattern encompasses the angular region $0 \leq \theta \lesssim \beta^{-1}$, which then continues smoothly into the so-called Fock transition region between the diffraction peak and strong illumination, $\beta^{-1} \lesssim \theta \lesssim \gamma$; this is precisely the semiclassical region.

Finally there is the intermediate, or Fresnel–Lommel region $a/\gamma \ll r \ll$

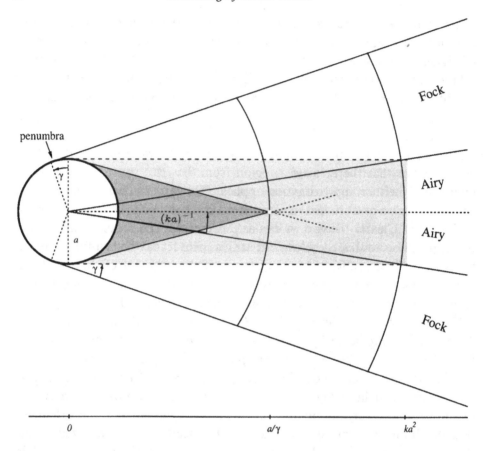

Fig. 2.6. An illustration of the qualitative behavior of the scattered wavefunction for an impenetrable sphere, in which the angles are greatly exaggerated. The physical significances of the various regions are discussed in the text.

ka^2, where $\theta \lesssim \theta_0$. Here one encounters the Poisson bright spot on the axis having the amplitude of the incident wave, which develops into the Poisson cone of apex angle β^{-1} (dotted lines). This cone is surrounded by diffraction rings, and all this merges into the Airy pattern in the far zone.

This brief analysis of what goes on physically in the neighborhood of the spherical surface leads to an understanding of what is seen in the scattering region. It would be quite pleasing, because of its simplicity, if one could describe the scattering completely in terms of rays and classical diffraction, but we can already see that this is not possible over a large range of ka. Short of the geometric optics limit, the major problem to be addressed is that of diffraction at a curved surface, and in Fig. 2.6 we see that corrections to classical diffraction theory originate in the penumbra and eventually manifest themselves in the Fock transition region.† This is an *edge effect* going beyond the simple terminator employed in Eq. (2.51). In addition, geometric optics describes the reflection of a ray by effectively replacing the curved surface by the tangent plane at the point of impact. However, short of the actual limit the curvature produces a noticeable effect and must be treated more carefully. This is again an edge effect and its emergence can be viewed as follows. At high frequencies an incident ray reflected between the backward direction and approximately 90° is described quite well by geometric optics. However, as the scattering angle moves forward the ray approaches grazing incidence and excites a number of more subtle physical phenomena. The wave–surface interaction attains crucial significance for a transparent sphere, which we describe very briefly here and study in considerable detail in the following chapters.

The transparent sphere

The preceding discussion is readily extended to the transparent sphere, Fig. 1.4, where one encounters a much richer array of physical phenomena. We presume the spherical medium to be homogeneous and isotropic, and to be described by an index of refraction n that for the moment is taken to be real and positive. Again, this problem might be applied to sound waves, or to quantum-mechanical scattering from a square well of the kind defined in Eq. (1.22). Our interests throughout the remaining chapters lead us to think of $\psi(r)$ as an electromagnetic wave of given polarization.

The only essential difference from the mathematics of the impenetrable sphere is that now we must require that *both* the wavefunction and its normal derivative be continuous at $r = a$. This is equivalent to matching logarithmic derivatives at the surface. Moreover, since $\psi(r, \theta)$ must be regular at the origin, only $j_\ell(kr)$ can appear in the interior radial wavefunction. When the logarithmic derivative of the latter is matched to that of the expression (2.17) at $r = a$ we obtain

$$nk\frac{j_\ell'(n\beta)}{j_\ell(n\beta)} = \frac{kA_\ell h_\ell^{(1)'}(\beta) + kB_\ell h_\ell^{(2)'}(\beta)}{A_\ell h_\ell^{(1)}(\beta) + B_\ell h_\ell^{(2)}(\beta)}, \tag{2.88}$$

where

$$\beta \equiv ka, \tag{2.89}$$

and in the external medium we set $N_0 = 1$. Once more we solve for the ratio

† Indeed, Fock (1965) appears to be the first to have carried out a careful study of diffraction at a curved edge.

of coefficients and introduce the notation of Eq. (2.18):

$$S_\ell \equiv \frac{A_\ell}{B_\ell} = -\frac{h_\ell^{(2)}(\beta)\,\ln' h_\ell^{(2)}(\beta) - n\ln' j_\ell(\alpha)}{h_\ell^{(1)}(\beta)\,\ln' h_\ell^{(1)}(\beta) - n\ln' j_\ell(\alpha)}, \tag{2.90}$$

where $\ln' f(z)$ denotes a logarithmic derivative with respect to z. Note that once again S_ℓ is a phase factor, since it has the form z^*/z for real n. All the scattering equations pertinent to the impenetrable sphere that do not depend on the explicit form of S_ℓ are still valid, including the partial-wave expansion (2.24).

Long- and short-wavelength limits of the cross sections are similar to those for the impenetrable sphere, and will be studied in detail in the next two chapters. It is already apparent, however, that S_ℓ possesses much more structure than the form in Eq. (2.18). Indeed, it can have poles, and therefore produce resonances in the scattering amplitude – these will be studied in Chapter 7.

2.3 The high-frequency reformulation

We have seen that, when $\beta = ka \gg 1$, the partial-wave expansions are not rapidly convergent, so that a large number of terms in the sums must be retained. Scattering functions are readily summed by computer, of course, and the results nicely displayed graphically. While this is a useful approach for a number of purposes, it is not by itself adequate for developing an understanding of the physical mechanisms underlying the far-zone phenomena. Ideally, it would be desirable to recast the series solutions into a form such that, instead of having to study many terms, the new form would consist of only a few terms containing the essential physics. There is, in fact, a lengthy history of such attempts, which have resulted in a satisfactory solution only relatively recently.

Not long after the series solution to the electromagnetic problem of scattering from a sphere became known (Chapter 3), various efforts to render the series more manageable at short and intermediate wavelengths were made. Both Poincaré (1910) and Nicholson (1910) replaced the series by an integral and then approximated that by means of the calculus of residues. Their analyses were rather elaborate and not valid for all values of θ. The main application of interest at that time was the propagation of radio waves on the Earth's surface, and in 1918 the essential mathematical solution was found by Watson (1918), who effectively introduced the notion of complex angular momentum. That is, he extended the index ℓ of the partial-wave series to complex values, thereby continuing the solution into

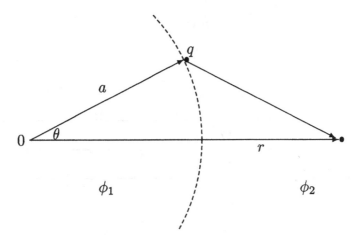

Fig. 2.7. A point charge q located a distance a from the origin. The electrostatic potential has different analytic expressions depending on whether $r < a$ or $r > a$.

the complex plane. The idea was also exploited by Sommerfeld (1949) in similar problems, and the method is best illustrated in an example used by Watson himself.

Consider a point charge located a distance a from the origin, as in Fig. 2.7. Near the origin, $r < a$, we know that expansion of Coulomb's law yields for the electrostatic potential the expression

$$\phi_1(r) = \frac{q}{a} \sum_{\ell=0}^{\infty} (r/a)^\ell P_\ell(\cos\theta), \qquad r < a. \tag{2.91}$$

This series in powers of r/a has a radius of convergence given by the unit circle, and therefore cannot represent the potential outside that circle, in the region $r > a$. Although one can easily find the exterior potential by reformulating the expansion procedure, a more elegant approach leading to a far more powerful tool is to continue ϕ_1 into the complex ℓ-plane.

Recall from Appendix D that

$$P_\ell(-\cos\theta) = (-1)^\ell P_\ell(\cos\theta), \tag{2.92}$$

which is valid *only* for integral values of ℓ (a point often overlooked in earlier treatments). This allows us to rewrite Eq. (2.91) as

$$\phi = \frac{q}{a} \sum_{\ell=0}^{\infty} (-1)^\ell (r/a)^\ell P_\ell(-\cos\theta)$$

$$= -\frac{q}{2ia} \int_C (r/a)^\lambda P_\lambda(-\cos\theta) \frac{d\lambda}{\sin(\pi\lambda)}, \tag{2.93}$$

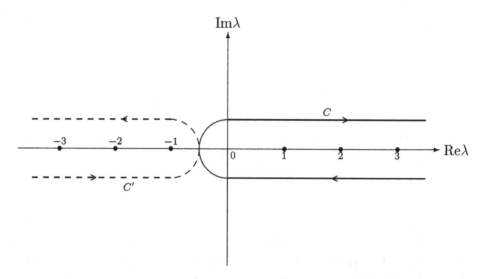

Fig. 2.8. An illustration of analytic continuation into the complex ℓ-plane. The contour C is appropriate for the integral in Eq (2.93), and C' is for that in Eq. (2.94).

where the contour C is that illustrated in Fig. 2.8, and the second line follows from Cauchy's residue theorem. Note that $P_\ell(z)$ is regular in the entire z-plane cut along the negative real axis $(-\infty, -1)$ and is an entire function of ℓ. For *any* ℓ, $P_{-\ell-1}(z) = P_\ell(z)$. Consequently, the contour C can be swung around through $180°$ by noting that convergence is guaranteed everywhere according to the asymptotic formula of Eq. (D.40b). We denote the new contour by C' and now have

$$-\frac{1}{2ia}\int_{C'}(r/a)^\lambda P_\lambda(-\cos\theta)\frac{d\lambda}{\sin(\pi\lambda)} = -\frac{1}{a}\sum_{\ell=1}^{\infty}(r/a)^{-\ell}(-1)^\ell P_{-\ell}(-\cos\theta)$$

$$= -\frac{1}{a}\sum_{\ell=0}^{\infty}(r/a)^{-\ell-1}(-1)^{\ell+1}P_{-\ell-1}(-\cos\theta), \quad (2.94)$$

with the help of the residue theorem. Using the above identities for the Legendre polynomials we find that

$$\phi_2(r) = \frac{q}{r}\sum_{\ell=0}^{\infty}(a/r)^\ell P_\ell(\cos\theta), \qquad r > a, \quad (2.95)$$

which is the desired analytic continuation of the potential function for large r.

With the extensive development of scattering theory in quantum mechanics in the 1950s interest in the Watson transformation was renewed, principally through the work of Regge (e.g., de Alfaro and Regge (1965)). Although we are primarily interested in classical scattering here, it is instructive to digress for a moment to review very briefly some of these developments. In the quantum theory of scattering $S_\ell(k) = \exp[2i\delta_\ell(k)]$ can be continued into the complex k-plane where it is an analytic function of k. For cutoff potentials, such as a hard sphere, it is meromorphic in k, and for exponentially decaying functions it is meromorphic in a cut plane. Under certain constraints the poles of $S_\ell(k)$ at $k = i\eta$ represent bound states of energy $-\eta^2/(2m)$. Complex poles – again under certain constraints – correspond to resonances, although they must be reasonably close to the real axis to be observable. These comments refer to a fixed value of ℓ, and we draw attention to them because a similar analysis of the S-function will prove fruitful in Chapter 7. In addition, this strongly suggests that new insights might arise from a continuation into the complex ℓ-plane as well.

Continuing with the quantum-mechanical discussion, we consider the partial amplitudes $f_\ell(E)$ as functions of energy, and for fixed E the poles of $f_\ell(E)$ as a function of ℓ are called *Regge poles*, located at $l = \eta_1(E), \eta_2(E), \cdots$. If E is now varied continuously each $\eta_i(E)$ traces out a path in the ℓ-plane, called a *Regge trajectory*. For $E = E_0 < 0$ suppose that there exists an s-wave bound state, which then corresponds to a Regge pole $\ell = \eta(E)$ at $\ell = 0$. Continue increasing E, keeping it negative, so that the pole moves to the right on the real ℓ-axis. During this movement $\eta(E)$ has no physical significance, but when it reaches $\ell = 1$ with $E < 0$ a p-wave bound state is inferred. A continuation of this process will pick out successive poles and bound states until E reaches a threshold, at which point $\eta(E)$ leaves the real axis. As long as it remains *near* the real axis it defines a resonance each time its real part takes on an integer value (Chapter 7), but eventually $\eta(E)$ will no longer describe observable phenomena. For cutoff potentials $\eta(E)$ moves off to infinity, whereas for exponentially decaying potentials it doubles back and moves off into the left half-plane.

In the quantum theory of scattering it becomes necessary at high energies to continue the amplitude into the complex angular-momentum plane to study the full scattering amplitude as a function of momentum transfer q^2. Following Watson, one rewrites the partial-wave series as a contour integral:

$$f(E, q^2) = \frac{1}{2i} \int_C \frac{(2\ell + 1)f_\ell(E)P_\ell(-\cos\theta)}{\sin(\pi\ell)}\, d\ell, \qquad (2.96)$$

where the contour C is identical with that in Fig. 2.8. Analytic continuation is effected by deforming the contour to the vertical and closing it to the left in two quarter circles to avoid the bound-state poles. In doing so the

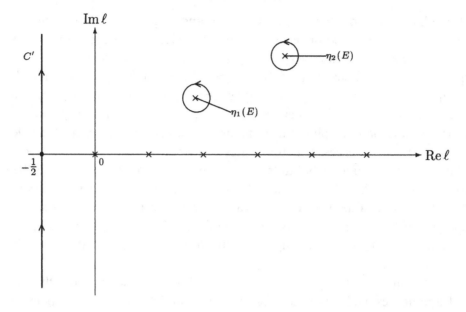

Fig. 2.9. The result of applying the Watson transformation to the scattering amplitude of Eq. (2.96), with correspoinding analytic expressions given by Eq. (2.97). The
significance of the abscissa $\ell = -\frac{1}{2}$ lies with the behavior of the radial function near
the origin. A portion of the Regge trajectory is indicated by the dashed line.

contour 'sweeps across' a number of Regge poles, thereby picking up the
corresponding residues. At least for an exponentially decaying potential, the
quarter circles do not contribute and there is only a finite number of Regge
poles, as shown in Fig. 2.9. The result of this continuation is the *Watson
transformation* of the scattering amplitude:

$$f(E, q^2) = \frac{1}{2i} \int_{C'} (2\ell + 1) f_\ell(E) \frac{P_\ell(-\cos\theta)}{\sin(\pi\ell)} \, d\ell + \pi \sum_{i=1}^{n} \beta_i(E) P_{\eta_i(E)}(-\cos\theta),$$

(2.97)

where the $\beta_i(E)$ are the residues of the integrand at the Regge poles with
the Legendre function separated out. The first term on the right-hand side
of Eq. (2.97) is referred to as the *background integral*, and the second is
the *residue series*. Together they provide an expression for $f(E, q^2)$ that is a
regular function of $\cos\theta$ in the cut plane $1 \le \cos\theta < \infty$.

These brief remarks suggest a powerful tool for studying the partial-wave
series of classical scattering as well. Application of the Watson transformation to the classical partial-wave series (2.24) takes the form

$$\sum_{\ell=0}^{\infty} g\left(\ell + \tfrac{1}{2}, \beta\right) = \frac{1}{2} \int_C g(\lambda, \beta) \frac{e^{-i\pi\lambda}}{\cos(\pi\lambda)} \, d\lambda,$$

(2.98)

where C is again the contour of Fig. 2.8, and the *interpolating function* $g(\lambda, \beta)$ reduces to $g(\ell + \frac{1}{2}, \beta)$ at the physical values $\lambda = \ell + \frac{1}{2}$ $(\ell = 0, 1, 2, \ldots)$. We also note that this is not the only way to represent the partial-wave series, for one could also employ the Poisson representation of Eq. (2.77). Indeed, this modification of the Watson transformation will become quite important in later chapters.

The equivalence of the two representations is easily demonstrated, so that (2.98) shows one way that the Poisson representation can be continued into the complex λ-plane. The denominator in Eq. (2.98) can be expanded into separate convergent series on each branch of the contour C:

$$\frac{1}{\cos(\pi\lambda)} = 2e^{\pm i\pi\lambda} \sum_{m=0}^{\infty} (-1)^m e^{\pm i2\pi m\lambda}, \qquad (2.99)$$

depending on whether λ is on the upper or lower branch. The separate integrals can now be brought onto the real axis, in opposite directions, for there are no longer any singularities in the integrands. Some judicious shifting and sign-changing of the indices yields precisely the result of Eq. (2.77). With these ideas in mind, we now return to a further analysis of scattering from an impenetrable sphere in the high-frequency domain $ka \gg 1$, which is analogous to the high-energy problem above.

The impenetrable sphere revisited

A dimensionless formulation is quite useful at this point, so we first rewrite the content of Eqs. (2.18) and (2.24) as

$$F(\beta, \theta) \equiv f(k, \theta)/a$$
$$= \frac{i}{\beta} \sum_{\ell=0}^{\infty} \left(\ell + \frac{1}{2}\right) [1 - S_\ell(\beta)] P_\ell(\cos\theta), \qquad (2.100)$$

and it is also convenient to express the optical theorem of Eq. (2.29) in terms of the scattering efficiency Q to be employed later for light scattering:

$$Q(\beta) \equiv \frac{\sigma_{tot}(\beta)}{\pi a^2} = \frac{4}{\beta} \operatorname{Im} F(\beta, 0). \qquad (2.101)$$

That is, the total cross section is normalized to the classical hard-sphere value.

It is already clear from Eq. (2.92) that a full calculation of $F(\beta, \theta)$ demands consideration of separate angular regions, owing to the singularity in $P_\nu(z)$ at $z = -1$. As noted there, the conventional choice of cut in the complex z-plane is $(-\infty, -1)$. Along with separate treatment of the forward and backward scattering directions, our calculational strategy involves a number of other points worth noting at this time. After moving into the complex λ-plane and locating the singularities of $S_\ell(\beta)$, which are all simple poles in this model, we then attempt to choose contours of integration prudently so as to take advantage of possible critical points and reduce the results to a handful of contributions. Because our interests lie with large $\beta = ka$ and $\ell = \lambda - \frac{1}{2}$, we look for saddle points, say. The calculations are heavily based on and guided by the asymptotic behavior of the Hankel and Legendre functions in the complex index plane, which is slightly complicated and leads to considerable algebra. Rather than detract from the physical aspects of the exposition, we shall refer constantly to the asymptotic formulas in Appendices A–D and thus provide the tools for the reader to verify as desired the details leading to the results. Much of the omitted mathematical detail can be found in the work of Nussenzveig (1965, 1988).

In the region $(0 < \theta \le \pi)$ we can apply the straightforward Watson transformation in the form of Eq. (2.98) and Fig. 2.8, but in doing so it is necessary to account for the Regge poles of the ensuing integrand. According to (2.18) these are located at the zeros in the λ-plane of the cylindrical Hankel function $H_\lambda^{(1)}(\beta)$, and their distribution has been determined for $\beta \gg 1$ by Streifer and Kodis (1964):†

$$\lambda_n = \beta + e^{i\pi/3}x_n\gamma^{-1} + e^{i2\pi/3}\frac{x_n^2}{60}\gamma + \frac{x_n^3}{1400}\left(1 - \frac{10}{x_n^3}\right)\gamma^3 + O(\gamma^5), \qquad (2.102)$$

where γ is the penumbra parameter $(2/\beta)^{1/3}$. We adopt the notations x_n and x_n' for the positive zeros of the Airy function and its derivative, respectively:

$$\text{Ai}(-x_n) = 0, \qquad \text{Ai}'(-x_n') = 0, \qquad (2.103)$$

with tables of x_n and x_n' provided in Appendix B. Figure 2.10 exhibits this pole distribution, along with a number of contours to be utilized presently.

Deformation of the contour C of Fig. 2.8 into that labeled Γ_1 in Fig. 2.10 results in the expected sum of residues in the first quadrant, along with a background integral,

$$F(\beta, \theta) = F_{\text{S}}(\beta, \theta) - \frac{1}{2\beta}\int_{\Gamma_1}[1 - S(\lambda, \beta)]P_{\lambda - 1/2}(-\cos\theta)\frac{\lambda\,d\lambda}{\cos(\pi\lambda)}, \qquad (2.104)$$

where $S(\lambda, \beta) = -H_\lambda^{(2)}(\beta)/H_\lambda^{(1)}(\beta)$ is the extension of $S_\ell(\beta)$ into the complex

---

† The order symbol 'big O' is defined loosely here as follows: $\phi(x) = O(\psi)$ means that there exists a constant A such that $\phi(x) \le A|\psi|$ as $x \to \infty$.

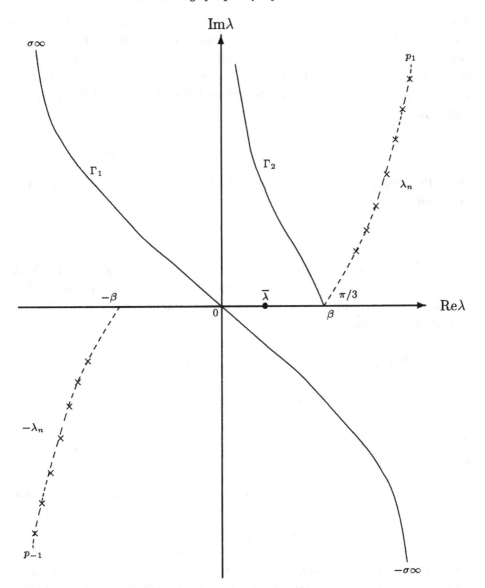

Fig. 2.10. The zeros of $H_\lambda^{(1)}(\beta)$, and hence the Regge poles $\times$ for the impenetrable sphere for $\lambda \gg 1$ are located on the curves p_1 (λ_n) and p_{-1} ($-\lambda_n$) The contour Γ_1 is symmetric about the origin and can also be deformed to pass through the saddle point $\bar\lambda$.

angular-momentum plane, we have employed the identity (D.6a), and the residue sum is

$$F_S(\beta,\theta) = -\frac{2\pi i}{\beta} \sum_{n=1}^{\infty} \lambda_n r_n \frac{e^{i\pi\lambda_n}}{1+e^{i2\pi\lambda_n}} P_{\lambda-1/2}(-\cos\theta). \qquad (2.105)$$

The quantity r_n is the residue of $S(\lambda, \beta)$ at λ_n in the first quadrant, and the sum is arranged in increasing values of $|\lambda_n|$. Since the contour Γ_1 has deliberately been chosen symmetric about the origin, the first contribution (unity) to the background integral vanishes by symmetry. In the second term we introduce one of the auxiliary functions $Q_\nu^{(1)}(\cos\theta)$ defined in Eq. (D.38) and write

$$P_{\lambda-1/2}(-\cos\theta) = ie^{-i\pi\lambda}P_{\lambda-1/2}(\cos\theta) - 2i\cos(\pi\lambda)Q_{\lambda-1/2}^{(1)}(\cos\theta). \quad (2.106)$$

From the reflection properties of the Hankel functions we obtain the symmetry relation

$$S(-\lambda, \beta) = e^{-2i\pi\lambda}S(\lambda, \beta), \quad (2.107)$$

which allows us to conclude that the first term on the right-hand side of (2.106) also leads to an odd integrand, according to the identity (D.31). The remaining integral can then be written in two possible ways and is called the reflection contribution:

$$\begin{aligned}
F_R(\beta, \theta) &= -\frac{i}{\beta}\int_{\sigma\infty}^{-\sigma\infty} S(\lambda, \beta)Q_{\lambda-1/2}^{(1)}(\cos\theta)\lambda\,d\lambda \\
&= \frac{i}{\beta}\int_{\sigma\infty}^{0} S(\lambda, \beta)P_{\lambda-1/2}(-\cos\theta)\tan(\pi\lambda)e^{-i\pi\lambda}\lambda\,d\lambda, \quad (2.108)
\end{aligned}$$

the contours being depicted in Fig. 2.10 as parts of Γ_1. Hence, the scattering amplitude in this region is

$$F(\beta, \theta) = F_R(\beta, \theta) + F_S(\beta, \theta), \quad 0 < \theta \leq \pi, \quad (2.109)$$

and the interpretation of F_S will be given presently.

In the forward direction ($0 \leq \theta < \pi$) it proves more useful to employ the Poisson sum formula, or modified Watson transformation (2.77), to obtain a representation manifestly regular at $\theta = 0$. The resulting amplitude will be written

$$F(\beta, \theta) = F_D(\beta, \theta) + F_E(\beta, \theta) + \tilde{F}_S(\beta, \theta), \quad (2.110)$$

whose interpretation will also be forthcoming presently. The importance of the term F_E is that it leads to a *uniform* approximation over the full range of scattering angles, providing a smooth transition from forward to backward regions.

With the aid of the identity (2.107) we write for the amplitude in the region $0 \leq \theta < \pi$

$$F(\beta,\theta) = \frac{i}{\beta} \sum_{m=-\infty}^{\infty} (-1)^m \int_0^{\infty} [1 - S(\lambda,\beta)] P_{\lambda-1/2}(\cos\theta) e^{i2m\pi\lambda} \lambda \, d\lambda$$

$$= \frac{i}{\beta} \sum_{m=0}^{\infty} (-1)^m \left(\int_{-\infty}^0 \left[e^{i2\pi\lambda} - S(\lambda,\beta) \right] P_{\lambda-1/2}(\cos\theta) e^{i2m\pi\lambda} \lambda \, d\lambda \right.$$

$$\left. + \int_0^{\infty} [1 - S(\lambda,\beta)] P_{\lambda-1/2}(\cos\theta) e^{i2m\pi\lambda} \lambda \, d\lambda \right). \quad (2.111)$$

For θ not too near 0 or π, the first line of this expression can be interpreted in the semiclassical limit as a superposition of contributions from 'pseudoclassical paths' associated with all values of λ (Berry and Mount 1972). Appealing to the asymptotic properties of the integrand described in Appendices C and D, we can shift the path of integration in the first integral of the second line to the positive imaginary axis, from $i\infty$ to 0. In doing so it is convenient to split the integrand into two terms,

$$\left[e^{i2\pi\lambda} - 1 \right] + [1 - S(\lambda,\beta)], \quad (2.112)$$

and then perform the m-sum over the first term. The integrals can now be split into integrals containing the S-function and those that do not, and, after some algebra, we make the identifications

$$F_{\mathrm{D}}(\beta,\theta) = \frac{i}{\beta} \int_0^{\beta} P_{\lambda-1/2}(\cos\theta) \lambda \, d\lambda - \frac{2i}{\beta} \int_0^{i\infty} \frac{e^{i2\pi\lambda}}{1 + e^{i2\pi\lambda}} P_{\lambda-1/2}(\cos\theta) \lambda \, d\lambda, \quad (2.113)$$

$$F_{\mathrm{E}}(\beta,\theta) = -\frac{i}{\beta} \int_{i\infty}^{\beta} S(\lambda,\beta) P_{\lambda-1/2}(\cos\theta) \lambda \, d\lambda + \frac{i}{\beta} \int_{\beta}^{\infty} [1 - S(\lambda,\beta)] P_{\lambda-1/2}(\cos\theta) \lambda \, d\lambda$$

$$\equiv F_{\mathrm{e}-}(\beta,\theta) + F_{\mathrm{e}+}(\beta,\theta), \quad (2.114)$$

$$\tilde{F}_{\mathrm{S}}(\beta,\theta) = \frac{i}{\beta} \sum_{m=1}^{\infty} (-1)^m \int_C [1 - S(\lambda,\beta)] P_{\lambda-1/2}(\cos\theta) e^{i2m\pi\lambda} \lambda \, d\lambda$$

$$= -\frac{2\pi}{\beta} \sum_{n=1}^{\infty} \lambda_n r_n \frac{e^{i2\pi\lambda_n}}{1 + e^{i2\pi\lambda_n}} P_{\lambda-1/2}(\cos\theta). \quad (2.115)$$

In $F_{\mathrm{e}-}$ the integration must stay to the left of the poles at λ_n, as on the contour Γ_2 in Fig. 2.10, whereas in Eq. (2.115) the contour C in the first line runs from $i\infty$ to 0 to ∞. Again the asymptotic properties of the integrand deduced from the appendices allow the path C to be closed and the integral evaluated as a sum of residues.

Before proceeding with the actual asymptotic evaluation of the scattering amplitude in the two regions let us briefly examine the physical interpretation of the residue series (2.105) and (2.115). In Eq. (D.39a) we express the

Legendre function in terms of two auxiliary functions as

$$P_{\lambda-1/2}(\cos\theta) = Q^{(1)}_{\lambda-1/2}(\cos\theta) + Q^{(2)}_{\lambda-1/2}(\cos\theta), \qquad (2.116a)$$

whose asymptotic behavior away from the forward and backward directions is

$$Q^{(1,2)}_{\lambda-1/2}(\cos\theta) \xrightarrow[|\lambda|\sin\theta\gg1]{} \frac{e^{\mp i(\lambda\theta-\pi/4)}}{\sqrt{2\pi\lambda\sin\theta}}. \qquad (2.116b)$$

In this limit, then, $Q^{(1)}_{\lambda-1/2}$ represents a clockwise angular traveling wave and $Q^{(2)}_{\lambda-1/2}$ is counter-clockwise. In the residue series they give rise to terms like $\exp(i\lambda_n\theta)$ and $\exp[i\lambda_n(2\pi-\theta)]$, and hence the Regge poles of the impenetrable sphere are associated with *surface wave* contributions. Moreover, from Eq. (2.102) we see that the amplitude of the surface waves is proportional to $\exp(-\frac{1}{2}\sqrt{3}x_n\theta/\gamma)$, so that the series converge rapidly for $\theta \gg \gamma$.

The residue contributions make contact with the geometric theory of diffraction proposed by Keller (1962) and exploited by Rubinow (1961) in this problem, in which, in addition to reflected and refracted rays, diffraction at high frequencies can take place via *diffracted rays* generated at the surface. That is, tangentially incident rays run along the surface for a way and, owing to exponential damping, shed energy tangentially as they go, in a kind of a pinwheel effect.† These diffracted rays are thus associated with surface waves, which have also been called *creeping modes* (Franz 1957) – their reality is strikingly evident in *Schlieren* photographs of sound pulses diffracted by cylinders (Neubauer 1973), as well as in photographs of backscattering of light from water droplets (Chapter 6). Once again the importance of edge phenomena becomes evident.

The generic approach in evaluating the above contour integrals is exemplified by setting $\lambda = \beta\cos\psi$ in the first line of Eq. (2.108), a substitution dictated entirely by the Debye asymptotic formulas for the Hankel functions, Eq. (C.16), in which we employ the notation of (C.23). We also require the expansion (D.40a) for $Q^{(1)}_{\lambda-1/2}(\cos\theta)$, and then a change of variables from λ to ψ is made. As a consequence the first integral in (2.108) is mapped into the form

$$F_R(\beta,\theta) \simeq e^{i\pi/4}\left(\frac{\beta}{2\pi\sin\theta}\right)^{1/2}\int_{\Gamma_\psi} A(\psi,\theta,\beta)e^{2i\beta\delta(\psi,\theta)}\,d\psi, \qquad (2.117)$$

† Indeed, Kepler advocated such an effect as being the mechanism producing the rainbow.

the quintessential form for which the method of steepest descents was designed. Indeed, this particular integral is used as an example in our brief exposition of the method in Appendix E, where the functions A and δ are defined. The exponential function contains a saddle point at $\psi_0 = \frac{1}{2}\theta$, which corresponds exactly to that of Eq. (2.82).

As explained in Appendix E, the procedure now is to deform the contour Γ_ψ so as to make it pass through ψ_0 along the steepest path, and thus extract the dominant contribution to F_R. Although $\beta \gg 1$ here, it is clear that $\lambda \lesssim \beta$ so that $\cos\psi$ is real. For the asymptotic expansions to be valid, however, λ cannot be too small, so that ψ is reasonably close to $\pi/2$. These expectations are met in the saddle-point evaluation, and the reflection amplitude is simply the WKB expansion:

$$F_R(\beta,\theta) \simeq -\frac{1}{2}e^{ik[(r-2a\sin(\theta/2)]}\left(1 + \frac{i}{2\beta\sin^3(\theta/2)} + \frac{2+3\cos^2(\theta/2)}{[2\beta\sin^3(\theta/2)]^2} + O(\beta^{-2})\right),$$

$$(2.118)$$

which is just a continuation of the series in (2.86). Just like for (2.86), the domain of usefulness is given by Eq. (2.87). Thus, this expression is valid in the interval $2\gamma \lesssim \theta \leq \pi$.

Equation (2.102) indicates that the poles of the S-function are $O(\beta)$, so that the residue contributions again correspond to $|\lambda| \gg 1$. From the asymptotic expansions (C.17) of the Hankel functions, along with the corresponding expansions for the Airy functions, the residues r_n in Eq. (2.105) are found to be

$$r_n = \left(\frac{H_\lambda^{(2)}(\beta)}{\partial H_\lambda^{(1)}(\beta)/\partial\lambda}\right)_{\lambda=\lambda_n}$$

$$\simeq \frac{e^{-i\pi/6}}{2\pi a_n'^2 \gamma}\left(1 + \frac{1}{30}x_n e^{i\pi/3}\gamma^2 - \frac{3}{1400}x_n^2 e^{2i\pi/3}\gamma^4 + O(\gamma^6)\right), \quad (2.119)$$

where $a_n' \equiv \mathrm{Ai}'(-x_n)$ and Ai' is the derivative of the Airy function. (Tables of a_n' and $a_n \equiv \mathrm{Ai}(-x_n')$ are provided in Appendix B.) We also employ the uniform asymptotic approximation $\mathscr{P}(\pi - \theta, \lambda_n)$ to the Legendre function given by (D.42), and for $\pi - \theta \gg \beta^{-1}$ extract the dominant contributions for the residue series:

$$F_S(\beta,\theta) \simeq \frac{1}{2}e^{-i5\pi/12}\left(\frac{\gamma}{\pi\sin\theta}\right)e^{i\beta\theta}\sum_n a_n'^{-2}\exp\left(-\frac{1}{2}(\sqrt{3}-i)x_n\theta/\gamma\right), \quad (2.120)$$

describing decaying surface waves as advertised above. That is, each surface wave propagates with wavevector

$$k_n \simeq k + \frac{x_n}{2\gamma a} + i\frac{\sqrt{3}}{2}\frac{x_n}{2\gamma a}, \tag{2.121}$$

extending the field forward into the shadow and lightening it (see Fig. 2.6). This is the phenomenon of *shadow scattering* that is often discussed in nuclear-scattering problems, and is just a euphemism for diffraction.

The preceding expressions for the scattering amplitude in the backward direction have been labeled the *outer representation* by Nussenzveig (1988), whereas the evaluation of (2.110) will constitute the *inner representation*. The diffraction amplitude (2.113) is independent of $S(\lambda, \beta)$ and refers to the sphere only through its radius a; it is basically Fresnel's *blocking amplitude*. Fresnel viewed diffraction as a blocking effect, in which the interference pattern of secondary waves emerging via Huygens' principle is perturbed owing to blocking of a portion of the incident wavefront by the obstacle.

The first integral in F_D has a structure that is elucidated by reference to the completeness relation (D.9) for Legendre polynomials. With $x' = 1$,

$$\sum_{\ell=0}^{\infty}(\ell + \tfrac{1}{2})P_\ell(x) = \delta(x - 1), \tag{2.122}$$

so that the sum is nonzero only in the forward direction. Now, for large β this is also the structure of our integral,

$$\begin{aligned}I(\theta, \beta) &\equiv \int_0^\beta \lambda P_{\lambda-1/2}(\cos\theta)\,d\lambda\\ &= \beta^2 \int_0^1 x P_{\beta x-1/2}(\cos\theta)\,dx,\end{aligned} \tag{2.123}$$

so that, for $\beta \gg 1$, we expect the dominant contributions to come from the region $\theta \simeq 0$. Indeed, for $\theta = 0$ the exact value of I is $\tfrac{1}{2}\beta^2$, so we are looking for corrections when θ is very small. If we split the integration interval into two at $x = \varepsilon \ll 1$, then the first piece will be of order ε^2, and the second can now be evaluated by substituting the uniform approximation $\mathscr{P}(\lambda, \theta)$ of Eq. (D.42). For $\beta \gg 1$, $\varepsilon \ll 1$, and $\theta \simeq 0$, we find for the diffraction amplitude the expression

$$\begin{aligned}F_D(\beta, \theta) &\simeq \frac{i}{\beta}\int_0^\beta P_{\lambda-1/2}(\cos\theta)\lambda\,d\lambda\\ &\approx i\beta^2\left(\frac{\theta}{\sin\theta}\right)^{1/2}\left(\frac{J_1(\beta\theta)}{\beta\theta} + \mathrm{O}(\theta)\right),\end{aligned} \tag{2.124}$$

which is just the Airy diffraction pattern.

The residue series (2.115) for $\tilde{F}_S$ is evaluated in much the same way as for F_S, but with the replacement of θ by $2\pi - \theta$. This contribution thus describes surface waves having already taken a half-turn or more around the sphere, so they are strongly damped and will not be considered further.

Finally we turn to Eq. (2.114) and the contribution $F_E(\beta, \theta)$ from the edge domain. This provides the important transition between diffraction and reflection and provides the key to a uniform description over the full range of scattering angles. We shall address the physical interpretation of the two separate contributions presently, but for now we simply note that F_{e_+} corresponds to impact parameters $\lambda/k \geq a$ and is thus an *above-edge* amplitude. Similarly, F_{e_-} refers to $\lambda/k \leq a$ and is a *below-edge* amplitude. In a manner similar to that leading to (2.118), we study F_{e_+} by first introducing new variables φ and z, as suggested by the uniform expansion parameters (C.23):

$$\lambda = \beta \cosh \varphi, \qquad z \equiv \left[\tfrac{3}{2} \beta (\varphi \cosh \varphi - \sinh \varphi) \right]^{2/3}. \qquad (2.125)$$

Now employ the uniform asymptotic forms of the Hankel functions from Eqs. (C.20) and (C.21) for $1 - S(\lambda, \beta)$, and again insert the asymptotic form $\mathscr{P}$ of Eq. (D.42) for the Legendre function. It turns out that the most convenient integration variable is z, so we next make that change. The above-edge amplitude can then be written

$$F_{e_+}(\beta, \theta) = -e^{-i\pi/6} \int_0^\infty H_+(z, \varphi) \mathscr{P}(\theta, \beta \cosh \varphi) \varphi^{-1} \cosh \varphi \sqrt{z} \, dz, \qquad (2.126)$$

where

$$H_+(z, \varphi) \equiv \frac{\mathrm{Ai}(z) - e^{i\pi/6} \sigma_+(z, \varphi) \, \mathrm{Ai}'(z)}{\mathrm{Ai}(z e^{i2\pi/3}) + e^{-i\pi/6} \sigma_+(z, \varphi) \, \mathrm{Ai}'(z e^{i2\pi/3})}, \qquad (2.127a)$$

in which

$$\sigma_+(z, \varphi) \equiv \frac{5}{24\beta} \frac{e^{-i\pi/6}}{\sqrt{z} \sinh \varphi} \left(\frac{3}{5} + [3(\varphi \coth \varphi - 1)]^{-1} - \coth^2 \varphi \right). \qquad (2.127b)$$

From the asymptotic expansions of the Airy functions in Appendix B we find the behavior

$$1 - S(\lambda, \beta) \simeq e^{i\pi/3} H_+(z, \varphi) \xrightarrow[z \gg 1]{} i \exp\left(-\tfrac{4}{3} z^{3/2} \right), \qquad (2.128)$$

which will prove significant presently; mathematically it provides an effective cutoff for the integral.

The below-edge amplitude can be evaluated in an almost-similar way by first defining a new variable ψ and re-defining z (because now $\lambda \leq \beta$):

$$\lambda = \beta \cos \psi, \qquad z \equiv e^{i\pi/3} \left[\tfrac{3}{2} \beta (\sin \psi - \psi \cos \psi) \right]^{2/3}. \qquad (2.129)$$

Thus,

$$F_{e-}(\beta, \theta) = -e^{-i\pi/3} \int_0^c H_-(z, \psi) \mathscr{P}(\theta, \beta \cos \psi) \psi^{-1} \cos \psi \sqrt{z} \, dz, \qquad (2.130)$$

where c is an effective cutoff, and

$$H_-(z, \psi) \equiv \frac{\mathrm{Ai}(z) - e^{-i\pi/6}\sigma_-(z, \psi)\,\mathrm{Ai}'(z)}{\mathrm{Ai}(ze^{-i2\pi/3}) + e^{i\pi/6}\sigma_-(z, \psi)\,\mathrm{Ai}'(ze^{-i2\pi/3})}, \qquad (2.131a)$$

in which

$$\sigma_-(z, \psi) \equiv -\frac{5}{24\beta} \frac{e^{-i\pi/3}}{\sqrt{z}\sin\psi} \left(\frac{3}{5} + [3(\psi \cot \psi - 1)]^{-1} + \cot^2 \psi \right). \qquad (2.131b)$$

In this case

$$S(\lambda, \beta) \simeq e^{-i\pi/3} H_-(z, \psi) \xrightarrow[z \gg 1]{} i \exp\left(-\tfrac{4}{3}z^{3/2}\right). \qquad (2.132)$$

We have noted here that the integrand is dominated by this behavior, so that the direction of fastest decrease is along the positive real z-axis. The result (2.130) is valid on $0 \leq \theta \leq \gamma$.

Nussenzveig (1988) has provided a careful proof that these asymptotic expressions of Eqs. (2.109) and (2.110), the outer and inner representations, yield a *uniform* description of the scattering from an impenetrable sphere over the range $0 \leq \theta \leq \pi$. The optical theorem (2.101), which requires only the inner representation, then gives us a very accurate expression for the normalized cross section:

$$\begin{aligned}
Q(\beta) \simeq 2 + \frac{1}{6\beta^2} &- \frac{8\pi}{\beta^2} \,\mathrm{Im}\left(\sum_n \lambda_n r_n \frac{e^{i2\pi\lambda_n}}{1 + e^{i2\pi\lambda_n}} \right) \\
&+ \frac{4}{\beta} \,\mathrm{Re}\left(e^{i\pi/6} \int_0^\infty H_-(z, \psi)\psi^{-1} \cos \psi \sqrt{z} \, dz \right. \\
&\left. + e^{i\pi/3} \int_0^\infty H_+(z, \varphi)\varphi^{-1} \cosh \varphi \sqrt{z} \, dz \right), \qquad (2.133)
\end{aligned}$$

verifying in no uncertain terms the limiting value for $\beta \to \infty$ mentioned earlier following Eq. (2.59). Owing to the behavior of the integrands for large z, we are able to extend the upper limits of integration to infinity. If one requires the high accuracy needed to demonstrate uniformity, then the integrals must be evaluated numerically in any event. However, for most discussions, further approximation is possible.

In particular, when β is very large and θ is very small one can invert the definitions of z and expand both φ and ψ in powers of z to reduce the complexity of the integrals and further approximate F_E. This process is slightly tedious, but has been carried through by Nussenzveig (1988) and leads to an expansion in powers of θ/γ with coefficients

$$M_m \equiv e^{i\pi/3} \int_0^\infty \frac{\mathrm{Ai}(z)}{\mathrm{Ai}(ze^{i2\pi/3})} z^m \, dz + e^{i(2m+1)\pi/3} \int_0^\infty \frac{\mathrm{Ai}(z)}{\mathrm{Ai}(ze^{-i2\pi/3})} z^m \, dz. \quad (2.134)$$

The first few coefficients are

$$\begin{aligned} M_0 &= (1.255\,13)e^{i\pi/3}, \\ M_1 &= (0.532\,291)e^{i2\pi/3}, \\ M_2 &= 0.067\,717. \end{aligned} \quad (2.135)$$

In this approximation the surface-wave contribution is negligible and, with $\gamma = (2/\beta)^{1/3}$, Eq. (2.133) reduces to

$$Q(\beta) \simeq 2 + 2\,\mathrm{Re}\left[M_0\gamma^2 + \frac{8}{15}M_1\gamma^4 + \left(\frac{4}{175}M_2 + \frac{23}{1680}\right)\gamma^6 + O(\gamma^8) \right]. \quad (2.136)$$

The physical interpretation

In the backward direction the reflection amplitude F_R is essentially the WKB approximation to geometric optics, and the interpretation of the residue contribution has already been discussed in terms of surface waves. In the forward direction the diffraction amplitude F_D is seen to comprise the classical Airy pattern, and again there is a surface-wave contribution. This sharp forward peak is an axial focusing effect arising from the azimuthal degeneracy due to spherical symmetry, and will become quite important in the backward direction as well (Chapter 6). In both regions the surface waves are rapidly damped at high frequencies.

The non-classical feature that is truly in need of interpretation is the edge amplitude F_E, which describes what is going on in the penumbra region of Fig. 2.6, and hence in the Fock transition regions in the far field. This contribution originates with glancing or near-glancing rays, yielding what might be called high-angle diffraction – and more. The physical picture is clarified considerably by appealing to some quantum mechanical ideas – though what we conclude has nothing to do with quantum theory.

Recall the relationship between the wave equation and a potential model as described by Eqs. (2.61) and (2.62). For the impenetrable sphere $V(r)$ is infinite at the surface, requiring the wavefunction to vanish there, and is zero for $r > a$. The effective potential, however, contains the centrifugal term

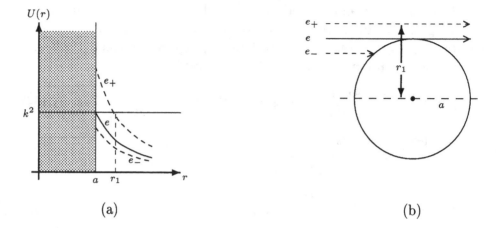

(a)   (b)

Fig. 2.11. A potential model of the edge effects. (a) The effective potential λ^2/r^2 for the impenetrable sphere for three different values of the angular momentum: above edge (e_+) with turning point r_1, tangential (e), and below edge (e_-). (b) The corresponding incident rays.

as well, and in the Langer sense of Eq. (2.73b) is λ^2/r^2 in the semiclassical regime. Because of the context we adopt units such that $\hbar = 2m = 1$, and thus the energy is just k^2. This is illustrated schematically in Fig. 2.11, along with the corresponding incident rays with impact parameters $b(\lambda) = \lambda/k$ near the edge of the sphere. The impact parameter $b = a$ defines the edge angular momentum $\lambda_e = \beta$, for which the incident *edge rays* are tangential to the surface. One sees that, for above-edge rays, $\lambda > \lambda_e$, a turning point develops at $r = r_1$. However, these are wave phenomena, so that, if the energy is sufficiently close to the top of the centrifugal barrier, the ray should be able to penetrate to the surface and produce a physical effect, effectively a nonlocal interaction. The propensity for such penetration is governed by the barrier-penetration factor of Eq. (2.71), keeping in mind Eq. (2.73b). In the present case the penetration factor is

$$T = \exp\left[-2\int_a^{r_1}\left(\frac{\lambda^2}{r^2} - k^2\right)^{1/2} dr\right] = \exp\left[-2\int_\beta^\lambda \left(\frac{\lambda^2}{y^2} - 1\right)^{1/2} dy\right]. \quad (2.137)$$

The integral $I(\lambda)$ in T can be done exactly,

$$-I(\lambda) = \sqrt{\lambda^2 - \beta^2} - \lambda \ln\left(\frac{\lambda + \sqrt{\lambda^2 - \beta^2}}{\beta}\right), \quad (2.138)$$

and, with Eq. (2.125) and the relations of hyperbolic functions to logarithms, we find that

$$T = \exp\left(-\tfrac{4}{3}z^{3/2}\right). \tag{2.139}$$

This is identical with the behavior (2.128) of H_+, and thus it is the tunneling of the above-edge rays through the barrier that accounts for the rapid decay of the integrand in the amplitude $F_{e_+}(x, \theta)$. Moreover, if we expand the hyperbolic functions in powers of $(\lambda - \beta)/\beta$, which is small in the high-frequency region, then one readily shows that

$$z \simeq (3/2)^{2/3}\gamma(\lambda - \beta), \qquad \lambda - \beta \ll \beta. \tag{2.140}$$

This result allows us to make an important connection with the localization principle. Namely, the transmissivity of the barrier is appreciable whenever the argument of the exponential becomes less than unity, or for

$$\beta \lesssim \lambda \lesssim \lambda_+ \equiv \beta + c_+\beta^{1/3}, \qquad c_+ = \tfrac{1}{2} = O(1). \tag{2.141}$$

Similarly, the reflected amplitude will be affected by the centrifugal barrier in the below-edge situation when

$$\lambda_- \equiv \beta - c_-\beta^{1/3} \lesssim \lambda \lesssim \beta, \qquad c_- = O(1). \tag{2.142}$$

These inequalities serve to define the *edge domain* $\lambda_- \lesssim \lambda \lesssim \lambda_+$, or, for physical values

$$\ell_- \lesssim \ell \lesssim \ell_+, \tag{2.143}$$

which provides the semiclassical transition region between geometric optics and pure wave phenomena.

In summary, the preceding equations provide a uniform description of scattering of a plane wave from an impenetrable sphere in terms both of classical ray reflection and of classical diffraction, along with correction terms and the connection between the two domains in the form of edge phenomena. The major effects of curvature of the surface are found in the amplitude F_E, which has been shown to have a satisfying physical interpretation. The above-edge contribution F_{e_+} arises from rays just above tunneling through the barrier to the surface, along complex paths. For $\theta \gtrsim \gamma$ it is the same order of magnitude as the blocking amplitude. The below-edge amplitude F_{e_-} describes the effects of curvature on reflection, and the near-glancing rays can be said to produce *anomalous reflection*. Indeed, this is the origin of the reflected wave as θ increases. In Fig. 2.12 we have plotted the exact partial-wave sum for Q along with the reduction (2.136) of the uniform asymptotic approximation, over a range of reasonably small values of β.

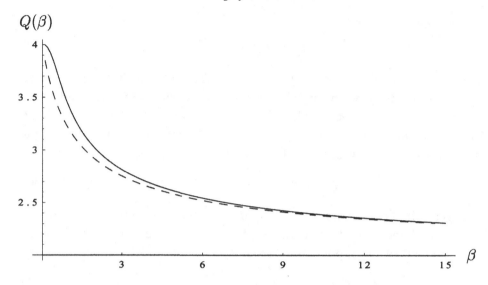

Fig. 2.12. The extinction efficiency $Q(\beta)$ over the range $0 < \beta \leq 15$. The exact result obtained numerically from the partial-wave (solid line) is compared with the approximation given by Eq. (2.136) (dashed line).

The good agreement down almost to $\beta = 10$ is truly impressive,† and makes one confident enough of the techniques to next extend them to a study of transparent spheres and the scattering of electromagnetic waves.

We remark in passing that these techniques of analytic continuation into the complex angular-momentum plane have also been applied to a scalar plane wave scattering from a transparent sphere (Nussenveig 1969a, b), which we shall not consider here because the electromagnetic generalization will be treated in great detail in Chapter 5. They have also been applied to the scattering of an acoustic plane wave from a fluid-loaded elastic sphere (Williams and Marston 1985). The analysis in the latter case differs significantly from the above in that the quantity of interest is the total pressure surrounding the sphere, which is the real part of

$$p(r,\theta,t) = e^{-i\omega t} \sum_{\ell=0}^{\infty} i^\ell (2\ell + 1) \left[j_\ell(kr) + A_\ell h_\ell^{(1)}(kr) \right] P_\ell(\cos\theta), \qquad (2.144)$$

where the A_ℓ are functions of the parameters of the problem.

---

† The uniform approximation (2.133) appears identical to the exact result down to $\beta \simeq 1$! Detailed numerical comparisons with the Mie theory have been carried out by Nussenzveig and Wiscombe (1991).

3

Scattering of electromagnetic waves from spherical targets

Up to this point our discussion has avoided any detailed reference to the underlying physical mechanisms producing scattered waves. One most often thinks of scattering in terms of particles undergoing elastic collision, either with other particles or with macroscopic objects such as walls. In the microscopic domain this view is even extended to light when the photon picture is appropriate. As we have seen, geometric optics permits a similar interpretation wherein rays mimic the scattering behavior of particles.

When one can no longer neglect the wave nature of light, however, this intuitive view of the scattering process is not entirely adequate and we are compelled to look beyond it for physical origins. On a microscopic level an electron, atom, or molecule will couple to an incident electromagnetic wave in an oscillating fashion, such that it re-radiates in all directions, producing a 'scattered' wave. It is tempting to amplify this mechanism to macroscopic targets, since they are certainly composed of these constituents, but to do so in detail would be a forbidding problem in many-body physics. A more appropriate macroscopic approach might be to envision the incident fields as inducing electric and magnetic multipoles that oscillate, and hence radiate, while maintaining definite phase relations with the incident wave. When the wavelength of the incident radiation is long compared with the dimensions of the scatterer, only the lowest-order multipoles will be important, and the re-radiation process can be approximated by invoking electric and magnetic dipoles. However, at the opposite extreme of short wavelengths the situation is not as simple, for many other varied physical phenomena can be present.

In this monograph we are primarily concerned with electromagnetic waves incident on macroscopic targets described by an index of refraction, so there is no need to address microscopic origins in any detail. Rather, we shall take the view that the energy incident upon a target in the form of a wave is thereby 'extinguished' by that object and reappears in the

63

form of a scattered wave less any energy absorbed by the target. Thus, our major goal is to describe and understand the process of extinction = scattering + absorption. By 'scattered wave' here we mean *all* incident radiation interacting with the scatterer and being re-radiated by the medium in the form of an electromagnetic field corresponding to a wave with speed c/n, where n is the appropriate index of refraction inside or outside the target. This is essentially the extinction theorem of Ewald (1916) and Oseen (1915). An exceptionally simple and clear derivation of this theorem that considers only macroscopic fields is provided by Ballenegger and Weber (1999).

The notion of extinction becomes more intuitive when it is applied to a system that itself constitutes a medium composed of nominally spherical particles, say. An example is visible light passing through a dust cloud. Absorption in such a system is then attributed both to scattering of light by the particles and to energy loss due to internal particle mechanisms. Such systems will be examined briefly in Chapter 8.

3.1 The electromagnetic formalism

For the most part we shall restrict ourselves to electromagnetic fields in homogeneous, isotropic, and nonmagnetic dielectric media, and it is most useful to adopt Gaussian units. The fields will then satisfy Maxwell's equations in the form

$$\nabla \times E = -\frac{1}{c}\frac{\partial B}{\partial t}, \quad \nabla \times H = \frac{4\pi}{c}J + \frac{1}{c}\frac{\partial D}{\partial t} \tag{3.1a}$$

$$\nabla \cdot B = 0, \quad \nabla \cdot E = 4\pi\rho, \tag{3.1b}$$

where ρ and J are the total charge and current densities, respectively. To these one must append the constitutive relations

$$J = \sigma E, \quad D = \varepsilon E, \tag{3.2}$$

with the electric conductivity σ and permitivity ε being the material parameters of the medium. The presumption that the medium is nonmagnetic means that there is no difference between H and B, so we shall choose the former notation unless the restriction is lifted.

As already discussed in the preceding chapter, it is sufficient for our purposes to presume that the general fields can be synthesized by means of Fourier transformation, and thus we employ a harmonic time dependence $e^{-i\omega t}$ for real frequencies ω. Equations (3.1a) thus reduce to

$$\nabla \times E = \kappa_2 H, \quad \nabla \times H = -\kappa_1 E, \tag{3.3}$$

where the magnitude of the propagation vector, or wavenumber, K in the medium is given by

$$K^2 \equiv -\kappa_1\kappa_2 = \frac{\omega^2}{c^2}\left(\varepsilon + i\frac{4\pi\sigma}{\omega}\right) = \frac{\omega^2}{c^2}n^2, \qquad (3.4)$$

and $n \equiv m + i\kappa$ is the complex index of refraction of the medium. Generally n is a function of the frequency and satisfies a set of dispersion relations, as developed at the beginning of Chapter 2. Conventionally the wavevector in vacuum is written as $k \equiv \omega/c$, in which case $K = nk$, $\kappa_1 = ikn^2$, and $\kappa_2 = ik$. Note that for infinite, or perfect, conductivity, $n^2 \rightarrow i\infty$. Implicit in our notation is that the dimensions characterizing the material specimen to be studied are well above the submicrometer range – say, $> 100\,\text{Å}$ ($= 10^{-8}\,\text{m} = 10^{-2}\,\mu\text{m}$) – so that a description in terms of bulk optical constants such as $n = \sqrt{\mu\varepsilon}$ is valid (e.g., Huffman (1988)).

Our major interest resides in materials with $m = \text{Re}\,n > 1$, and when m is very large the material is optically very dense. Complex n implies an absorbing material, as already noted in conjunction with Eq. (2.4). However, even for strong absorption κ is not exceedingly large; with our choice of sign in the harmonic time variation, absorption of power corresponds to $\kappa > 0$. As examples, $n = 1.27 + 1.37i$ describes iron at a wavelength of $0.42\,\mu\text{m}$; platinum at $\lambda = 10\,\mu\text{m}$ has the refractive index $n = 37 + 41i$; and, at $\lambda = 10\,\text{cm}$ the refractive index of water is $n = 8.9 + 0.69i$. At optical frequencies, however, water has an almost-real index of $n = m \simeq 1.33$, or about $\frac{4}{3}$. Clearly most materials are dispersive to some extent.

Multipole expansions

It follows directly from the source-free Maxwell equations that, at the interface between two media, the tangential components of E and H are continuous, as are the normal components of H and $D = \varepsilon E$. In Chapter 1 we extracted several consequences of these boundary conditions, such as the laws of reflection and refraction, and shall not repeat the exercise here. Rather, we concentrate for the time being on empty space and note that, if we eliminate E from Maxwell's equations in this case, we obtain the equivalent set

$$(\nabla^2 + k^2)H = 0, \qquad \nabla \cdot H = 0,$$

$$E = \frac{i}{k}\nabla \times H. \qquad (3.5)$$

Alternatively, elimination of H yields the same set with E and H interchanged and $i \rightarrow -i$. In either case the Cartesian components of E and H

all satisfy the scalar Helmholtz equation, and it remains only to ensure that the solutions satisfy the divergence equations.

General solutions are constructed with the aid of the vector identity

$$\nabla^2(r \cdot A) = r \cdot (\nabla^2 A) + 2\nabla \cdot A. \tag{3.6}$$

This implies that the scalars $r \cdot E$ and $r \cdot H$ also satisfy the Helmholtz equation:

$$(\nabla^2 + k^2)(r \cdot E) = 0,$$
$$(\nabla^2 + k^2)(r \cdot H) = 0. \tag{3.7}$$

With these expressions one readily constructs multipole expansions from solutions to the scalar wave equation.

We follow Jackson (1975) in defining pure electric and magnetic multipole fields of order (ℓ, m), where $\ell = 0, 1, 2, \ldots,$ $-\ell \leq m \leq \ell$, although we presume magnetic monopoles to be nonexistent. As we have seen in Chapter 2, the scalar equations (3.7) in spherical coordinates separate into radial and angular equations, with solutions in terms of spherical Bessel functions and spherical harmonics, respectively. Our subsequent interest lies with the scattering problem, so we adopt solutions to the radial equation as linear combinations of the form suggested by Eq. (2.13):

$$g_\ell(kr) = A_\ell^{(1)} h_\ell^{(1)}(kr) + A_\ell^{(2)} h_\ell^{(2)}(kr), \tag{3.8}$$

a mixture of incoming and outgoing spherical waves. A magnetic multipole field (TE) of order (ℓ, m) is defined as

$$E_{\ell m}^{(M)} \equiv g_\ell(kr) \, LY_\ell^m(\theta, \phi),$$
$$H_{\ell m}^{(M)} \equiv -\frac{i}{k}\nabla \times E_{\ell m}^{(M)}, \tag{3.9}$$

where $L \equiv -i(r \times \nabla)$ is related to the angular part of ∇^2 and is effectively an angular-momentum operator. In like manner, let $f_\ell(kr)$ be another linear combination of spherical Hankel functions similar to that of (3.8) and define an electric multipole field (TM) of order (ℓ, m) as

$$H_{\ell m}^{(E)} \equiv f_\ell(kr) \, LY_\ell^m(\theta, \phi),$$
$$E_{\ell m}^{(E)} \equiv \frac{i}{k}\nabla \times H_{\ell m}^{(E)}. \tag{3.10}$$

The electric field $E_{\ell m}^{(M)}$ has only transverse components with respect to r, as does the magnetic field $H_{\ell m}^{(E)}$, because $r \cdot L = 0$ – hence the appellations TE and TM, respectively.

Because these multipole fields form complete sets, they can be combined into general solutions to Maxwell's equations. To do this, define vector spherical harmonics

$$X_{\ell m}(\theta, \phi) \equiv \frac{1}{\sqrt{\ell(\ell+1)}} LY_\ell^m(\theta, \phi), \qquad (3.11)$$

for $\ell = 0, 1, 2, \ldots, -\ell \leq m \leq \ell$, which is taken to be identically equal to zero if $\ell = 0$. Various properties of the spherical Bessel functions and the various angular functions, including vector spherical harmonics, are to be found in Appendices A and D, respectively.

With this notation the general solutions are†

$$E = \sum_{\ell,m} \left(\frac{i}{k} a_E(\ell, m) \nabla \times f_\ell(kr) X_{\ell m} + a_M(\ell, m) g_\ell(kr) X_{\ell m} \right), \qquad (3.12a)$$

$$H = \sum_{\ell,m} \left(a_E(\ell, m) f_\ell(kr) X_{\ell m} - \frac{i}{k} a_M(\ell, m) \nabla \times g_\ell(kr) X_{\ell m} \right), \qquad (3.12b)$$

where f_ℓ and g_ℓ are the linear combinations of spherical Hankel functions introduced above. The coefficients $a_E(\ell, m)$ and $a_M(\ell, m)$ specify the amounts of electric and magnetic multipole fields in a given multipole, respectively, and are determined from boundary conditions. In this respect the following identities are useful:

$$a_M(\ell, m) g_\ell(kr) = \frac{k}{\sqrt{\ell(\ell+1)}} \int Y_\ell^{m*}(\theta, \phi) r \cdot H \, d\Omega, \qquad (3.13a)$$

$$a_E(\ell, m) f_\ell(kr) = -\frac{k}{\sqrt{\ell(\ell+1)}} \int Y_\ell^{m*}(\theta, \phi) r \cdot E \, d\Omega. \qquad (3.13b)$$

They are obtained by means of the orthogonality relations found in Appendix D. Explicit expressions for the first few vector spherical harmonics can also be found there.

Plane waves

We recall that a plane wave is partially described by its wavevector k, defining the direction of propagation. An expansion in spherical coordinates for a scalar wave is provided by Bauer's formula, Eq. (2.16), which can also be written in the form

$$e^{ik \cdot r} = \sum_{\ell=0}^{\infty} i^\ell \sqrt{4\pi(2\ell+1)} j_\ell(kr) Y_\ell^0(\theta, \phi). \qquad (3.14)$$

---

† Similar expansions when sources are present have been developed by Lambert (1978). In addition, there are also other forms in which one can write the solutions to the vector wave equation (e.g., Stratton (1941)), and one such is employed in Chapter 8 below.

In the same way one can write a similar expansion for an electromagnetic plane wave, after first accounting for its polarization.

By definition a transverse plane wave satisfying Maxwell's equations has fields

$$E(r,t) = E_0\epsilon_1 e^{ik\cdot r - i\omega t},$$
$$H(r,t) = H_0\epsilon_2 e^{ik\cdot r - i\omega t},$$
$$= \frac{c}{\omega} k \times E(r,t), \tag{3.15}$$

such that $(\epsilon_1, \epsilon_2, k)$ form a complete orthonormal triad.† That is, the set (ϵ_1, ϵ_2) is comprised of *polarization vectors* spanning a plane perpendicular to the propagation direction of the transverse wave. More generally, suppose that

$$E = (\epsilon_1 E_1 + \epsilon_2 E_2) e^{i(k\cdot r - \omega t)}, \tag{3.16}$$

where E_1 and E_2 are complex numbers. If E_1 and E_2 have the same phase the wave is *linearly polarized* in a fixed direction, making an angle $\theta \equiv \tan^{-1}(E_2/E_1)$ with ϵ_1, and with amplitude $(E_1^2 + E_2^2)^{1/2}$. Should E_1 and E_2 have different phases the wave is *elliptically polarized*, and if this phase difference is $\pi/2$ it is *circularly polarized* with $|E_1| = |E_2| = E_0$. In the latter case

$$E = E_0(\epsilon_1 + i\epsilon_2)e^{i(k\cdot r - \omega t)}, \tag{3.17}$$

and the constant-magnitude E at a fixed point in space rotates in a circle at frequency ω. The rotation is either counter-clockwise (left) or clockwise (right), often referred to as positive and negative *helicity*, respectively. For circular polarization we can equally well employ the circular basis

$$\epsilon_\pm \equiv \frac{1}{\sqrt{2}}(\epsilon_1 \pm i\epsilon_2). \tag{3.18}$$

Now let the circularly polarized plane wave in (3.17) propagate along the positive z-axis. From the orthogonality relations in Appendix D one verifies that Eqs. (3.13) yield

$$a_{\mathrm{M}}^\pm(\ell, m) = i^\ell [4\pi(2\ell + 1)]^{1/2} \delta_{m,\pm 1}, \qquad a_{\mathrm{E}}^\pm(\ell, m) = \mp i a_{\mathrm{M}}^\pm(\ell, m), \tag{3.19}$$

for the two polarizations. Because the plane wave is everywhere finite we have taken $f_\ell(kr) = g_\ell(kr) = j_\ell(kr)$. The general solutions (3.12) now provide the

---

† Note that, in a general medium of refractive index n, the wavenumber $k = n\omega/c$.

multipole expansion of the circularly polarized electromagnetic plane wave:

$$E = E_0 \sum_{\ell=1}^{\infty} i^\ell \sqrt{4\pi(2\ell+1)} \left(j_\ell(kr)X_{\ell,\pm 1} \pm \frac{1}{k}\nabla \times j_\ell(kr)X_{\ell,\pm 1} \right), \qquad (3.20a)$$

$$H = E_0 \sum_{\ell=1}^{\infty} i^\ell \sqrt{4\pi(2\ell+1)} \left(-\frac{i}{k}\nabla \times j_\ell(kr)X_{\ell,\pm 1} \mp i j_\ell(kr)X_{\ell,\pm 1} \right). \qquad (3.20b)$$

Note that only the values $m = \pm 1$, $\ell \geq 1$ occur, corresponding to positive and negative helicity, respectively.

Of particular interest for the scattering problem, as a convention, is the plane wave propagating along the positive z-axis with electric vector linearly polarized in the positive x-direction. In this scenario the identities (3.13) yield

$$a_M(\ell,m) = \tfrac{1}{2} i^\ell \sqrt{4\pi(2\ell+1)}\,(\delta_{m,1} + \delta_{m,-1}),$$
$$a_E(\ell,m) = \tfrac{1}{2i} i^\ell \sqrt{4\pi(2\ell+1)}\,(\delta_{m,-1} - \delta_{m,1}), \qquad (3.21)$$

and the fields are

$$E = \hat{x}E_0 e^{ikz} = \frac{1}{2}E_0 \sum_{\ell=1}^{\infty} i^\ell \sqrt{4\pi(2\ell+1)}$$
$$\times \left(j_\ell(kr)(X_{\ell,1} + X_{\ell,-1}) + \frac{1}{k}\nabla \times j_\ell(kr)(X_{\ell,-1} - X_{\ell,1}) \right), \qquad (3.22a)$$

$$H = \hat{y}H_0 e^{ikz} = -\frac{i}{2}H_0 \sum_{\ell=1}^{\infty} i^\ell \sqrt{4\pi(2\ell+1)}$$
$$\times \left(j_\ell(kr)(X_{\ell,-1} - X_{\ell,1}) + \frac{1}{k}\nabla \times j_\ell(kr)(X_{\ell,-1} + X_{\ell,1}) \right). \qquad (3.22b)$$

These multipole fields are also called partial wave expansions.

Scattered fields

Let a circularly polarized plane wave propagating along the positive z-axis be incident upon a localized target. The fields outside the object are then sums of incident and scattered fields:

$$E = E_{in} + E_{sc},$$
$$H = H_{in} + H_{sc}. \qquad (3.23)$$

If the target geometry is arbitrary the complete solution for the scattered fields is very difficult to obtain, which is one reason why we focus generally on spherical targets. For the moment, however, we shall seek a number of results independent of any particular symmetry; the difficulty with nonspherical targets comes in determining the coefficients from boundary conditions. In addition, we presume the frequency of scattered radiation to be the same as that of the incident fields; this excludes Raman scattering, for example.

Equations (3.20) describe the incident fields, and suggest the form of the scattered fields as well. However, now the radial functions must be spherical Hankel functions of the first kind so as to describe outgoing spherical waves, and the scattered fields can be written as

$$E_{sc} = \sum_{\ell,m} i^\ell \sqrt{4\pi(2\ell+1)} \left(\frac{1}{k} a(\ell,m) \nabla \times h_\ell^{(1)}(kr) + b(\ell,m) h_\ell^{(1)}(kr) \right) X_{\ell m}, \quad (3.24a)$$

$$H_{sc} = \sum_{\ell,m} i^\ell \sqrt{4\pi(2\ell+1)} \left(-\frac{i}{k} b(\ell,m) \nabla \times h_\ell^{(1)}(kr) - ia(\ell,m) h_\ell^{(1)}(kr) \right) X_{\ell m}. \quad (3.24b)$$

We now denote by $a(\ell,m)$ and $b(\ell,m)$ the amounts of electric (TM) and magnetic (TE) multipole, respectively. These coefficients are to be determined by consideration of boundary conditions on the surface of the scatterer.

The incident plane wave has a time-averaged energy flux given by the Poynting vector

$$S = \frac{1}{2}\frac{c}{4\pi} E \times H^* = \frac{c}{8\pi k} \sqrt{\varepsilon} |E_0|^2 k, \quad (3.25a)$$

and the energy density

$$u = \frac{\varepsilon}{8\pi} |E_0|^2. \quad (3.25b)$$

In general the target can have a complex refractive index, and thus will absorb some of this incident energy. The scattered and absorbed powers are calculated by integrating the appropriate components of the appropriate Poynting vector over a spherical surface of radius r_0 surrounding the target. That is,

$$P_{sca} = -\frac{cr_0^2}{8\pi} \text{Re} \int E_{sca} \cdot (\hat{r} \times H^*_{sca}) \, d\Omega, \quad (3.26a)$$

$$P_{abs} = \frac{cr_0^2}{8\pi} \text{Re} \int E_{tot} \cdot (\hat{r} \times H^*_{tot}) \, d\Omega, \quad (3.26b)$$

where the scattered fields are given by Eqs. (3.24) and the total fields are the sums of these and the incident fields (3.20). Upon substitution of these fields into (3.26) one can evaluate the integrals with the aid of the orthogonality relations for the vector spherical harmonics (Appendix D) and the Wronskian for the spherical Bessel functions (Appendix A). The total scattering cross section is defined as P_{sca} divided by the incident flux $c/(4\pi)$, and we find that

$$\sigma_{sca} = \frac{2\pi}{k^2} \sum_{\ell,m} (2\ell + 1) \left[|a(\ell,m)|^2 + |b(\ell,m)|^2 \right], \qquad (3.27)$$

with a similar expression for σ_{abs}. The total, or *extinction cross section* is defined as the sum of the two,

$$\sigma_{ext} \equiv \sigma_{sca} + \sigma_{abs}, \qquad (3.28)$$

and calculation yields

$$\sigma_{ext} = -\frac{2\pi}{k^2} \sum_{\ell,m} (2\ell + 1) \,\mathrm{Re}[a(\ell,m) + b(\ell,m)], \qquad (3.29)$$

depending on the state of circular polarization. The absorption cross section is obtained from (3.28).

In the special case that the target possesses spherical symmetry, only the values $m = \pm 1$ contribute to the scattered fields (3.24). The cross sections then simplify to

$$\sigma_{sca} = \frac{2\pi}{k^2} \sum_{\ell} (2\ell + 1) \left[|a(\ell)|^2 + |b(\ell)|^2 \right], \qquad (3.30a)$$

$$\sigma_{abs} = \frac{2\pi}{k^2} \sum_{\ell} (2\ell + 1) \left[\tfrac{1}{2} - |a(\ell) - \tfrac{1}{2}|^2 - |b(\ell) - \tfrac{1}{2}|^2 \right], \qquad (3.30b)$$

$$\sigma_{ext} = \frac{2\pi}{k^2} \sum_{\ell} (2\ell + 1) \,\mathrm{Re}[a(\ell) + b(\ell)]. \qquad (3.30c)$$

Generally the scattered radiation is elliptically polarized; only if $a(\ell,\pm 1) = b(\ell,\pm 1)$ for all ℓ is it circularly polarized. These points will become clearer as we develop the electromagnetic scattering theory further and specialize completely to spherical scattering.

3.2 Vector scattering theory

Envision a plane wave propagating in direction k_{in} with electric field E_{in} and incident upon an arbitrary but localized target. The wave is scattered by the target and in the far field is described by k_{sc} and E_{sc}. Owing to the wave nature of the process the propagation vector k_{sc} is not fixed, but varies with angles (θ and ϕ) with respect to a fixed reference system. For given angles the pair of vectors (k_{in}, k_{sc}) defines the *scattering plane*, and Θ denotes the *scattering angle* between k_{in} and k_{sc} in this plane. The geometry is depicted in Fig. 3.1, and in what follows there should be no confusion if we write

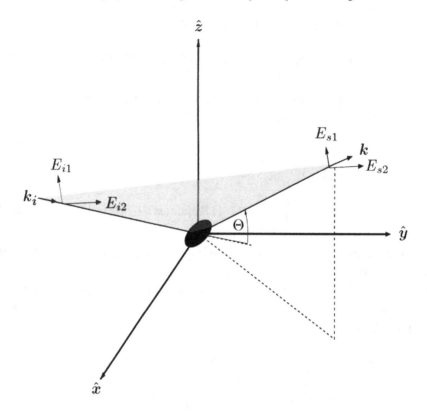

Fig. 3.1. Scattering of a linearly polarized plane wave from an arbitrary target at scattering angle Θ.

$k = k_{sc}$. Note, however, that the scattering plane is undefined for $\theta = 0$ and π, when the two vectors are collinear.

To describe the process in more detail we introduce unit vectors $\hat{e}_1$ and $\hat{e}_2$ perpendicular and parallel to the scattering plane, respectively, such that $\hat{e}_1 \times \hat{e}_2 = k/k$ both for incident and for scattered waves. Thus, $\hat{e}_1$ is the same for both waves, whereas $\hat{e}_2$ differs between the two. As indicated in Fig. 3.1, one can decompose E_{in} and E_{sc} into components along $\hat{e}_1$ and $\hat{e}_2$, with a similar decomposition for the magnetic field vectors.

The linearity of Maxwell's equations implies that the scattering process mixes the components E_1 and E_2 of Eq. (3.16) linearly. Such a process is described most conveniently in a matrix context, allowing us to write for the far field amplitudes

$$\begin{pmatrix} E_2 \\ E_1 \end{pmatrix}_{sc} = \frac{e^{ikr}}{ikr} \begin{pmatrix} S_2 & S_3 \\ S_4 & S_1 \end{pmatrix} \begin{pmatrix} E_2 \\ E_1 \end{pmatrix}_{in}, \tag{3.31}$$

defining the *scattering–amplitude matrix* S. Note that this expression incorporates the intrinsic asymptotic behavior of the spherical Hankel function $h_\ell^{(1)}(kr)$, and that the harmonic time dependence of the incident wave is transferred directly to the scattered wave.†

It is apparent that the scenario outlined here is inherently based on the notion of linear polarization, in that the unit vectors ($\hat{e}_1$ and $\hat{e}_2$) are just the polarization vectors (ϵ_1, ϵ_2) of the previous section. In actual scattering experiments it is often more useful to introduce a description in terms of the *Stokes parameters* (Stokes 1852). If we write

$$E_1 \equiv a_1 e^{i\delta_1}, \qquad E_2 \equiv a_2 e^{i\delta_2}, \tag{3.32}$$

these parameters are defined as follows:

$$t_1 \equiv E_1 E_1^* = |E_1|^2, \tag{3.33a}$$

$$t_2 \equiv E_2 E_2^* = |E_2|^2, \tag{3.33b}$$

$$t_3 \equiv E_1 E_2^* + E_1^* E_2 = 2\,\mathrm{Re}(E_1 E_2^*), \tag{3.33c}$$

$$t_4 \equiv i(E_1 E_2^* - E_1^* E_2) = -2\,\mathrm{Im}(E_1 E_2^*). \tag{3.33d}$$

Substitution of (3.32) into (3.33) then yields the identifications

$$t_1 = a_1^2, \tag{3.34a}$$

$$t_2 = a_2^2, \tag{3.34b}$$

$$t_3 = 2a_1 a_2 \cos\delta, \tag{3.34c}$$

$$t_4 = 2a_1 a_2 \sin\delta, \tag{3.34d}$$

with $\delta \equiv \delta_1 - \delta_2$. These are not all independent, however, as one readily verifies:

$$4t_1 t_2 = t_3^2 + t_4^2. \tag{3.35}$$

According to the discussion following (3.16), $t_3 = 0$ for circular polarization, and $t_4 = 0$ for linear polarization.

Clearly t_1 and t_2 describe the intensity of each polarization, while δ is the phase difference between the two. In fact, these three parameters completely describe the electromagnetic wave, an observation known as the *principle of optical equivalence*. The physical significance of t_3 and t_4 is encompassed in that of δ, which is related to the polarization itself.

The usefulness of the Stokes parameters is suggested by the linear relation (3.31). We first define Stokes vectors $(t) \equiv (t_1, t_2, t_3, t_4)$, along with the *transfer*

---

† S is not what is usually termed the 'scattering matrix', but rather, as will be seen later, a matrix of scattering amplitudes.

matrix T, in terms of the linear transformation

$$(t)_{\text{sc}} = T\,(t)_{\text{in}}. \tag{3.36}$$

We also define

$$M_k \equiv S_k S_k^*, \tag{3.37a}$$

$$N_{kj} \equiv \tfrac{1}{2}\left(S_j S_k^* + S_k S_j^*\right) = N_{jk}, \tag{3.37b}$$

$$-D_{kj} \equiv (i/2)\left(S_j S_k^* - S_k S_j^*\right) = D_{jk}. \tag{3.37c}$$

After some algebraic reduction we find that[†]

$$T = \begin{pmatrix} M_2 & M_3 & N_{23} & -D_{23} \\ M_4 & M_1 & N_{41} & -D_{41} \\ 2N_{24} & 2N_{31} & N_{21}+N_{34} & -D_{21}+D_{34} \\ 2D_{24} & 2D_{31} & D_{21}+D_{34} & N_{21}-N_{34} \end{pmatrix}. \tag{3.38}$$

Thus, by calculating the scattering amplitudes S_i one has also effected a solution of the scattering problem through Eq. (3.36).

To explicate the meaning of the phase parameter δ we note that there also exists a geometric representation equivalent to that in terms of Stokes parameters. An electric vector with the most general type of elliptic polarization can be written

$$E = a\boldsymbol{n}\cos\varphi\sin(kz - \omega t + \alpha) + a\boldsymbol{m}\sin\varphi\cos(kz - \omega t + \alpha), \tag{3.39}$$

where $\boldsymbol{n}$ and $\boldsymbol{m}$ are unit vectors along the long and short axes of the ellipse, respectively, a^2 is the intensity, and α is an arbitrary phase angle. The ellipticity is measured by $\tan\varphi$, such that it is zero for linear polarization, and ± 1 for right- or left-handed circular polarization. The geometry is depicted in Fig. 3.2, where χ is the orientation angle of the ellipse. With reference to Eq. (3.32), one finds that

$$a_1^2 = a^2(\cos^2\varphi\cos^2\chi + \sin^2\varphi\sin^2\chi),$$
$$a_2^2 = a^2(\cos^2\varphi\sin^2\chi + \sin^2\varphi\cos^2\chi), \tag{3.40a}$$

$$\tan(\alpha + \delta_1) = -\cot\chi\cot\varphi, \quad \tan(\alpha + \delta_2) = \tan\chi\cot\varphi, \tag{3.40b}$$

and as a consequence

$$\tan\delta = \frac{\tan(2\varphi)}{\sin(2\chi)}. \tag{3.41}$$

Therefore, δ provides a measure of the ellipticity of the polarization.

[†] Depending on the context, T is sometimes called a *Mueller matrix*, or a *scattering matrix*. An excellent review of its utility is provided by Bickel and Bailey (1985).

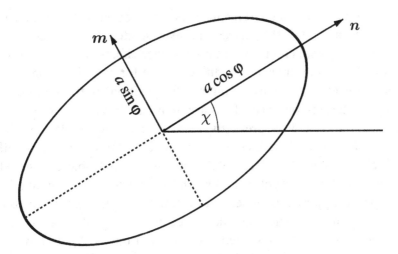

Fig. 3.2. The polarization ellipse corresponding to the electric field of Eq. (3.39). The plane of the ellipse is perpendicular to the direction of propagation, which is into the page, and the angle χ describes the orientation of the major axis with respect to the arbitrary horizontal axis.

The preceding formulation is not, of course, unique. For example, one could as well describe a wave in terms of circular polarization by means of the basis vectors (3.18) and write the electric field as $E = E_+\epsilon_+ + E_-\epsilon_-$. If we define a unitary matrix

$$\mathcal{U} \equiv \frac{1}{\sqrt{2}}\begin{pmatrix} i & 1 \\ -i & 1 \end{pmatrix}, \tag{3.42}$$

it is readily verified that the two sets of basis vectors are related by

$$\begin{pmatrix} \epsilon_+ \\ \epsilon_- \end{pmatrix} = \mathcal{U}\begin{pmatrix} \epsilon_2 \\ \epsilon_1 \end{pmatrix} = \frac{1}{\sqrt{2}}\begin{pmatrix} \epsilon_1 + i\epsilon_2 \\ \epsilon_1 - i\epsilon_2 \end{pmatrix}, \tag{3.43a}$$

and the fields by

$$\begin{pmatrix} E_+ \\ E_- \end{pmatrix} = \mathcal{U}\begin{pmatrix} E_2 \\ E_1 \end{pmatrix}. \tag{3.43b}$$

Thus, up to a factor of $(ik)^{-1}$, the scattering-amplitude matrix in the circular representation is

$$\begin{aligned} S^C &= \mathcal{U}S^L\mathcal{U}^\dagger \\ &= \frac{1}{2}\begin{pmatrix} S_1 - iS_4 + iS_3 + S_2 & S_1 + iS_4 + iS_3 - S_2 \\ S_1 - iS_4 - iS_3 - S_2 & S_1 + iS_4 - iS_3 + S_2 \end{pmatrix}. \end{aligned} \tag{3.44}$$

In general, the scattering process is considerably more complicated than just that of a single wave impinging upon a single target, although that can be done. A more common arrangement is a collection of targets, such as a gas. Except in a crystal, the phases of waves scattered in a direction other than forward are incoherent, so that interference effects are averaged to zero. Hence, for $\theta \neq 0$ intensities are additive. In the forward direction no phase differences are introduced by changing the position of the scatterer, even randomly, so that interference effects are manifest. In this case one must add amplitudes. As we shall see presently, these observations imply that all cross sections are additive. If the incident wavelength is large compared with the interparticle spacing, the effect on the medium as a whole can sometimes be described in terms of polarizability and the well-known Lorentz–Lorenz formula. At the opposite extreme, and for low densities, it is often possible to treat the medium as a single target described by an index of refraction derived from the properties of its constituent particles. For the most part, however, we shall consider the collection of identical targets to be independent scatterers, which is effectively achieved for interparticle spacings greater than a few radii.

Multiple scattering is important if much of the incident beam is actually extinguished and the incident waves on many particles are predominantly those scattered from other particles. For example, it is estimated that only about 10% of the original sunlight actually emerges from clouds after single scattering. These phenomena have been studied extensively by van de Hulst (1980) and will not be considered further here.

Another feature of actual scattering that must be assessed is that most often the incident radiation consists of an incoherent beam. This is a super-position of many waves propagating in the same direction, but with different phases and polarizations. Such incoherence implies that the intensity of the beam is just the sum of the individual intensities. We can study this scenario by first noting that what one defines as Stokes parameters is by no means a unique scheme. A very useful set as an alternative to Eqs. (3.33) is the following (e.g., van de Hulst (1957)):

$$
\begin{aligned}
I &= |E_1|^2 + |E_2|^2, \\
Q &= |E_2|^2 - |E_1|^2, \\
U &= 2\,\mathrm{Re}(E_1 E_2^*), \\
V &= -2\,\mathrm{Im}(E_1 E_2^*).
\end{aligned}
\tag{3.45}
$$

Not surprisingly, these are not all independent:

$$
I^2 = Q^2 + U^2 + V^2.
\tag{3.46}
$$

Upon comparison with (3.33) we find the relations

$$I = t_1 + t_2,$$
$$Q = t_1 - t_2,$$
$$U = t_3,$$
$$V = t_4, \tag{3.47a}$$

which in the geometric picture of Fig. 3.2 and Eqs. (3.40a) become

$$I = a^2,$$
$$Q = a^2 \cos(2\beta) \cos(2\chi),$$
$$U = a^2 \cos(2\beta) \sin(2\chi),$$
$$V = a^2 \sin(2\beta). \tag{3.47b}$$

The quantities I, $Q^2 + U^2$, and V are clearly invariant under rotation of the reference line in Fig. 3.2, since they are independent of χ. A similar scheme can also be constructed in the circular representation.

For an incoherent beam the new Stokes parameters are just the sums for all the individual wavelets:

$$I = \sum_i I_i, \qquad Q = \sum_i Q_i,$$
$$U = \sum_i U_i, \qquad V = \sum_i V_i, \tag{3.48}$$

and in this case Eq. (3.46) implies the triangle inequality

$$I^2 \geq Q^2 + U^2 + V^2. \tag{3.49}$$

The equality sign holds if and only if the beam is fully polarized; if it is completely unpolarized then $Q = U = V = 0$. The latter case is often referred to as 'natural' light. It is equally natural to define the *degree of polarization* as

$$P \equiv \frac{(Q^2 + U^2 + V^2)^{1/2}}{I}. \tag{3.50}$$

Owing to their invariant character the quantities $(Q^2 + U^2)^{1/2}/I$ and V/I can also be considered as measures of the degrees of linear and circular polarization, respectively. (See Eq. (3.47a) and the discussion following (3.35)).

In terms of the scheme we have adopted in Eqs. (3.33) and (3.34) we can write the degree of polarization explicitly in the form

$$P^2 = \frac{(t_1 - t_2)^2 + t_3^2 + t_4^2}{(t_1 + t_2)^2}. \tag{3.51}$$

Note that we are not entitled to reduce this further at this point by means of

(3.35), because that has already effectively been used in the derivation from (3.46). Equation (3.51) will subsequently prove very useful in determining the polarization of scattered radiation by means of (3.36).

Cross sections

An elegant alternative to the description of polarization in terms of Stokes parameters, which also provides an expeditious means for calculating scattering cross sections, is to introduce a *density matrix* (e.g., Newton (1982)):

$$\rho_{ij} \equiv \frac{1}{I} \left(E_{j2}^* \ E_{j1}^* \right) \begin{pmatrix} E_{i2} \\ E_{i1} \end{pmatrix}. \tag{3.52}$$

In the representation of linear polarization,

$$\rho^{\mathrm{L}} = \frac{1}{I} \begin{pmatrix} |E_2|^2 & E_2 E_1^* \\ E_1 E_2^* & |E_1|^2 \end{pmatrix}, \tag{3.53a}$$

whereas for circular polarization

$$\rho^{\mathrm{C}} = \frac{1}{I} \begin{pmatrix} |E_+|^2 & E_+ E_-^* \\ E_- E_+^* & |E_-|^2 \end{pmatrix}, \tag{3.53b}$$

and in both cases the normalization is $\mathrm{Tr}\,\rho = 1$. In either case ρ is idempotent for a fully polarized beam – i.e., $\rho^2 = \rho$. If the beam is only partially polarized and ρ is written in terms of the Stokes parameters of Eq. (3.45), one readily shows that

$$\rho^2 - \rho = \frac{1}{4}\left(P^2 - 1 \right) \begin{pmatrix} 1 & 0 \\ 0 & 1 \end{pmatrix}. \tag{3.54}$$

Finally, if the beam is completely unpolarized ρ is just half the unit matrix.

The present relevance of this formulation lies with its relation to the differential scattering cross section, which we recall is defined as the ratio of the scattered flux to the incident flux per unit area:

$$\frac{d\sigma}{d\Omega} \equiv \frac{I_{\mathrm{sc}}}{I_{\mathrm{in}}}. \tag{3.55}$$

From Eqs. (3.52) and (3.31) we find that

$$I_{\mathrm{sc}} (\rho_{\mathrm{sc}})_{ij} = (E_{\mathrm{sc}})_i (E_{\mathrm{sc}}^*)_j$$
$$= \sum_{k,\ell} (\rho_{\mathrm{in}})_{k\ell} \ S_{ik} S_{j\ell}^* I_{\mathrm{in}},$$

or

$$\left(\frac{d\sigma}{d\Omega} \right) \rho_{\mathrm{sc}} = S \rho_{\mathrm{in}} S^\dagger. \tag{3.56}$$

By taking traces and noting that $\text{Tr}\,\rho = 1$ we see that

$$\frac{d\sigma}{d\Omega} = \text{Tr}\left(S\rho_{\text{in}}S^\dagger\right), \tag{3.57}$$

and calculation of the differential scattering cross section is reduced to simple algebra once the scattering amplitudes are known. By definition σ_{sca} is then given by integration over all solid angles. It is of some value to notice that if the incident beam is completely unpolarized, we have

$$\frac{d\sigma}{d\Omega} = \frac{1}{2}\,\text{Tr}(SS^\dagger). \tag{3.58}$$

The total field received by a detector situated at large distances from the target is just $E = E_{\text{in}} + E_{\text{sc}}$. If this detector is centered on the forward direction, Eq. (3.31) implies that the field components received there are, in matrix notation,

$$E \simeq \left(1 + \frac{e^{ik(r-z)}}{ikr}S(0)\right)E_{\text{in}}, \tag{3.59}$$

for plane waves incident along the z-axis. (Here 1 represents the unit matrix and E is a column matrix.) Presume that the dimensions of the detector are small compared with its distance from the target, so that at the detector $x^2 + y^2 \ll z^2$, and that the dimensions are much larger than a wavelength. If L is a representative dimension of the detector, these presumptions imply that $(z\lambda)^{1/2} \ll L \ll z$, and we can then write $r \simeq z + (x^2 + y^2)/(2z)$. Consequently, the energy flux at a point (x, y) of the detecting screen is just

$$I(x, y) \simeq I_{\text{in}} + \frac{2}{z}\,\text{Re}\left(\frac{e^{ik(x^2+y^2)/2z}}{ik}E_{\text{in}}^\dagger S(0)E_{\text{in}}\right) + O(z^{-2}). \tag{3.60}$$

We can calculate the total energy received at the detector by integrating the intensity there over its area A. The above restrictions on relative dimensions lead to the evaluation

$$\int_A\!\!\int dx\,dy\,e^{ik(x^2+y^2)/2z} \simeq \int_{-\infty}^{\infty} dx \int_{-\infty}^{\infty} dy\,e^{ik(x^2+y^2)/2z} = 2\pi iz/k. \tag{3.61}$$

The total energy received at the detector is then

$$AI \simeq AI_{\text{in}} - \frac{4\pi}{k}\,\text{Re}\left(E_{\text{in}}^\dagger \frac{S(0)}{k}E_{\text{in}}\right), \tag{3.62}$$

so that the second term on the right-hand side represents the energy removed from the incident beam per unit incident intensity; it describes the diminution of energy owing to the presence of the scatterer, and (3.62) is merely a statement of the conservation of energy.

These results and remarks serve to define the *extinction cross section*

$$\sigma_{\text{ext}} \equiv \frac{4\pi}{Ik^2} \operatorname{Re}\left(E_{\text{in}}^{\dagger} S(0) E_{\text{in}}\right)$$

$$= \frac{4\pi}{Ik^2} \operatorname{Re}\left(S_2(0)|E_2|^2 + S_1(0)|E_1|^2 + S_4(0)E_1^* E_2 + S_3(0)E_1 E_2^*\right), \quad (3.63)$$

the complete electromagnetic form of the optical theorem. One could also perform the calculation in the circular representation and obtain the same result. In either case we note again that conservation of energy implies that $\sigma_{\text{ext}} = \sigma_{\text{sca}} + \sigma_{\text{abs}}$, representing the total energy removed from the incident wave by the target. For an unpolarized incident beam $E_1^* E_2 = 0$ and $\frac{1}{2}I = |E_1|^2 = |E_2|^2$, so that (3.63) reduces to

$$\sigma_{\text{ext}} = \frac{2\pi}{k^2} \operatorname{Re}\left(S_2(0) + S_1(0)\right). \quad (3.64)$$

As they must be, all these results are completely equivalent to those obtained by integration of the Poynting vector in the preceding section. It is worth noting, however, that the optical theorem is strictly valid only for incident *plane waves*. In the presence of a Gaussian beam, for example, $S(0)$ provides only a first approximation to the extinction cross section (e.g., Lock *et al.* (1995)), a scenario that will be considered further in Chapter 8.

The foregoing derivation of the optical theorem follows that of van de Hulst (1957), who also discusses how σ_{ext} tends to *twice* the geometric cross section of the target as its dimensions become very much larger than the incident wavelength. This observation describes the 'extinction paradox' that we have already noted in previous chapters, and is a result of residual diffraction effects even in the geometric-optics limit. A rigorous mathematical derivation of this phenomenon will emerge in Chapter 4, but was already discussed in some detail by Brillouin (1949).

3.3 Spherically symmetric scatterers

For various reasons already enumerated, the main application of the scattering theory developed in the preceding sections is to targets possessing spherical symmetry. Only for spheres and a few other shapes is it easy to match solutions at the boundary surface and obtain exact solutions. When the targets are either very small or very large compared with the incident wavelength there are approximation methods that are quite effective, giving essentially the same results as do spheres under similar conditions. If the shape possesses a definite asymmetry, one also has the problem of specifying the target's orientation with respect to the incoming beam. We shall discuss nonspherical targets to some extent in Chapter 8, but it is important above all else to first acquire a deep appreciation of the theory in the case of spherical symmetry. In addition, application to water drops in the atmosphere, which are usually spherical to a very good approximation, and to man-made spheres provides a strong incentive for considering this scenario in some detail.

A first simplification arising from spherical symmetry is that the scattering-amplitude matrix of Eq. (3.31) is now diagonal:

$$\begin{pmatrix} E_2 \\ E_1 \end{pmatrix}_{sc} = \frac{e^{ikr}}{ikr} \begin{pmatrix} S_2 & 0 \\ 0 & S_1 \end{pmatrix} \begin{pmatrix} E_2 \\ E_1 \end{pmatrix}_{in}. \tag{3.65}$$

We verify this from (3.31) as follows: reflection of the scatterer through the scattering plane is equivalent to reflecting everything but the scatterer; then $(E_1)_{in}$ and $(E_1)_{sc}$ change sign, but $(E_2)_{in}$ and $(E_2)_{sc}$ do not; hence, S_3 and S_4 must change sign; but, if the scatterer is completely symmetric under reflection, S_3 and S_4 can *not* change sign, and the only conclusion is that $S_3 = S_4 = 0$.

An immediate consequence of this result is that, with a single exception to be discussed later, there is no direct mixing of polarization in the scattered fields. If I_{in} is the initial flux, then in terms of the different polarizations, the scattered fluxes are

$$I_1 = \frac{i_1}{k^2 r^2} I_{in}, \tag{3.66a}$$

$$I_2 = \frac{i_2}{k^2 r^2} I_{in}, \tag{3.66b}$$

$$I = \frac{i_1 + i_2}{2k^2 r^2} I_{in} \qquad \text{(for an unpolarized beam)}, \tag{3.66c}$$

where we find it convenient here and subsequently to define the scattered polarized intensities

$$i_1 \equiv |S_1(\theta)|^2, \qquad i_2 \equiv |S_2(\theta)|^2, \tag{3.67}$$

noting that now the amplitudes are independent of the azimuthal angle ϕ. Phases are defined similarly by writing

$$S_1 = \sqrt{i_1}\, e^{i\delta_1}, \qquad S_2 = \sqrt{i_2}\, e^{i\delta_2},$$
$$\delta \equiv \delta_1 - \delta_2, \tag{3.68}$$

a definition necessary to describe scattering of incident light with arbitrary polarization. As for nomenclature, the flux I is simply the magnitude of the Poynting vector, sometimes called the irradiance. We take the intensity to be proportional to the ratio of scattered to incident irradiance.

As expected, the description in terms of Stokes parameters also simplifies considerably, the transfer matrix (3.38) becoming explicitly

$$
T = \begin{pmatrix} |S_2|^2 & 0 & 0 & 0 \\ 0 & |S_1|^2 & 0 & 0 \\ 0 & 0 & \frac{1}{2}(S_1 S_2^* + S_2 S_1^*) & (i/2)(S_1 S_2^* - S_2 S_1^*) \\ 0 & 0 & -(i/2)(S_1 S_2^* - S_2 S_1^*) & \frac{1}{2}(S_1 S_2^* + S_2 S_1^*) \end{pmatrix} \tag{3.69}
$$

and this contains only three independent parameters. The Stokes parameters of the scattered wave are now given explicitly by

$$
(t_1)_{sc} = \frac{i_1}{k^2 r^2}(t_1)_{in}, \qquad (t_2)_{sc} = \frac{i_2}{k^2 r^2}(t_2)_{in}, \tag{3.70a}
$$

$$
(t_3)_{sc} = \frac{\sqrt{i_1 i_2}}{k^2 r^2}[(t_3)_{in} \cos \delta - (t_4)_{in} \sin \delta], \tag{3.70b}
$$

$$
(t_4)_{sc} = \frac{\sqrt{i_1 i_2}}{k^2 r^2}[(t_3)_{in} \sin \delta + (t_4)_{in} \cos \delta]. \tag{3.70c}
$$

That is, both the intensity and the polarization of the scattered wave are completely determined by measuring i_1, i_2, and the phase difference δ between S_1 and S_2. With reference to (3.66), it should be quite clear that both the polarization and the angular distribution of the scattered radiation depend intimately on the polarization of the incident wave. If the latter is polarized completely either parallel or perpendicular to the scattering plane, so is the scattered wave. For an unpolarized incident beam the degree of polarization of the scattered wave follows from (3.51), along with the observation that $(t_1)_{in} = (t_2)_{in}$ and $(t_3)_{in} = (t_4)_{in} = 0$:

$$
P = \frac{i_1 - i_2}{i_1 + i_2}. \tag{3.71}
$$

We could, of course, have developed the preceding expressions in the circular basis as well. Although we shall not pursue that course explicitly, it is nevertheless worth noting from (3.44) that the scattering-amplitude matrix in the circular representation for spherical symmetry is

$$
S^C = \frac{1}{2}\begin{pmatrix} S_1 + S_2 & S_1 - S_2 \\ S_1 - S_2 & S_1 + S_2 \end{pmatrix}. \tag{3.72}
$$

The optical theorem of Eq. (3.63) becomes particularly simple for spherical targets, not only because we have $S_3 = S_4 = 0$, but also because the remaining scattering amplitudes degenerate into a single quantity in the forward direction. This follows because the amplitudes are independent of angles in this case and the scattering plane is not defined. If we let

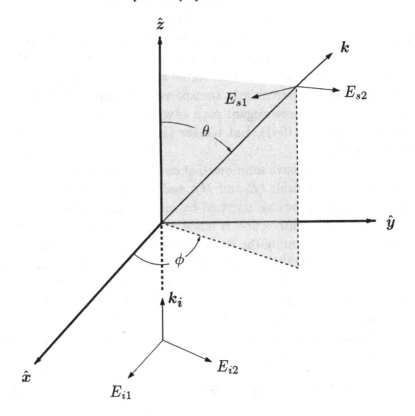

Fig. 3.3. Scattering of a linearly polarized plane wave from a spherical target located at the origin.

$S(0) \equiv S_1(0) = S_2(0)$, then (3.64) yields

$$\sigma_{\text{ext}} = \frac{4\pi}{k^2} \operatorname{Re} S(0). \tag{3.73}$$

The homogeneous dielectric sphere

We now focus specifically on a spherical target that is a homogeneous and isotropic dielectric with complex index of refraction $n = m + i\kappa$, although our principal interest lies with real n. Let a transverse plane wave be incident along the positive z-axis with linear polarization, along which we align the positive x-axis, as illustrated in Fig. 3.3. This is usually a typical member of an incident unpolarized beam whose electric vectors may point in any direction in the xy-plane. Explicit expressions for the incident fields are given by Eqs. (3.22), and a straightforward approach is to adopt forms for the scattered fields and those within the sphere suggested by these multipole expansions; the coefficients, and, therefore, the scattering amplitudes, are determined by applying boundary conditions at the surface of the sphere. Although this procedure is correct and workable (e.g., Bohren and Huffman (1983)), owing to the appearance of vector spherical harmonics it is rather cumbersome, and not particularly transparent physically. Rather, we shall follow a simpler and more elegant path advocated in the past by Born and Wolf (1975), Stratton (1941), and Kerker (1969), which will serve us very well in the following.

Maxwell's equations have solutions that can be constructed as two linearly independent pairs of fields (E and H), and owing to the arbitrariness in polarization, each can be characterized by zero radial component for one or the other field. This construction is made more transparent, and ultimately more useful, by representing the fields in terms of two linearly independent Hertz vectors (Hertz 1889). The vector ${}^{e}\mathbf{\Pi}$ has its source generally in a distribution of electric polarization P and is called the *electric Hertz vector*. Similarly, the *magnetic Hertz vector* ${}^{m}\mathbf{\Pi}$ has its source in a distribution of magnetic moment M. What is most useful in the current context is that both quantities satisfy the vector wave equation, and in a non-magnetic medium free of sources, the fields are

$$E = \nabla \times (\nabla \times {}^{e}\mathbf{\Pi}) - \frac{1}{c}\nabla \times \frac{\partial\,({}^{m}\mathbf{\Pi})}{\partial t}, \tag{3.74a}$$

$$H = \nabla \times (\nabla \times {}^{m}\mathbf{\Pi}) + \nabla \times \left(\varepsilon\frac{1}{c^2}\frac{\partial\,({}^{e}\mathbf{\Pi})}{\partial t} + \frac{4\pi\sigma}{c^2}\,{}^{e}\mathbf{\Pi}\right). \tag{3.74b}$$

Magnetic medium or not, the speed of light is $v = c/\sqrt{\mu\varepsilon}$, which in free space is simply c, so that when μ and ε appear explicitly they are to be considered *relative* to the free-space quantities.

Major simplification emerges upon realization that the Hertz vectors can be derived from scalar potentials (Debye 1909b). These so-called *Debye potentials* are introduced as radial Hertz vectors,

$${}^{e}\mathbf{\Pi} \equiv r\Pi_1, \qquad {}^{m}\mathbf{\Pi} \equiv r\Pi_2, \tag{3.75}$$

and both satisfy the *scalar* wave equation:

$$\nabla^2\Pi + k^2\Pi = 0. \tag{3.76}$$

With harmonic time dependence the fields take the forms

$$E = \nabla \times \nabla \times (r\Pi_1) + ik\nabla \times (r\Pi_2), \tag{3.77a}$$

$$H = \nabla \times \nabla \times (r\Pi_2) - ikn^2\nabla \times (r\Pi_1). \tag{3.77b}$$

The electric potential Π_1 is *transverse magnetic* ($H_r = 0$), and the magnetic potential Π_2 is *transverse electric* ($E_r = 0$). An excellent review of Debye potentials, as well as a general proof that the fields can always be represented by two scalar potentials of types TE and TM, is given by Nisbet (1955).

Our primary interest lies with the scattering of electromagnetic plane waves from a dielectric sphere of radius a and refractive index $m = \mathrm{Re}\, n > 1$. In terms of the Debye potentials the explicit field components are verified to be

$$E_r = \partial_{rr}^2(r\Pi_1) + K^2(r\Pi_2),$$

$$E_\theta = \frac{1}{r}\partial_{r\theta}^2(r\Pi_1) + \kappa_2\frac{1}{r\sin\theta}\partial_\phi(r\Pi_2),$$

$$E_\phi = \frac{1}{r\sin\theta}\partial_{r\phi}^2(r\Pi_1) - \kappa_2\frac{1}{r}\partial_\theta(r\Pi_2); \tag{3.78a}$$

$$H_r = \partial_{rr}^2(r\Pi_2) + K^2(r\Pi_2),$$

$$H_\theta = \frac{1}{r}\partial_{r\theta}^2(r\Pi_2) - \kappa_1\frac{1}{r\sin\theta}\partial_\phi(r\Pi_1),$$

$$H_\phi = \frac{1}{r\sin\theta}\partial_{r\phi}^2(r\Pi_2) + \kappa_1\frac{1}{r}\partial_\theta(r\Pi_1), \tag{3.78b}$$

in the notation of Eq. (3.4). Note that E_r is independent of Π_2 and H_r is independent of Π_1. Boundary conditions are obtained from the requirement that the tangential field components are continuous across the boundary at $r = a$, and this translates here into continuity of

$$\kappa_1\Pi_1, \quad \frac{\partial}{\partial r}(r\Pi_1), \quad \kappa_2\Pi_2, \quad \frac{\partial}{\partial r}(r\Pi_2). \tag{3.79}$$

Solutions to the wave equation (3.76) are found in the usual way, by separation of variables: $\Pi(r) = R(r)\Theta(\theta)\Phi(\phi)$. As always, the azimuthal functions are $\exp(\pm i\phi m)$, and the polar functions are $P_\ell^m(\cos\theta)$; the radial functions are solutions of

$$\frac{d^2(rR)}{dr^2} + \left(k^2 - \frac{\ell(\ell+1)}{r^2}\right)(rR) = 0. \tag{3.80}$$

The incident wave, as well as these solutions, can be expanded in partial waves, and, just like in (3.22), we know that only the azimuthal eigenvalue $m = 1$ occurs. For the incident wave

$$\Pi_1^{\text{inc}} = -i\frac{\cos\phi}{kN_0}\sum_{\ell=1}^{\infty}\varepsilon_\ell\, j_\ell(N_0 kr)P_\ell^1(\cos\theta), \tag{3.81a}$$

$$\Pi_2^{\text{inc}} = -i\frac{\sin\phi}{k}\sum_{\ell=1}^{\infty}\varepsilon_\ell\, j_\ell(N_0 kr)P_\ell^1(\cos\theta), \tag{3.81b}$$

where it is convenient to define

$$\varepsilon_\ell \equiv i^\ell \frac{2\ell+1}{\ell(\ell+1)}, \tag{3.82}$$

and N_0 is the refractive index of the external medium in which the sphere is embedded – it is usually taken as unity. By analogy, introduction of undetermined coefficients yields for the external fields after scattering

$$\Pi_1^{\text{ext}} = i\frac{\cos\phi}{kN_0} \sum_{\ell=1}^{\infty} \varepsilon_\ell a_\ell\, h_\ell^{(1)}(N_0kr) P_\ell^1(\cos\theta), \tag{3.83a}$$

$$\Pi_2^{\text{ext}} = i\frac{\sin\phi}{k} \sum_{\ell=1}^{\infty} \varepsilon_\ell b_\ell\, h_\ell^{(1)}(N_0kr) P_\ell^1(\cos\theta), \tag{3.83b}$$

and inside the sphere

$$\Pi_1^{\text{int}} = -i\frac{\cos\phi}{kN} \sum_{\ell=1}^{\infty} \varepsilon_\ell\, c_\ell\, j_\ell(Nkr) P_\ell^1(\cos\theta), \tag{3.84a}$$

$$\Pi_2^{\text{int}} = -i\frac{\sin\phi}{k} \sum_{\ell=1}^{\infty} \varepsilon_\ell\, d_\ell\, j_\ell(Nkr) P_\ell^1(\cos\theta), \tag{3.84b}$$

where N is the refractive index of the sphere. The factors of N_0 and N have been inserted for later convenience, and capital letters have been used so that we can now introduce the *relative refractive index* $n \equiv N/N_0$, which will appear everywhere in what follows. In conjunction with this we also define the *size parameter* β as follows:

$$\beta \equiv N_0 ka = \frac{2\pi a}{\lambda_0/N_0}, \tag{3.85}$$

so that λ_0 is the vacuum wavelength.

Equation (3.80) indicates that the natural radial solutions for this problem are Ricatti–Bessel functions (Appendix A), so it is convenient to adopt the conventional optics notation† and define three new radial functions:

$$\psi_\ell(z) \equiv z j_\ell(z), \qquad \xi_\ell(z) \equiv z y_\ell(z),$$
$$\zeta_\ell^{(1,2)}(z) \equiv z h_\ell^{(1,2)}(z) = \psi_\ell(z) + i\xi_\ell(z). \tag{3.86}$$

Matching the solutions (3.81)–(3.84) leads to the following set of linear equations determining the coefficients:

$$c_\ell \psi_\ell'(n\beta) = n\left[\psi_\ell'(\beta) - a_\ell \zeta_\ell^{(1)'}(\beta)\right],$$

† Note that our notation differs from that of Kerker (1969), as well as from that of Bohren and Huffman (1983).

$$d_\ell \psi'_\ell(n\beta) = \psi'_\ell(\beta) - b_\ell \zeta_\ell^{(1)'}(\beta),$$
$$c_\ell \psi(n\beta) = \psi_\ell(\beta) - a_\ell \zeta_\ell^{(1)}(\beta),$$
$$d_\ell \psi_\ell(n\beta) = n\left[\psi_\ell(\beta) - b_\ell \zeta_\ell^{(1)}(\beta)\right], \tag{3.87}$$

where the primes denote differentiation with respect to the argument. We find that the *partial-wave scattering coefficients* are

$$a_\ell(\beta) = \frac{\psi_\ell(\beta)\psi'_\ell(n\beta) - n\psi_\ell(n\beta)\psi'_\ell(\beta)}{\zeta_\ell^{(1)}(\beta)\psi'_\ell(n\beta) - n\psi_\ell(n\beta)\zeta_\ell^{(1)'}(\beta)}, \tag{3.88a}$$

$$b_\ell(\beta) = \frac{\psi_\ell(n\beta)\psi'_\ell(\beta) - n\psi'_\ell(n\beta)\psi_\ell(\beta)}{\zeta_\ell^{(1)'}(\beta)\psi_\ell(n\beta) - n\zeta_\ell^{(1)}(\beta)\psi'_\ell(n\beta)}. \tag{3.88b}$$

With the aid of the Wronskian (A.20) for the Ricatti–Bessel functions the internal coefficients can be written as

$$c_\ell(\beta) = \frac{-in}{\zeta_\ell^{(1)}(\beta)\psi'_\ell(n\beta) - n\psi_\ell(n\beta)\zeta_\ell^{(1)'}(\beta)}, \tag{3.89a}$$

$$d_\ell(\beta) = \frac{in}{\zeta_\ell^{(1)'}(\beta)\psi_\ell(n\beta) - n\zeta_\ell^{(1)}(\beta)\psi'_\ell(n\beta)}. \tag{3.89b}$$

Note that the denominators in a_ℓ and c_ℓ are identical, as are those in b_ℓ and d_ℓ.

Equations (3.78) provide us with the external scattered fields, both near and far:

$$E_r = iE_0 \cos\phi \sum_\ell i^\ell (2\ell+1) a_\ell(\beta) \frac{\zeta_\ell^{(1)}(K_0 r)}{(K_0 r)^2} P_\ell^1(\cos\theta),$$

$$E_\theta = -iE_0 \cos\phi \sum_\ell \varepsilon_\ell \left(a_\ell(\beta) \frac{\zeta_\ell^{(1)'}(K_0 r)}{K_0 r} \tau_\ell(\cos\theta) + b_\ell(\beta) \frac{i\zeta_\ell^{(1)}(K_0 r)}{K_0 r} \pi_\ell(\cos\theta) \right),$$

$$E_\phi = iE_0 \sin\phi \sum_\ell \varepsilon_\ell \left(a_\ell(\beta) \frac{\zeta_\ell^{(1)'}(K_0 r)}{K_0 r} \pi_\ell(\cos\theta) + b_\ell(\beta) \frac{i\zeta_\ell^{(1)}(K_0 r)}{K_0 r} \tau_\ell(\cos\theta) \right); \tag{3.90a}$$

$$H_r = iE_0 \sin\phi \sum_\ell i^\ell (2\ell+1) b_\ell(\beta) \frac{\zeta_\ell^{(1)}(K_0 r)}{(K_0 r)^2} P_\ell^1(\cos\theta),$$

$$H_\theta = -iE_0 N_0 \sin\phi \sum_\ell \varepsilon_\ell \left(a_\ell(\beta) \frac{i\zeta_\ell^{(1)}(K_0 r)}{K_0 r} \pi_\ell(\cos\theta) + b_\ell(\beta) \frac{\zeta_\ell^{(1)'}(K_0 r)}{K_0 r} \tau_\ell(\cos\theta) \right),$$

$$H_\phi = -iE_0 N_0 \cos\phi \sum_\ell \varepsilon_\ell \left(a_\ell(\beta) \frac{i\zeta_\ell^{(1)}(K_0 r)}{K_0 r} \tau_\ell(\cos\theta) + b_\ell(\beta) \frac{\zeta_\ell^{(1)'}(K_0 r)}{K_0 r} \pi_\ell(\cos\theta) \right), \tag{3.90b}$$

where $K_0 \equiv kN_0$, E_0 is the amplitude of the incident wave, ε_ℓ is defined in (3.82), and we have defined angular functions

$$\pi_\ell(\cos\theta) \equiv \frac{P_\ell^1(\cos\theta)}{\sin\theta}, \qquad \tau_\ell(\cos\theta) \equiv \frac{dP_\ell^1(\cos\theta)}{d\theta}, \qquad (3.91)$$

whose properties are studied in Appendix D. The reader may wish to verify that the internal fields are obtained from these equations by the following replacements: $N_0 \to N$, $K_0 \to K$, $a_\ell \to c_\ell$, $b_\ell \to d_\ell$, and $\zeta_\ell \to \psi_\ell$.

Explicit expressions for the scattered fields in the radiation, or far, zone can be identified from the asymptotic behavior of the spherical Hankel functions (Appendix A):

$$\zeta_\ell^{(1)}(x) \xrightarrow[x \gg 1]{} (-1)^{\ell+1} e^{ix}, \qquad \zeta_\ell^{(1)'}(x) \xrightarrow[x \gg 1]{} (-1)^\ell e^{ix}. \qquad (3.92)$$

The longitudinal fields drop off like r^{-2} and can be neglected relative to the transverse fields. Hence, the far fields are

$$E_\theta = -iE_0 \cos\phi \frac{e^{iK_0 r}}{K_0 r} \sum_\ell \frac{2\ell+1}{\ell(\ell+1)} (a_\ell \tau_\ell + b_\ell \pi_\ell) = \frac{H_\phi}{N_0}, \qquad (3.93a)$$

$$E_\phi = iE_0 \sin\phi \frac{e^{iK_0 r}}{K_0 r} \sum_\ell \frac{2\ell+1}{\ell(\ell+1)} (a_\ell \pi_\ell + b_\ell \tau_\ell) = -\frac{H_\theta}{N_0}. \qquad (3.93b)$$

Note that the orthogonal components of E and H are related to each other properly, so that these are indeed outgoing spherical electromagnetic waves in a medium of refractive index N_0.

Reference to Eq. (3.77a) allows us to express the scattered electric field as

$$E_{sc} = \frac{i}{K_0 r} e^{iK_0 r} E_0 \left[S_2(\beta,\theta) \cos\phi \, \hat{\theta} + S_1(\beta,\theta) \sin\phi \, (-\hat{\phi}) \right], \qquad (3.94)$$

where $\hat{\theta}$ and $\hat{\phi}$ are unit vectors associated with the coordinate system of Fig. 3.3. The scattering amplitudes are identified as

$$S_1(\beta,\theta) \equiv \sum_{\ell=1}^{\infty} \frac{2\ell+1}{\ell(\ell+1)} [a_\ell(\beta)\pi_\ell(\cos\theta) + b_\ell(\beta)\tau_\ell(\cos\theta)], \qquad (3.95a)$$

$$S_2(\beta,\theta) \equiv \sum_{\ell=1}^{\infty} \frac{2\ell+1}{\ell(\ell+1)} [b_\ell(\beta)\pi_\ell(\cos\theta) + a_\ell(\beta)\tau_\ell(\cos\theta)], \qquad (3.95b)$$

and, unlike the scalar case, there are no $\ell = 0$ terms. Thus, $S_1(\beta,\theta)$ describes polarization perpendicular, and $S_2(\beta,\theta)$ polarization parallel, to the scattering plane. Both amplitudes contain contributions from electric and magnetic multipoles, such that b_ℓ measures the strength of magnetic multipole radiation and a_ℓ that of electric multipole radiation. The scattered intensities

(3.67) follow directly from these results, and the Stokes parameters can be obtained from Eqs. (3.33), (3.36), and (3.69). Note that in writing (3.94) we have referred to (3.65) and noticed from Fig. 3.3 that, relative to the scattering plane,

$$(E_1)_{in} = E_0 \sin \phi, \qquad (E_2)_{in} = E_0 \cos \phi. \qquad (3.96)$$

Equations (3.88)–(3.95) comprise the exact solution to the problem of electromagnetic plane waves scattering from a homogeneous dielectric sphere, which is generally known as the *Mie solution*. Although it is traditionally attributed to Mie (1908), and to Debye (1909b), there is a considerable pre-history. Clebsch (1863) had actually solved the vector wave equation for elastic waves much earlier, as did Lamb (1881) somewhat later. An exact solution to the scalar problem of scattering of sound waves from a sphere was also found by Rayleigh (Strutt 1872), which he later extended to various aspects of the electromagnetic case (e.g., Strutt (1910)). This problem was solved completely in 1890 by Lorenz, who wrote down precisely the above results (Lorenz 1890), along with the expression given below for the scattering cross section.† Detailed surveys of the earlier work on plane-wave scattering from spherical targets have been provided both by Logan (1965), in a delightful essay, and by Kerker (1969).

While the Mie solution is exact, it is nevertheless a *series* solution within which many of the desired physical insights lie buried. Much of the ensuing discussion will be devoted to extracting the detailed characteristics of the scattered radiation from these series, either analytically or computationally. Our next step toward understanding the physics of the model is to incorporate these solutions into a number of expressions for the experimental observables, which will, initially at least, also be partial-wave expansions.

3.4 Measurable quantities

From our earlier work it is clear that by 'measurable quantities' describing the scattering of a linearly–polarized plane wave from a dielectric sphere we mean primarily cross sections. At first glance the forward and backward directions appear to be somewhat awkward, given that the scattering plane is not defined in these directions – as luck would have it, however, they actually

---

† On the one hand there is certainly good reason to refer to the 'Lorenz solution', and subsequently to the 'Lorenz theory' – and, indeed, reference to the *Lorenz–Mie theory* is often encountered. As we shall see subsequently, the extensive development of the theory by Debye also argues for inclusion of his name. On the other hand, it is difficult to get too worked up about this historical point, for usually we do not name various applications of a general theory after the appliers, though it is definitely in order to assign priority as done above. Thus, both as a convenience and to avoid awkward prose, we shall herein refer simply to the Mie solution and Mie theory, as is common.

become simpler! From Appendix D we note that the angular functions in these directions are

$$\pi_\ell(1) = \tau_\ell(1) = \frac{\ell(\ell+1)}{2},$$ (3.97a)

$$\pi_\ell(-1) = -\tau_\ell(-1) = (-1)^{\ell-1}\frac{\ell(\ell+1)}{2}.$$ (3.97b)

In the forward direction

$$S_1(\beta,0) = \frac{1}{2}\sum_{\ell=1}^{\infty}(2\ell+1)(a_\ell+b_\ell) = S_2(\beta,0) \equiv S(\beta,0)$$ (3.98)

and thus Eq. (3.73) yields the extinction cross section:

$$\sigma_{\text{ext}} = \frac{2\pi}{k^2}\sum_{\ell=1}^{\infty}(2\ell+1)\,\text{Re}(a_\ell+b_\ell).$$ (3.99a)

As an aside, it is not unreasonable to ask whether the *imaginary* part of the forward scattering amplitude also has observable physical significance. It does, indeed, and we shall address this point at the end of this chapter, and again in Chapter 8.

Conventional discussions of light scattering from spheres introduce dimensionless *efficiencies*, which are the cross sections normalized by the geometric cross section of the sphere, πa^2. Thus, (3.99a) can be re-expressed as the *extinction efficiency*† $Q_{\text{ext}} \equiv \sigma_{\text{ex}}/(\pi a^2)$:

$$Q_{\text{ext}}(\beta) = \frac{2}{\beta^2}\sum_{\ell=1}^{\infty}(2\ell+1)\,\text{Re}(a_\ell+b_\ell),$$ (3.99b)

and in Fig. 3.4 we have plotted it as a function of the size parameter $\beta = N_0 ka$ for an index of refraction close to that of water. There are several notable features of this curve, the first of which is that it exhibits a tendency to approach the value 2 as β becomes large. This so-called 'extinction paradox' has been noted several times earlier, particularly in the discussion following Eq. (3.64). Although this asymptotic behavior will be derived rigorously in the following chapter, it is worth posting a *caveat* at this point regarding the meaning to be attributed to the variation of the size parameter. From its definition β can be changed by varying either the radius of the sphere or the wavelength, their ratio being the operative parameter. However, all media are dispersive to some extent, so that the refractive index varies with wavelength. There is a tacit presumption in Fig. 3.4 that n is held constant as β varies, which can only be true to some degree of approximation. Although

---

† An unfortunate misnomer, since this and other 'efficiencies' can certainly exceed unity.

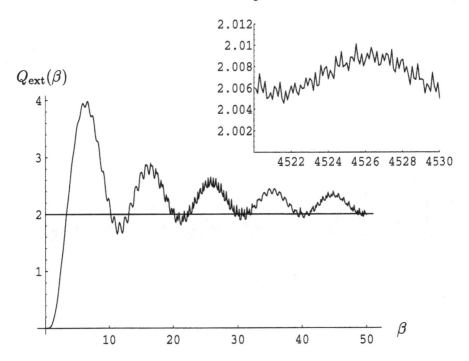

Fig. 3.4. The extinction efficiency given by Eq. (3.99) over a broad range of size parameters for a refractive index approximating that for water at optical wavelengths. The horizontal line at $Q_{ext} = 2$ is the geometric-optics limit, and the inset illustrates the persistence of the ripple at very large size parameters.

this is usually a small detail, it is an observation worth keeping in mind when drawing far-ranging conclusions from such figures.

The dominant structural features of $Q_{ext}(\beta)$ will need, and receive, careful explanation subsequently, but one can understand them qualitatively already. The slowly varying background oscillation of the curve will be seen to arise from interference between the diffracted waves and rays transmitted directly through the sphere. The latter correspond to small impact parameters, and from Eq. (2.39) these occur for $\beta \gg \ell$. In this limit the numerators of the coefficients a_ℓ and b_ℓ in (3.88) can be approximated by appeal to the asymptotic behavior of the Riccati–Bessel functions given in Appendix A, and we see that in both there is an ℓ-independent contribution given by $\sin[\beta(n-1)]$, for real n. In this approximation there is thus a series of maxima and minima in Q_{ext} at

$$\beta(n-1) = \begin{cases} (2p+1)(\pi/2), & \text{maxima} \\ p\pi, & \text{minima} \end{cases} \qquad (3.100)$$

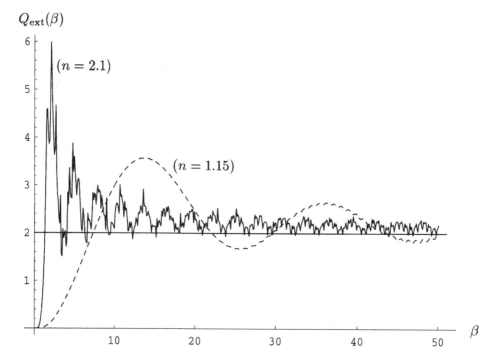

Fig. 3.5. The extinction efficiency for $n = 2.1$ (solid line) and $n = 1.15$ (dashed line), illustrating the variation in the ripple structure with the refractive index.

for integers p. This interpretation is reinforced by recalling that such oscillatory behavior is lost when there is no direct transmission term, as for the corresponding hard-sphere extinction efficiency in Fig. 2.12.

There is also a rapidly varying irregular structure superimposed on the background variation of Q_{ext} that is fundamentally a resonance phenomenon, to be discussed in great detail in Chapter 7. Its origin can again be surmised from the expressions for the partial-wave coefficients in (3.88). Although the denominators cannot vanish for real β, they indeed have zeros in the complex β-plane. Resonance effects are manifest when the imaginary parts of these zeros are small, and will be shown to underlie the *ripple structure* evident in Fig. 3.4. Although this ripple has the appearance of noise, and in some applications can be considered as such, it will come to be of much deeper consequence both theoretically and experimentally. The inset in Fig. 3.4 demonstrates that the ripple persists for very large size parameters. However, in Fig. 3.5 we see that it becomes more erratic and is present at smaller values of β for larger refractive index, while smoothing out altogether as n decreases toward unity.

In the presence of absorption the scattering cross section differs from σ_{ext}, so to calculate it we return to Eqs. (3.58) and (3.65) and begin with the differential cross section:

$$\frac{d\sigma}{d\Omega} = \frac{1}{k^2}\left(|S_2|^2 \cos^2\phi + |S_1|^2 \sin^2\phi\right). \tag{3.101}$$

The total scattering cross section is found by integrating over all solid angles, but the integrals turn out to be rather difficult. Recall, though, that matrix traces are invariant under unitary transformation, so that perhaps the calculation may simplify in the circular basis. From (3.42), (3.53), and (3.72) we find an expression alternative to (3.101):

$$\frac{d\sigma}{d\Omega} = \frac{1}{2k^2}\left[|S_+|^2 + |S_-|^2 + (S_+S_-^* + S_-S_+^*)(\cos^2\phi - \sin^2\phi)\right], \tag{3.102}$$

where

$$S_+ \equiv S_1 + S_2 = \sum_{\ell=1}^{\infty} \frac{2\ell+1}{\ell(\ell+1)}(a_\ell + b_\ell)(\pi_\ell + \tau_\ell), \tag{3.103a}$$

$$S_- \equiv S_1 - S_2 = \sum_{\ell=1}^{\infty} \frac{2\ell+1}{\ell(\ell+1)}(a_\ell - b_\ell)(\pi_\ell - \tau_\ell). \tag{3.103b}$$

Then,

$$\sigma_{\text{sca}} = \int\left(\frac{d\sigma}{d\Omega}\right)d\Omega = \frac{\pi}{2k^2}\int_{-1}^{+1}\left(|S_+|^2 + |S_-|^2\right)dx, \tag{3.104}$$

where $x \equiv \cos\theta$. Whereas (3.102) provides both intensity and phase information, σ_{sca} does not – it is also independent of of the incident polarization.

Absorption

Substitution of (3.103) into (3.104) leads to integrals of products of angular functions that are evaluated in Appendix D, and the *scattering efficiency* is obtained as

$$Q_{\text{sca}}(\beta) = \frac{2}{\beta^2}\sum_{\ell=1}^{\infty}(2\ell+1)\left(|a_\ell|^2 + |b_\ell|^2\right). \tag{3.105}$$

Conservation of energy then provides the *absorption efficiency*:

$$\begin{aligned}
Q_{\text{abs}}(\beta) &= Q_{\text{ext}}(\beta) - Q_{\text{sca}}(\beta) \\
&= \frac{2}{\beta^2}\sum_{\ell=1}^{\infty}(2\ell+1)\left[\left(\text{Re}\,a_\ell - |a_\ell|^2\right) + \left(\text{Re}\,b_\ell - |b_\ell|^2\right)\right] \\
&= \frac{1}{2\beta^2}\sum_{\ell=1}^{\infty}(2\ell+1)\left(2 - |1 - 2a_\ell|^2 - |1 - 2b_\ell|^2\right).
\end{aligned} \tag{3.106}$$

These efficiencies agree with the previous results (3.30) obtained from direct integration of the Poynting vectors, as they must. Furthermore, they can be combined into another observable, the *albedo*: Q_{sca}/Q_{ext}.

While energy conservation is a perfectly valid way to compute the absorption, it provides no direct insight into the dissipative mechanisms at work – neither does it reveal a direct dependence on the internal coefficients c_ℓ and d_ℓ, as it should. Indeed, with a little more effort we can unravel exactly how σ_{abs} depends directly on the internal fields and complex refractive index. To do this we first note that the intensity of the electric field anywhere relative to that of the incident plane wave is

$$I(r, \theta, \phi) \equiv \frac{E(r) \cdot E^*(r)}{|E_0|^2}. \tag{3.107}$$

Although this intensity is often studied on its own merits to extract interesting details of the scattering process, attention is usually focused on various integrals. For example, the angle-averaged intensity is called the *source function*,

$$\mathscr{S}(r) \equiv \frac{1}{4\pi} \int I(r) \, d\Omega, \tag{3.108}$$

and it is the internal fields that are generally of interest.† Explicit expressions for these fields are readily found from Eqs. (3.90a) by means of the replacements noted following (3.91), which are then substituted into (3.108). The angular integrations are carried out with the help of results found in Appendix D, and after some algebra we find that, with $K = nk = (m + i\kappa)k$,

$$\mathscr{S}(r) = \frac{1}{2} \sum_{\ell=1}^{\infty} (2\ell + 1) \left[|c_\ell|^2 \left(\ell(\ell+1) \left| \frac{\psi_\ell(Kr)}{K^2 r^2} \right|^2 + \left| \frac{\psi_\ell'(Kr)}{Kr} \right|^2 \right) \right.$$
$$\left. + |d_\ell|^2 \left| \frac{\psi_\ell(Kr)}{Kr} \right|^2 \right] \tag{3.109a}$$

$$= \frac{1}{2} \sum_{\ell=1}^{\infty} (2\ell + 1) \left(\frac{|c_\ell|^2}{2\ell + 1} \left[\ell \, |j_{\ell+1}(Kr)|^2 + (\ell + 1) |j_{\ell-1}(Kr)|^2 \right] \right.$$
$$\left. + |d_\ell|^2 \, |j_\ell(Kr)|^2 \right). \tag{3.109b}$$

The second line is obtained from the first by means of the recurrence relations for the spherical Bessel functions in Appendix A, and is often more useful computationally because it contains no derivatives. We shall return to a study of the internal energy distribution in the following chapters.

---

† Often I is referred to as the source function, but the term relative intensity seems more appropriate here.

Of more immediate interest here is the integral over the complete volume of the sphere, which gives us the total field intensity within:

$$\mathcal{S} \equiv \int_0^a 4\pi r^2 \mathcal{S}(r)\, dr = \frac{2\pi}{|K|^2} \sum_{\ell=1}^{\infty} (2\ell + 1)\left(\alpha_\ell\, |c_\ell|^2 + \beta_\ell\, |d_\ell|^2\right), \qquad (3.110)$$

where

$$\alpha_\ell \equiv \int_0^a \left(|\psi_\ell'(Kr)|^2 + \ell(\ell + 1)\left|\frac{\psi_\ell(Kr)}{Kr}\right|^2\right) dr, \qquad (3.111a)$$

$$\beta_\ell \equiv \int_0^a |\psi_\ell(Kr)|^2\, dr. \qquad (3.111b)$$

Remarkably, these integrals can be evaluated exactly in closed form (Karam and Fung 1993).

In the second-order equation for the Ricatti–Bessel functions, (3.80) with $k \to K$, multiply by $r^2 \psi_\ell^*(Kr)$ and integrate over $(0, a)$. After an integration by parts one can identify the integrals α_ℓ and β_ℓ, and in the resulting equation define the evaluated logarithmic derivative as

$$D_\ell(Ka) \equiv \psi_\ell'(Ka)/\psi_\ell(Ka). \qquad (3.112)$$

This and the corresponding complex-conjugate expression constitute two equations in two unknowns and, with the observation that $z\, \mathrm{Im}\, w - w\, \mathrm{Im}\, z = \mathrm{Im}(z^* w)$, the solutions are[†]

$$\alpha_\ell = -|\psi_\ell(Ka)|^2 \frac{\mathrm{Im}[K^* D_\ell(Ka)]}{\mathrm{Im}(K^2)}, \qquad (3.113a)$$

$$\beta_\ell = -|\psi_\ell(Ka)|^2 \frac{\mathrm{Im}[K D_\ell(Ka)]}{\mathrm{Im}(K^2)}. \qquad (3.113b)$$

Finally, write the partial-wave scattering coefficients in a form that will take on deeper meaning presently,

$$a_\ell = \tfrac{1}{2}(1 - A_\ell), \qquad b_\ell = \tfrac{1}{2}(1 - B_\ell), \qquad (3.114)$$

and by identifying A_ℓ and B_ℓ show that

$$\mathrm{Re}\, a_\ell - |a_\ell|^2 = \frac{1}{4}(1 - |A_\ell|^2) = \frac{|c_\ell|^2}{k|n|^2} \alpha_\ell\, \mathrm{Im}(K^2), \qquad (3.115a)$$

$$\mathrm{Re}\, b_\ell - |b_\ell|^2 = \frac{1}{4}(1 - |B_\ell|^2) = \frac{|d_\ell|^2}{k|n|^2} \beta_\ell\, \mathrm{Im}(K^2). \qquad (3.115b)$$

[†] Although this derivation depends on ψ_ℓ being complex, the integrals certainly have finite values for a real index of refraction as well. The reader may wish to find these values by taking a limit.

Hence, the absorption cross section is

$$\sigma_{\text{abs}} = \frac{2\pi}{|K|^2} \frac{\text{Im}(K^2)}{k} \sum_{\ell=1}^{\infty} (2\ell + 1)\left(\alpha_\ell |c_\ell|^2 + \beta_\ell |d_\ell|^2\right), \qquad (3.116)$$

which is proportional to $\mathscr{S}$, (3.110), but necessarily vanishes if n is real. This result is easily checked physically by recalling that the time-averaged rate of dissipation of energy per unit volume in ohmic losses is $\frac{1}{2} J \cdot E^*$, where here $J = \sigma E$ is the induced current density in the sphere. The rate of absorption of energy by the sphere is thus

$$W = \frac{1}{2}\sigma \int E \cdot E^* \, d^3r, \qquad (3.117)$$

and from Eq. (3.4) we can write $\sigma = c\,\text{Im}(K^2)/(4\pi k)$. The integral has essentially been done above, so that, when we normalize W by the incident time-averaged energy flux to compute

$$\sigma_{\text{abs}} = \frac{W}{c|E_0|^2/(8\pi)}, \qquad (3.118)$$

we regain (3.116).

For future application it is useful to rewrite Eq. (3.116) in a cleaner form in terms of the *absorption efficiency*:

$$Q_{\text{abs}}(\beta) = \frac{2}{\beta^2 |n|^2} \sum_{\ell=1}^{\infty} (2\ell + 1)\left(p_\ell |c_\ell|^2 + q_\ell |d_\ell|^2\right), \qquad (3.119a)$$

with

$$p_\ell(\beta) \equiv -\text{Im}\left[n^* \psi_\ell'(n\beta)\psi_\ell^*(n\beta)\right] \xrightarrow[n \text{ real}]{} 0, \qquad (3.119b)$$

$$q_\ell(\beta) \equiv -\text{Im}\left[n\psi_\ell'(n\beta)\psi_\ell^*(n\beta)\right] \xrightarrow[n \text{ real}]{} 0. \qquad (3.119c)$$

These expressions can also be verified directly, though tediously, from Eqs. (3.106) and (3.88).

Backscattering

In analogy with (3.98), Eqs. (3.95) and (3.97b) give for the amplitudes in the backward direction

$$S_1(\pi) = \frac{1}{2} \sum_{\ell=1}^{\infty} (-1)^\ell (2\ell + 1)(a_\ell - b_\ell) = -S_2(\pi). \qquad (3.120)$$

We define the *backscattering cross section* as the differential cross section at $\theta = \pi$ averaged over ϕ, which is tantamount to summing over all scattered polarizations:

$$\sigma_b \equiv \left(\frac{d\sigma}{d\Omega}\right)_{\text{av}}\Bigg|_{\theta=\pi} = \frac{1}{k^2}|S_1(\pi)|^2, \tag{3.121}$$

and is often called the *radar cross section* as well. Hence, the corresponding efficiency follows from (3.120):

$$Q_b = \frac{1}{4\pi\beta^2}\left|\sum_{\ell=1}^{\infty}(-1)^{\ell}(2\ell+1)(a_\ell - b_\ell)\right|^2. \tag{3.122}$$

This definition is a factor of 4π less than that usually employed by most other writers, and the discrepancy has an interesting origin. Backscattering studies often employ another function related to σ_b, a special case of the *normalized phase function*

$$p(\beta,\theta) \equiv \frac{1}{\sigma_{\text{sca}}}\left(\frac{d\sigma}{d\Omega}\right)_{\text{av}} = \frac{1}{2\pi\beta^2 Q_{\text{sca}}}\left(|S_2|^2 + |S_1|^2\right). \tag{3.123}$$

which integrates to unity over all solid angles. Chandrasekhar (1950) defined a similar function in his studies of radiative transfer, but normalized it by 4π so that the *average* phase function over the entire solid angle is unity. Somehow this factor of 4π has crept into the optics literature and persisted, where it continues to cause mischief. With the definition (3.123) we find for the backscattered phase function

$$p(\beta,\pi) = \frac{1}{\pi\beta^2 Q_{\text{sca}}}|S_1(\pi)|^2 = \frac{Q_b}{Q_{\text{sca}}}. \tag{3.124}$$

Another descriptive measure of the scattering is the *gain*, defined by van de Hulst (1957) as the ratio of the polarized intensity to the limiting values for an ideal isotropic scatterer (a totally reflecting sphere):

$$G_1 \equiv \frac{4|S_1|^2}{\beta^2}, \qquad G_2 \equiv \frac{4|S_2|^2}{\beta^2}, \tag{3.125}$$

which are also normalized by 4π. Then the average gain $\frac{1}{2}(G_1 + G_2)$ is just $4\pi p(\beta,\theta)$. One reason for taking note of these various descriptive measures is that detailed analysis of the backscattering is rather complicated, as will become clear in the following two chapters. The origin of this complexity is apparent in Fig. 3.6, in which we have plotted $p(\beta,\pi)$ over a small range of size parameters common in atmospheric scattering.

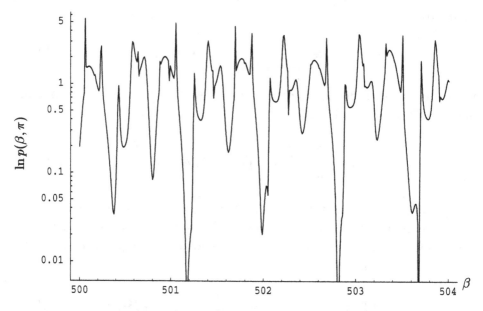

Fig. 3.6. The logarithm of the normalized phase function for backscattering over a small range of size parameters and for a refractive index $n = 1.333$.

Radiation pressure

The degree of polarization and the ellipticity can be found from Eqs. (3.71) and (3.68), respectively,

$$P = \frac{|S_1|^2 - |S_2|^2}{|S_1|^2 + |S_2|^2}, \qquad \tan\delta = \frac{\mathrm{Im}(S_1 S_2^*)}{\mathrm{Re}(S_1 S_2^*)}, \tag{3.126}$$

but neither is simply expressed in terms of the partial-wave coefficients. One last observable that can be so expressed is related to a measure of the momentum in the scattered radiation field. Observe that the total momentum removed from the incident beam in the forward direction is proportional to σ_{ext}, whereas that replaced is proportional to σ_{sca}. The total forward component scattered in any direction will be proportional to $\cos\theta\,(d\sigma/d\Omega)$, so that the total forward momentum of the scattered radiation is then

$$\langle\cos\theta\rangle\sigma_{\mathrm{sca}} \equiv \int \cos\theta \left(\frac{d\sigma}{d\Omega}\right) d\Omega. \tag{3.127}$$

It thus seems natural to define the *radiation-pressure cross section*

$$\sigma_{\mathrm{pr}} \equiv \sigma_{\mathrm{ext}} - \langle\cos\theta\rangle\sigma_{\mathrm{sca}} \tag{3.128}$$

as a measure of the forward momentum transferred to the target, and the resulting force on, or rate of transfer of momentum to, the target particle is just $F_{pr} = I_{in}\sigma_{pr}/c$. Debye (1909b) provided a rigorous derivation of this expression in his initial paper on this subject.

To evaluate σ_{pr}, given the other cross sections, one need only calculate the *asymmetry factor*

$$\langle \cos \theta \rangle = \frac{\pi}{2k^2\sigma_{sca}} \int_{-1}^{+1} \left(|S_+|^2 + |S_-|^2 \right) x \, dx, \tag{3.129}$$

a calculation very similar to that of the scattering cross section, (3.104). The requisite integrals over products of angular functions are again to be found in Appendix D, and the result is

$$\sigma_{sca}\langle \cos \theta \rangle = \frac{4\pi}{k^2} \sum_{\ell=1}^{\infty} \left(\frac{\ell(\ell+2)}{\ell+1} \operatorname{Re}(a_\ell a_{\ell+1}^* + b_\ell b_{\ell+1}^*) + \frac{2\ell+1}{\ell(\ell+1)} \operatorname{Re}(a_\ell b_\ell^*) \right). \tag{3.130}$$

Forces generated by laser light are quite capable of levitating micrometer-sized neutral particles, so the effects of radiation pressure have become amenable to study over the past few decades (e.g., Ashkin (1980)). The power required for levitation is inversely proportional to Q_{pr}. Further applications of the asymmetry parameter are plentiful in the field of atmospheric modeling (e.g., Liou (1980)) and, along with Q_{abs}, it will be studied further in Chapter 6.

One can, of course, calculate the intensity in any other fixed direction, as well as in angular sectors (Chýlek 1973; Pendleton 1982). For example, in Appendix D we find that

$$\pi_\ell(0) = \ell P_{\ell-1}(0), \qquad \tau_\ell(0) = \ell(\ell+1)P_\ell(0), \tag{3.131a}$$

with

$$P_\ell(0) = \frac{\cos\left(\dfrac{\ell\pi}{2}\right)}{\sqrt{\pi}} \frac{\Gamma\left(\dfrac{\ell+1}{2}\right)}{\Gamma\left(\dfrac{\ell}{2}+1\right)}, \tag{3.131b}$$

from which one can compute the intensities at $\theta = \pi/2$. We shall not pursue this here, but merely observe that π_ℓ and τ_ℓ vanish alternately as ℓ varies, so that the contribution to the total intensity is alternately completely electric or completely magnetic. In directions other than $0, \pi/2$, and π, as ℓ becomes large τ_ℓ tends to dominate π_ℓ (but see Appendix D), so, for spheres large compared with the wavelength, S_1 is dominated by contributions from magnetic multipoles and S_2 by electric multipoles. However, as can be seen from (3.97), there is an ambiguity in the direction of polarization when one views the scattered radiation along the axis of incidence, either forward or backward. In both cases electric and magnetic multipoles contribute with equal intensity and so there is an interference between them that must be kept in mind. This has been referred to as a *cross polarization effect* by van de Hulst (1957).

The partial-wave coefficients

With most of the measurable quantities expressed in terms of the partial–wave coefficients it now behooves us to understand the behavior of these coefficients in some detail. We shall make a start at that here and continue to investigate them further as circumstances demand. To begin, it will be very convenient to re-express (3.88) and (3.89) more compactly by defining new quantities

$$
\begin{aligned}
P_\ell^e &\equiv \psi_\ell(\beta)\psi_\ell'(n\beta) - n\psi_\ell(n\beta)\psi_\ell'(\beta), \\
Q_\ell^e &\equiv \xi_\ell(\beta)\psi_\ell'(n\beta) - n\psi_\ell(n\beta)\xi_\ell'(\beta), \\
P_\ell^m &\equiv \psi_\ell(n\beta)\psi_\ell'(\beta) - n\psi_\ell'(n\beta)\psi_\ell(\beta), \\
Q_\ell^m &\equiv \xi_\ell'(\beta)\psi_\ell(n\beta) - n\psi_\ell'(n\beta)\xi_\ell(\beta),
\end{aligned}
\tag{3.132}
$$

with the notation of Eq. (3.86). These are real *if the relative refractive index n is real.* The partial-wave coefficients are then equivalent to

External coefficients:

$$
a_\ell = \frac{P_\ell^e}{P_\ell^e + iQ_\ell^e}, \qquad b_\ell = \frac{P_\ell^m}{P_\ell^m + iQ_\ell^m};
\tag{3.133a}
$$

Internal coefficients:

$$
c_\ell = \frac{-in}{P_\ell^e + iQ_\ell^e}, \qquad d_\ell = \frac{in}{P_\ell^m + iQ_\ell^m}.
\tag{3.133b}
$$

This notation allows us not only to take advantage of the earlier observation that the external and internal denominators are similar, but also to gain additional insight into the physical meaning of the coefficients. To that end, we take n to be real for the moment and define real *phase shifts δ_ℓ* as follows:

$$
\tan\delta_\ell^e \equiv \frac{P_\ell^e}{Q_\ell^e}, \qquad \tan\delta_\ell^m \equiv \frac{P_\ell^m}{Q_\ell^m},
\tag{3.134}
$$

so that

$$
\begin{aligned}
a_\ell &= \frac{\tan\delta_\ell^e}{\tan\delta_\ell^e + i} = \frac{1}{2}\left(1 - e^{2i\delta_\ell^e}\right), \\
b_\ell &= \frac{\tan\delta_\ell^m}{\tan\delta_\ell^m + i} = \frac{1}{2}\left(1 - e^{2i\delta_\ell^m}\right).
\end{aligned}
\tag{3.135a}
$$

The coefficient a_ℓ in (3.133a) can be expanded into

$$a_\ell = \frac{P_\ell^{e2}}{P_\ell^{e2} + Q_\ell^{e2}} - i\frac{P_\ell^e Q_\ell^e}{P_\ell^{e2} + Q_\ell^{e2}},$$

(3.136)

with a similar expression for b_ℓ. Clearly, the real part of a_ℓ is identical to its squared magnitude and ranges over $(0, 1)$. Additionally, the magnitude of the imaginary part is bounded above by $\frac{1}{2}$, since its maximum is attained when $Q_\ell^e = P_\ell^e$. These conclusions follow also from (3.135a):

$$\operatorname{Re} a_\ell = |a_\ell|^2 = \tfrac{1}{2}[1 - \cos(2\delta_\ell^e)] = \sin^2 \delta_\ell^e,$$
$$\operatorname{Im} a_\ell = \tfrac{1}{2}\sin 2\delta_\ell^e,$$

(3.137)

with a similar set for b_ℓ. In S-matrix language, reaching these upper bounds saturates the unitarity limit, and cannot happen if there is any absorption. Thus, in the absence of absorption the efficiencies can be written as

$$Q_{\text{ext}} = Q_{\text{sca}} = \frac{2}{\beta^2}\sum_{\ell=1}^{\infty}(2\ell + 1)\left(\sin^2 \delta_\ell^m + \sin^2 \delta_\ell^e\right),$$

(3.138)

which is reminiscent of the scalar expression (2.28).

Return to the scalar theory for a moment and recall the expression (2.22) for the partial scattering amplitude,

$$f_\ell(k) = \frac{e^{-i\ell\pi/2}}{2ik}\,[S_\ell(ka) - 1],$$

and $S_\ell(ka) \equiv e^{2i\delta_\ell}$ is the partial-wave 'scattering matrix' in terms of the phase shifts. Similarly, we define here the matrix

$$S_\ell \equiv \begin{pmatrix} S_\ell^E & 0 \\ 0 & S_\ell^M \end{pmatrix},$$

(3.139)

and conservation of energy again requires that S_ℓ be unitary for real n. Thus, these quantities must be phase factors and can only be those introduced in (3.135):

$$a_\ell = \tfrac{1}{2}\left[1 - S_\ell^E(\beta)\right], \qquad b_\ell = \tfrac{1}{2}\left[1 - S_\ell^M(\beta)\right],$$

(3.140)

with obvious replacements in S_1 and S_2. Substitution into (3.88) leads to explicit expressions similar to that of Eq. (2.88):

$$S_\ell^E(\beta) = -\frac{\zeta_\ell^{(2)}(\beta)}{\zeta_\ell^{(1)}(\beta)}\left(\frac{\ln' \zeta_\ell^{(2)}(\beta) - n^{-1}\ln' \psi_\ell(n\beta)}{\ln' \zeta_\ell^{(1)}(\beta) - n^{-1}\ln' \psi_\ell(n\beta)}\right),$$

$$S_\ell^M(\beta) = -\frac{\zeta_\ell^{(2)}(\beta)}{\zeta_\ell^{(1)}(\beta)}\left(\frac{\ln' \zeta_\ell^{(2)}(\beta) - n\ln' \psi_\ell(n\beta)}{\ln' \zeta_\ell^{(1)}(\beta) - n\ln' \psi_\ell(n\beta)}\right),$$

(3.141)

where once again, and subsequently, it becomes useful to define the logarithmic derivative as $\ln' f(x) \equiv d \ln f(x)/dx = f'(x)/f(x)$. For real n the observation that z^*/z is just a phase factor confirms the identification of these quantities with those in (3.134).

As in the scalar case, Eqs. (2.30)–(2.32), the presence of absorption implies that the S-matrix is no longer unitary, which is now expressed by the inequalities

$$\left| S_\ell^E \right| \leq 1, \qquad \left| S_\ell^M \right| \leq 1, \qquad (3.142)$$

as expressed by Eqs. (3.115). Alternatively, $\operatorname{Re} a_\ell \leq 1$, $|\operatorname{Im} a_\ell| \leq \frac{1}{2}$, and similarly for b_ℓ.

The limiting behavior of the partial-wave coefficients can be adduced from the asymptotic properties of the Bessel functions in Appendix A. For $\beta \ll 1$,

$$a_\ell = -i\beta^{2\ell+1} \frac{(2\ell+1)(\ell+1)(n^2-1)}{[(2\ell+1)!!]^2 (\ell n^2 + \ell + 1)} \left[1 + O\left(\beta^2\right) \right] \simeq -i\delta_\ell^e, \quad (3.143a)$$

$$b_\ell = -\frac{i}{2} \frac{\beta^{2\ell+3}(\ell+1)}{[(2\ell+1)!!]^2} \frac{n^2-1}{2\ell+3} \left[1 + O\left(\beta^2\right) \right] \simeq -i\delta_\ell^m, \qquad (3.143b)$$

which will be quite useful presently. In the opposite extreme, $\beta \gg 1$, the coefficients approach no definite limit, but rather oscillate. Indeed, it is just this limit that will occupy us completely in Chapter 5.

Figure 3.7 illustrates the oscillatory character of the partial-wave coefficients as functions of ℓ, as well as their rapid decay to zero for $\ell \gtrsim \beta$. Such behavior guarantees convergence of the Mie series while at the same time emphasizing the slow rate of that convergence. This cutoff can be attributed to a rapid decay of the phase shifts at $\ell \sim \beta$, and clearly supports the principle of localization.

Dispersion relations and $\operatorname{Im} S(\beta, 0)$

For our purposes the *causality condition* on the scattering process can be stated as follows: the outgoing wave cannot appear before the incoming wave has reached the target. Often this is expressed in terms of the Fourier transform of an output time signal $g(t)$; if we require $g(t)$ to vanish for $t < 0$ this can be written as

$$G(\omega) = \int_0^\infty g(t) e^{i\omega t} \, dt. \qquad (3.144)$$

Under broad physical conditions both $g(t)$ and $G(\omega)$ can be presumed square integrable; i.e., both $|g(t)|^2$ and $|G(\omega)|^2$ have finite integrals over the real

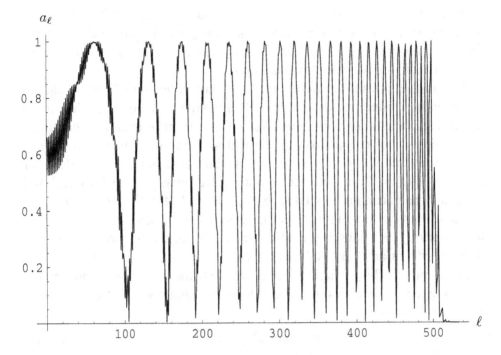

Fig. 3.7. The partial-wave coefficient $a_l(\beta)$ for $\beta = 500.5$ and $n = 1.3$. The individual points have been joined in a continuous curve as an aid to the eye.

line. The key theorem in analyzing these processes, due to Titchmarsh (1948), imposes a slightly stronger requirement.

Titchmarsh's Theorem:

If a square integrable function $G(\omega)$ satisfies any one of the following four conditions, then it satisfies all of them.

(i) The inverse Fourier transform $g(t)$ of $G(\omega)$ vanishes for $t < 0$.

(ii) $G(\omega)$ is, for all ω other than a set of measure zero, the limit as $\gamma \to 0$ of an analytic function $G(\omega + i\gamma)$ that is regular in the upper half-plane and square integrable over any line parallel to the real axis.

(iii) The real and imaginary parts of G verify the first Plemelj formula:

$$\operatorname{Re} G(\omega) = \frac{1}{\pi} P \int_{-\infty}^{\infty} \frac{\operatorname{Im} G(\omega')}{\omega' - \omega} \, d\omega. \qquad (3.145a)$$

(iv) The real and imaginary parts of G verify the second Plemelj formula:

$$\operatorname{Im} G(\omega) = -\frac{1}{\pi} P \int_{-\infty}^{\infty} \frac{\operatorname{Re} G(\omega')}{\omega' - \omega} \, d\omega. \qquad (3.145b)$$

A function $G(\omega)$ satisfying any one (and therefore all) of the conditions of Titchmarsh's theorem is called a *causal transform*, and for such functions the *dispersion relations* (3.145) are both necessary and sufficient conditions for the validity of the causality condition.

We must emphasize, however, that a square integrable function that is the boundary value on the real line of an analytic function regular in the upper half-plane is *not necessarily* a causal transform; it must also be square integrable on any line parallel to the real axis, as in condition (ii). It may also happen that $G(\omega)$ is not square integrable at all, but only bounded. For example, $G(\omega) = C$, a constant, is not square integrable, and its real and imaginary parts cannot be related through Eqs. (3.145). Rather, if $g(t)$ is the causal output corresponding to an input signal $f(t)$, then $g(t) + bf(t)$, with b an arbitrary real constant, is also a causal output. Thus, in this case one must specify an arbitrary additive constant to relate Re G and Im G.

Let us pursue this scenario of constant Re G for a moment. Convenience alone suggests we specify the constant as some definite value of G, say $G(\omega_0)$. Then the function

$$D(\omega) \equiv \frac{G(\omega) - G(\omega_0)}{\omega - \omega_0} \qquad (3.146)$$

is a causal transform vanishing on any line parallel to the real axis as $|\omega| \to \infty$. It also satisfies the dispersion relation

$$G(\omega) = G(\omega_0) + \frac{\omega - \omega_0}{i\pi} P \int_{-\infty}^{\infty} \frac{G(\omega') - G(\omega)}{\omega' - \omega_0} \frac{d\omega'}{\omega' - \omega}, \qquad (3.147)$$

similar to that of Eq. (2.7). This is generally called a dispersion relation with one subtraction, and most often one chooses $\omega_0 = 0$ so that the constant is just the static value of G.

An explicit example of this subtraction procedure is provided by the propagation of light in a homogeneous dielectric medium. In such a system the polarization vector $\boldsymbol{P}$ is related to the electric field $\boldsymbol{E}$ by the constitutive relation $\boldsymbol{P} = \chi \boldsymbol{E}$, where the electric susceptibility in the optical range is given by (e.g., Jackson (1975))

$$\chi(\omega) = \frac{\epsilon(\omega) - 1}{4\pi} = \frac{n^2(\omega) - 1}{4\pi} \simeq \frac{Ne^2}{m\omega^2}, \qquad (3.148)$$

in terms of N molecular dipoles. Thus, while $n(\omega)$ is not square integrable, $n^2 - 1$ is so and satisfies Titchmarsh's theorem. Indeed, $n + 1 \to 2$ for large ω and $n - 1 \to 0$, so that $n(\omega) - 1$ itself satisfies the theorem, is a causal transform, and verifies the dispersion (or Kramers–Kronig) relations (2.9).

If $G(\omega)$ diverges linearly as $|\omega| \to \infty$ we need two subtractions to obtain a square integrable function. The additional constant can be either the derivative $G'(\omega_0)$, or the value of the function itself at a second point, $G(\omega_1)$, and one proceeds as before to construct dispersion relations. In general, if $G(\omega) = O(\omega^n)$ as $|\omega| \to \infty$, dispersion relations will require $n+1$ subtractions. Numerous examples are given by Nussenzveig (1972).

With this brief background we can return specifically to light scattering from a homogeneous sphere. The causality condition here is readily found by reference to Fig. 2.3, along with the observation that the shortest path connecting an incident wavefront along the z-axis with a distant detection point via the target is that taken by a ray undergoing specular reflection in direction θ at the surface. Causality implies that the scattered wave cannot arrive at a point r far removed from the scatterer prior to time

$$t_0 = \frac{1}{c}[r - 2a\sin(\theta/2)]. \tag{3.149}$$

Hence, if the incident wave vanishes for $t < z/c$ the scattered wave in direction θ must vanish for $t < 0$.

Now consider the scattering amplitudes (3.95) and note that, for real n, they satisfy the symmetry relations

$$S_j(-\beta, \theta) = S_j^*(\beta, \theta), \qquad j = 1, 2. \tag{3.150}$$

These follow readily from the realization that, if the time-dependent electric field is to be real, its Fourier transform must satisfy $E(-\omega) = E^*(\omega)$. Or, one can prove them directly for $a_\ell(\beta)$ and $b_\ell(\beta)$ by employing the relations (A.14). With (3.150) and the identities

$$P\int_{-\infty}^{\infty} \frac{d\omega'}{\omega'(\omega' - \omega)} = 2P\int_0^{\infty} \frac{d\omega'}{\omega'^2 - \omega^2} = 0, \tag{3.151}$$

it is straightforward to construct a dispersion relation for the single forward-scattering amplitude $S(\beta, 0)$. Since $\beta = \omega a/c$, we shall here write this as $S(\omega, 0)$ and note from (3.73) and the asymptotic behavior implied by Fig. 3.4 (and proved in Chapter 5) that

$$\sigma_{\text{ext}} = \frac{4\pi c^2}{\omega^2} \operatorname{Re} S(\omega a/c, 0) \xrightarrow[\omega \to \infty]{} 2\pi a^2. \tag{3.152}$$

That is,

$$\frac{\operatorname{Re} S(\omega, 0)}{\omega^2} \xrightarrow[\omega \to \infty]{} \frac{a^2}{2c^2}, \tag{3.153}$$

suggesting that

$$D(\omega) \equiv \frac{S(\omega, 0)}{\omega^2} - \frac{a^2}{2c^2} \tag{3.154}$$

is a causal transform. (The symmetry condition (3.150) confirms that $\operatorname{Im} S(\omega, 0)$ vanishes as $\omega \to \infty$.)

These considerations guarantee the existence of dispersion relations, for a real index of refraction, for $D(\omega)$. Just like with Eqs. (2.7) and (2.9), it is convenient to express these as one-sided integrals (Kramers–Kronig relations):

$$\operatorname{Re} D(\omega) = \frac{2}{\pi} P \int_0^\infty \frac{\omega' \operatorname{Im} D(\omega')}{\omega'^2 - \omega^2} \, d\omega', \qquad (3.155a)$$

$$\operatorname{Im} D(\omega) = -\frac{2\omega}{\pi} P \int_0^\infty \frac{\operatorname{Re} D(\omega')}{\omega'^2 - \omega^2} \, d\omega'. \qquad (3.155b)$$

In turn, the identities (3.151) and the optical theorem allow us to rewrite (3.155b) as

$$2\pi c^2 \frac{\operatorname{Im} S(\omega, 0)}{\omega^3} = -P \int_0^\infty \frac{\sigma_{\text{ext}}(\omega')}{\omega'^2 - \omega^2} \, d\omega'. \qquad (3.156)$$

The left-hand side of this expression is readily evaluated for $\omega \to 0$, either directly from Eqs. (3.95) and (3.143), or from the result (4.1) in the following chapter. Either way, we can now express the integral in (3.156) in terms of the *static* dielectric function, and it is tidier to change variables from frequency to wavelength:

$$\int_0^\infty \sigma_{\text{ext}}(\lambda) \, d\lambda = 4\pi a^3 \frac{\varepsilon(0) - 1}{\varepsilon(0) + 2}. \qquad (3.157)$$

Thus, the role of the imaginary part of the forward scattering amplitude apparently is to provide a *sum rule* for the extinction. Note that the integrated extinction cross section is bounded above by the particle volume, independent of the particle's material.

One can actually obtain dispersion relations for $\theta \neq 0$, in which $D(\omega)$ in (3.155) acquires a factor $e^{2ika \sin(\theta/2)}$, $k = \omega/c$. This is of less physical interest, however, if for no other reason than that there is no optical theorem to reduce $\operatorname{Re} D(\omega, \theta)$ to a simple experimental observable. Finally, we note that Purcell (1969) derived the expression (3.157) in a slightly different way in an application to a *medium* of interstellar grains.

4

First applications of the Mie solution

In this chapter we first examine some models for which only the leading-order terms in the partial-wave series are required, thereby providing relatively simple analytic descriptions. These models in turn suggest further approximations and extrapolations away from very small size parameters. The extension to large size parameters begins here and continues full blown in the following chapter.

4.1 Small size parameters

Perhaps the simplest application of the Mie theory is to dielectric spheres with real refractive indices in the long-wavelength limit: $\beta = N_0 ka \ll 1$. In this limit the expansions (3.143) are valid and the electric–multipole contributions a_ℓ are dominant. However, these expansions assert that $n\beta \ll 1$ as well as the size parameter itself being small, and it is this set of conditions which we shall adopt as our definition of *Rayleigh scattering* (Strutt 1871). An illuminating history of this phenomenon has been provided by Young (1982). To leading order, then, we consider only $a_1(\beta)$ in the expressions (3.95), and from Appendix D the associated angular functions are $\pi_1 = 1$ and $\tau_1 = \cos\theta$. Substitution yields

$$S_1(\theta) \simeq \tfrac{3}{2}(a_1\pi_1 + b_1\tau_1) \simeq -i\beta^3 \frac{n^2 - 1}{n^2 + 2}, \tag{4.1a}$$

$$S_2(\theta) \simeq \tfrac{3}{2}(b_1\pi_1 + a_1\tau_1) \simeq -i\beta^3 \frac{n^2 - 1}{n^2 + 2} \cos\theta, \tag{4.1b}$$

and the transfer matrix (3.69) reduces to

$$T = \beta^6 \left| \frac{n^2 - 1}{n^2 + 2} \right|^2 \begin{pmatrix} \cos^2\theta & 0 & 0 & 0 \\ 0 & 1 & 0 & 0 \\ 0 & 0 & \cos\theta & 0 \\ 0 & 0 & 0 & \cos\theta \end{pmatrix}. \tag{4.2}$$

With the presumption of linear incident polarization the differential cross section (3.101) reduces to

$$\frac{d\sigma}{d\Omega} \simeq k^4 a^6 \left|\frac{n^2-1}{n^2+2}\right|^2 (\cos^2\theta\cos^2\phi + \sin^2\phi). \tag{4.3}$$

The total cross section follows from angular integration, both extinction and scattering cross sections becoming

$$\sigma_{\text{sca}} \simeq \frac{2\pi}{k^2} 3|a_1|^2 \simeq \frac{8\pi}{3} k^4 a^6 \left|\frac{n^2-1}{n^2+1}\right|^2. \tag{4.4}$$

Note that the structure of σ_{sca} in (4.4) is just that of the Lorentz–Lorenz formula for the molecular electric polarizability; with $n^2 = \varepsilon$ it is the Clausius–Mossotti formula (e.g., Jackson (1975)).

Equations (4.1) and (3.73) yield the leading-order expression

$$Q_{\text{ext}}(\beta) \simeq 4\beta \,\text{Im}\left(\frac{n^2-1}{n^2+2}\right) = 12\beta \frac{\text{Im}\,\varepsilon}{|\varepsilon+2|^2}, \tag{4.5}$$

which vanishes for real n or ε. Thus, if n is complex the amplitudes must be calculated at least through order β^4 to compute the absorption efficiency from $Q_{\text{ext}} - Q_{\text{sca}}$. However, then the leading-order contribution to Q_{abs} must be just that of (4.5), so that, to the extent that there is any significant absorption at all, it will clearly dominate the scattering for very small size parameters. Recall, however, that $n = n(\omega)$, so that (4.5) only implicitly provides the correct frequency dependence of the cross section.

It is customary to sum the differential cross section over scattered polarizations, which is equivalent to averaging over ϕ. That is, we focus attention on the phase function (3.123), which reduces here to

$$p(\beta,\theta) \simeq \frac{3}{16\pi}(1 + \cos^2\theta). \tag{4.6}$$

This expression is identical to the angular distribution of radiation from crossed dipoles oscillating along the x- and y-axes, Table D.1. From (3.71) and (4.1) we also find for the degree of polarization

$$P(\theta) \simeq \frac{\sin^2\theta}{1 + \cos^2\theta}, \tag{4.7}$$

which is completely symmetric about $\theta = \pi/2$. Indeed, at this angle the scattered light is completely perpendicularly polarized. Both the differential cross section and the degree of polarization for an unpolarized incident beam are exhibited in Fig. 4.1.

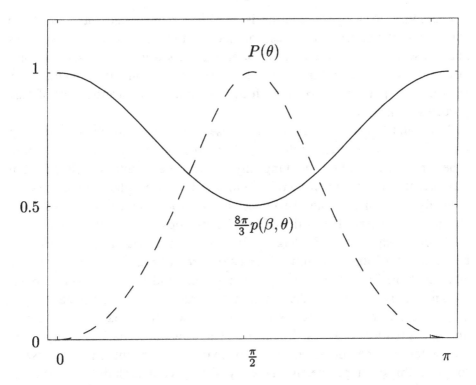

Fig. 4.1. The normalized phase function for Rayleigh scattering (solid line), along with the corresponding degree of polarization (dashed line), in the case of an unpolarized incident beam.

According to (4.4) the scattered intensity is proportional to λ^{-4}, being thus dominated by the blue end of the spectrum. It was therefore proposed by Lord Rayleigh that this result explains the blue of the sky as well as its polarization (Strutt 1899), which it does following some additional comment.

If one argues that these phenomena arise when sunlight is scattered from molecules in the upper atmosphere, a counter argument that scattering can take place only from inhomogeneities can be made. In an effectively random homogeneous medium the phases of scattered radiation tend to cancel out in all but the forward direction, so that the incident wave merely propagates forward through the medium with a speed different from c. In the atmosphere, however, small density fluctuations destroy the homogeneity, thereby rendering the Rayleigh argument correct. Evidently the Earth's atmosphere is dilute enough that summing the scattering from molecules is equivalent to considering scattering from fluctuations. Leonardo da Vinci apparently recognized some of the essential features of the phenomenon in wood smoke as long ago as 1500!

This long-wavelength scattering is also the origin of the wanness of the winter sun at northern latitudes and the predominant redness of sunrises and sunsets. The blue end of the spectrum is scattered out at low solar angles and the directly transmitted red wavelengths dominate. An effectively thicker atmosphere enhances the effect, as do unusual concentrations of dust particles in the air.

We should hasten to point out that observation of atmospheric scattering is not in *exact* agreement with the picture presented above – neither should it be expected to be. For one thing, any significant absorption in the medium can alter the spectrum considerably. For another, atmospheric molecules are certainly not spheres, and the molecules can rotate and vibrate. Rayleigh was indeed concerned with the effect of anisotropy in the actual data.

Figure 4.1 shows that Rayleigh scattering is symmetric in the forward and backward directions, exhibiting the characteristic radiation pattern of an electric dipole for parallel incident polarization, and that of crossed dipoles for an unpolarized beam. As the target sphere becomes somewhat larger, $\beta \gtrsim 1$, significantly more terms must be retained in the partial-wave series and the situation changes drastically. The general scenario of large size parameters will be addressed in the following chapter, but it may be useful to place things in perspective here by simply contemplating the angular distribution of scattered radiation for slightly larger spheres. In Fig. 4.2 we have constructed polar diagrams of the scattered intensities for two values of the size parameter just greater than unity. Even for these relatively small particles the dominance of forward scattering is striking, as is the difference between the two patterns. In Chapter 2 we saw how the forward diffraction peak dominated the scattering for impenetrable spheres, and the same can surely be expected for transparent targets. What is most instructive in Fig. 4.2, however, is that the concept of a 'large' size parameter is very relative!

An interesting variant of the above scenario arises when n decreases toward unity while β is not too large. Rather than require the particle size to be much less than a wavelength, we presume that

$$|n - 1| \ll 1, \qquad 2\beta|n - 1| \ll 1. \tag{4.8}$$

The first condition implies that the optical properties of the particle and medium are almost identical, so that we expect the incident and internal fields not to differ greatly. The second inequality, not entirely unrelated to the first, places a restriction on the *phase shift* $2\beta|n-1|$ experienced by the central light ray in passing through the sphere – recall the somewhat analogous shift exhibited in Fig. 2.3 for the impenetrable sphere. This condition of small phase shifts corresponds precisely to the Born approximation in quantum

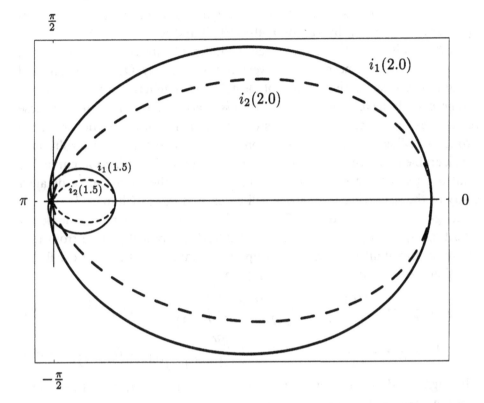

Fig. 4.2. Polar diagrms for scattering from a small dielectric sphere at two different size parameters; $\beta = 1.5$ (thin lines), and $\beta = 2.0$ (thick lines). The intensities are plotted on the same scale, indicating how rapidly the forward scattering grows as β increases from unity.

mechanics, and here is referred to as *Rayleigh–Gans theory* (Strutt 1881; Gans 1925). The model can be quite useful in studying small particles of arbitrary shape, for one approach to its formulation is to begin with the Rayleigh results of Eqs. (4.1) and normalize them by the particle volume V. The consequent amplitudes per unit particle volume can then be defined in the limit of vanishing particle radius and thus be independent of shape:

$$s_1 \equiv \lim_{a \to 0} \frac{S_1}{V} = -i\frac{3k^3}{4\pi}\frac{n^2 - 1}{n^2 + 2},$$

$$s_2 \equiv \lim_{a \to 0} \frac{S_2}{V} = -i\frac{3k^3}{4\pi}\frac{n^2 - 1}{n^2 + 2}\cos\theta. \qquad (4.9)$$

An immediate consequence of these expressions is that Q_{sca}, for example, is proportional to the square of the phase shift, and hence is always very small. Extensive discussions of the Rayleigh–Gans theory and its various applications are given by van de Hulst (1957) and Kerker (1969).

As β remains small, but the refractive index is allowed to become large, optical resonances emerge. These were first studied by Debye (1909b) and are almost-self-sustaining modes of electric and magnetic vibration that do not necessarily have anything to do with scattering. The conditions for these resonances are exactly those discussed in Chapter 3 in connection with the extinction efficiency, and one solution for $\beta \ll 1$ and large n results in a considerable enhancement of the ripple structure in the extinction curve. We shall not pursue these large-n phenomena any further here, but a detailed study of the scattering resonances will be carried out in Chapter 7.

Long-wavelength scattering from small perfectly conducting (completely reflecting) spheres presents a somewhat different physical picture from that above. In this limit $n \to i\infty$ the asymptotic behavior of the spherical Bessel functions (Appendix A) reduces Eqs. (3.88) to

$$a_\ell \xrightarrow[n \to i\infty]{} \frac{\psi'_\ell(\beta)}{\zeta'_\ell(\beta)} \xrightarrow[\beta \ll 1]{} i\frac{(\ell + 1)\beta^{2\ell+1}}{\ell(2\ell - 1)!!(2\ell + 1)!!}\left[1 + O\left(\beta^2\right)\right], \quad (4.10a)$$

$$b_\ell \xrightarrow[n \to i\infty]{} \frac{\psi_\ell(\beta)}{\zeta_\ell(\beta)} \xrightarrow[\beta \ll 1]{} -i\frac{\beta^{2\ell+1}}{(2\ell - 1)!!(2\ell + 1)!!}\left[1 + O\left(\beta^2\right)\right]. \quad (4.10b)$$

Although the dominant terms in the series are still $\ell = 1$, in this case both types of coefficient contribute:

$$a_1 \simeq \frac{2i}{3}\beta^3, \qquad b_1 \simeq -\frac{i}{3}\beta^3. \quad (4.11)$$

Whereas long-wavelength scattering from a small dielectric sphere is dominated by a single electric dipole, in the present case the source consists of orthogonal magnetic and electric dipoles. The equal importance of a_1 and b_1 can be attributed to induced surface currents on the conducting sphere. Individual intensities are found to be $i_1 \simeq \beta^6(1 - \frac{1}{2}\cos\theta)^2$ and $i_2 \simeq \beta^6(\frac{1}{2} - \cos\theta)^2$, and summation over final polarizations provides the differential and total cross sections, as well as the degree of polarization:

$$\left(\frac{d\sigma}{d\Omega}\right)_{\text{av}} \simeq k^4 a^6\left[\tfrac{5}{8}(1 + \cos^2\theta) - \cos\theta\right], \quad (4.12a)$$

$$\sigma_{\text{sca}} \simeq \frac{10\pi}{3}k^4 a^6, \quad (4.12b)$$

$$P(\theta) \simeq \frac{3\sin^2\theta}{5(1 + \cos^2\theta) - 8\cos\theta}. \quad (4.12c)$$

The angular distributions are exhibited in Fig. 4.3, demonstrating that there is a large departure from those of Fig. 4.1.

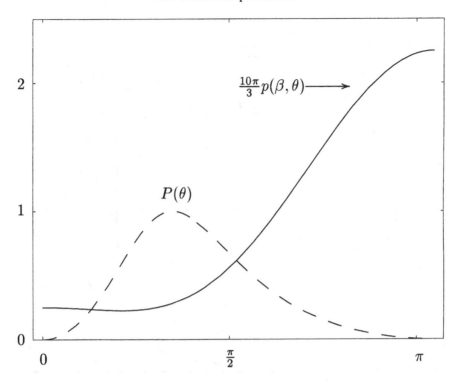

Fig. 4.3. The normalized phase function for scattering from a small perfectly conducting sphere (solid line), along with the corresponding degree of polarization (dashed line), in the case of an unpolarized incident beam.

A more revealing picture of the scattering is provided by the polar diagram of the intensities in Fig. 4.4. The process is strongly dominated by backscattering, differing considerably from the results of Rayleigh scattering. This is to be expected, of course, because an infinitely conducting surface provides an entirely different physical process in which the Rayleigh condition $n\beta \ll 1$ is strongly violated, prohibiting the fields from entering the particle. For Rayleigh scattering one assumes that full and instantaneous penetration of the fields into the sphere occurs.

Further results for scattering from very small, not necessarily homogeneous, spherical particles are given by van de Hulst (1957), Kerker (1969), and Bohren and Huffman (1983), and as a last example here involving small particles we consider magnetic spheres, for which the relative magnetic permeability $\mu \neq 1$. In this case one must lift the restriction against nonmagnetic materials in the derivation of the scattering coefficients. This translates into an additional factor of μ in κ_2 in Eq. (3.4), which finally emerges as the replacement $n \to n/\mu$ in the coefficients of Eqs. (3.88) and (3.89). An imme-

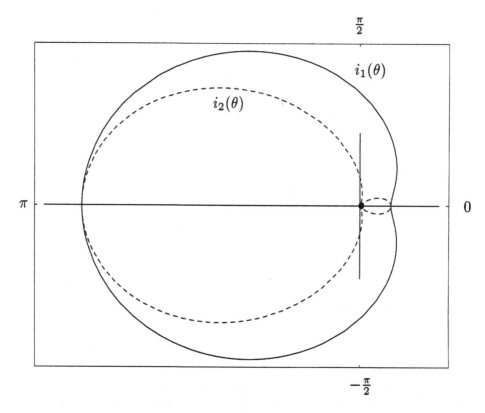

Fig. 4.4. A polar diagram for scattering from a small perfectly conducting sphere located at the origin, illustrating the dominance of backscattering. The intensities i_1 and i_2 are the squares of the absolute values of the scattering amplitudes for the two polarizations.

diate consequence of this is that if $\mu = \varepsilon$, then $a_\ell = b_\ell$ and from (3.120) the backscattering cross section vanishes for any size of particle.

For small magnetic particles one must re-examine the asymptotic behavior of the external coefficients, and we find the following forms:

$$a_\ell \simeq i\beta^{2\ell+1} \frac{\ell+1}{(2\ell-1)!!(2\ell+1)!!} \frac{\mu - n^2}{n^2\ell + \mu(\ell+1)}$$

$$\xrightarrow[\ell=1]{} -i\tfrac{2}{3}\beta^3 \frac{\varepsilon - 1}{\varepsilon + 2};$$

(4.13a)

$$b_\ell \simeq i\beta^{2\ell+1} \frac{(\ell+1)(\mu-1)}{(2\ell-1)!!(2\ell+1)!!(\ell\mu+\ell+1)}$$

$$\xrightarrow[\ell=1]{} i\tfrac{2}{3}\beta^3 \frac{\mu - 1}{\mu + 2},$$

(4.13b)

which agree with the Rayleigh expressions for $\mu = 1$. However, for magnetic spheres the magnetic and electric dipoles make identical contributions in terms of their respective polarizabilities. We leave it to the reader to extract the relation between ε and μ leading to zero forward scattering, in which case the polarization of the scattered radiation is the same as that of the incident radiation (Kerker *et al.* 1983).

4.2 Larger size parameters

In his classic monograph van de Hulst (1957) classified light-scattering processes in terms of regions in the n-β plane. Rayleigh scattering, for example, corresponds to small β and $n > 1$, whereas geometric optics has similar refractive indices but $\beta \gg 1$. This is a very useful classification scheme because it permits ready construction of various models through systematic approximations in n and β, as was done above for the Rayleigh–Gans regime defined by (4.8). In this spirit, van de Hulst also observed that that model could be extended to larger particles that were somewhat tenuous, or 'soft', and he defined the region of *anomalous diffraction* through the restrictions

$$|n - 1| \ll 1, \qquad \beta \gg 1, \tag{4.14}$$

while maintaining the phase shift $2\beta(n - 1)$ approximately constant.

Envision a large sphere with real $n \simeq 1$, behind which is positioned a (fictitious) screen normal to the incident rays, as in Fig. 4.5. Because β is large we can follow a ray through the sphere and, owing to (4.14), it will deviate very little from a straight line. The Fresnel coefficients tend to vanish as $n \to 1$, so the amplitude is essentially unchanged at the point P on the screen, though the phase of the field suffers a delay $\psi \equiv kx(n - 1)$, where $x = 2a \sin \varphi$ is the path length through the sphere. Under the given constraints we expect a sharp shadow on the screen, and, if the field outside the shadow circle is taken as unity the field to be added to the incident plane wave is $e^{i\psi} - 1$, so that the sum of the original and scattered fields in the shadow area is $e^{i\psi}$.

Now employ Babinet's principle in the manner leading to Eq. (2.54), which yields for the forward scattering amplitude

$$S(0) = \frac{k^2}{2\pi} \int \int_A \left(1 - e^{i2\beta(n-1)\sin\varphi}\right) dA, \tag{4.15}$$

where A is the area of the shadow projected onto the screen.† In polar

---

† The sign in the exponential is dictated by the choice of sign conventions at the beginning of Chapter 2, and differs from that of van de Hulst (1957), say.

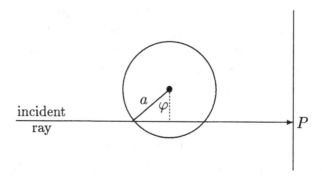

Fig. 4.5. An incident beam passes almost unperturbed through a sphere of refractive index $n \simeq 1$, other than suffering a phase lag $2\beta(n-1)\sin\varphi$.

coordinates,

$$S(0) = \frac{k^2}{2\pi} \int_0^{\pi/2} \left(1 - e^{i2\beta(n-1)\sin\varphi}\right) \cos\varphi \sin\varphi \, d\varphi, \qquad (4.16)$$

which is readily evaluated by changing variables to $y = \sin\varphi$ and integrating by parts. The optical theorem then provides the extinction efficiency for anomalous diffraction:

$$Q_{\text{ext}}^{\text{ad}}(\rho) = 4\left(\frac{1}{2} - \frac{\sin\rho}{\rho} + \frac{1-\cos\rho}{\rho^2}\right), \qquad (4.17)$$

where $\rho \equiv 2\beta(n-1)$ is the phase-shift parameter defined earlier.

This is a remarkable expression, for it summarizes succinctly numerous salient features of the extinction curve in the Mie theory. Perhaps the most striking of these is shown in Fig. 4.6, in which we compare, for n not even too close to unity, the full Mie curve and $Q_{\text{ext}}^{\text{ad}}$ as functions of β and ρ, respectively. Although the latter has a slightly smaller amplitude, it matches the periodicity stated in Eq. (3.100) exactly. Furthermore, as stated in connection with that equation, the underlying reason for the periodicity is seen quite clearly to originate from interference between transmitted and diffracted light. Equation (4.17) cannot reproduce subtler features such as the ripple, of course, but that was already clear from Fig. 3.5 for n approaching unity. Very probably the expression (4.17) remains valid for $|n-1| \ll |n+1|$ (Sharma 1992).

Other features evident from (4.17) are the correct limiting value of 2 for $\beta \gg 1$, and the reduction to the Rayleigh–Gans expression $\frac{1}{2}\rho^2$ for $\rho \ll 1$. In addition, if the refractive index is complex, $n = m + i\kappa$, the exponential in (4.16) yields an attenuation factor $e^{-2\beta\kappa\sin\varphi}$ that leads to a definite expression for the absorption efficiency. This is in complete agreement with the more

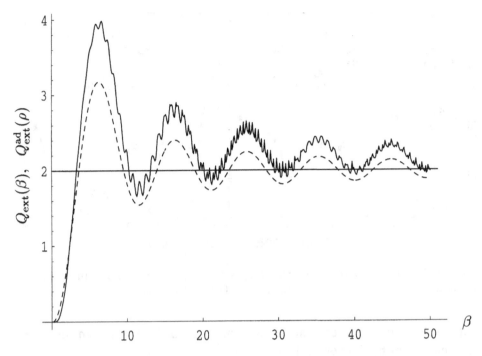

Fig. 4.6. A comparison of the full Mie extinction curve (solid line) with that for $Q_{\text{ext}}^{\text{ad}}(\rho)$ (dashed line) given by Eq. (4.17). The refractive index is that for water at optical wavelengths.

general form in geometric optics that we shall discuss presently, so we shall not pursue it here.

This model of anomalous diffraction has some utility in studying nonspherical particles, under the stated conditions. The integral in (4.15) depends on the spherical nature of the target only through the path length x of the ray through the particle, and the cross sectional area A can be anything an irregular shape might present. Thus, if one can compute x and evaluate the integral, any geometry can be treated in this region. Several other regular shapes have been treated in this way by Chýlek and Klett (1991).

Van de Hulst also defined the region intermediate between Rayleigh–Gans and anomalous diffraction as

$$\beta|n-1| \ll 1, \qquad \beta \gg 1. \tag{4.18}$$

This is seen to be the $\beta \gg 1$ limit of the Rayleigh–Gans region and the $\beta|n-1| \ll 1$ limit of the anomalous diffraction region, and thus provides a transition region between the two. In this spirit one can enlarge the range of

First applications of the Mie solution

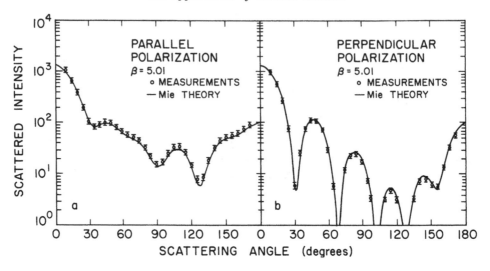

Fig. 4.7. Experimental verification of Mie scattering theory over the full range of scattering angles for $\beta = 5.01$. (Pinnick *et al.* (1976), with permission.)

applicable models in a continuous way, and the method has been discussed to some extent by Chýlek and Li (1995).

Very large size parameters

Almost all the results of this chapter have been relatively easy to establish because they involved only the leading terms in the partial-wave series. As β increases toward unity it becomes necessary to retain more terms in the sums in order to achieve an accurate description of the physics, and when $\beta \gtrsim 1$ the analytic expressions become very unwieldy, calling into question the mathematical utility of the series expansions. At this point one turns to the computer to carry out partial numerical summations of the series, and to the power of computer graphics to visualize partially the physical processes. In Fig. 4.7 we present plots of typical measurements of the intensities of scattered radiation from a dielectric sphere for both polarizations and for a representative size parameter. It is evident that agreement with the Mie theory is excellent at this value of β.

With ever increasing β, however, extraction of the physics embedded in the Mie solution becomes more tedious and the origins of many physical phenomena become more difficult to locate. Convergence of the series is not at issue, but rather the rate at which the sums converge. Recall, for example, the discussion in Chapter 2 of the localization principle, which applies equally to the transparent sphere, and the expression (2.39) providing a cutoff $\ell + \frac{1}{2} \gtrsim \beta$ for the summations. Some perspective emerges when it is noted that visible light incident upon a water drop about $\simeq 1 \text{mm}$ in diameter corresponds to $\beta \simeq 6000$. Thus, the problem is not so much summing the series numerically, for that has become almost trivial with a desktop computer, but in envisioning the physics that produces the resulting curves. A simple picture akin to the radiating electric dipole of Rayleigh scattering is not as easy to create for all the physics contained in that water drop.

At the upper extreme sits geometric optics, which is the limit of electromagnetic theory as $\beta \rightarrow \infty$. More specifically, a ray description becomes appropriate when $\beta \gg 1$ and when the phase shift $2\beta(n-1) \gg 1$. Under these conditions illumination from an incident light wave can be localized to a small area of the target surface – by a screen with a very small hole in it, say, or a laser – and thus treated as a pencil of rays. We have already seen that forward scattering dominates as size parameters become greater than unity, eventually becoming compressed into an intense narrow lobe centered on $\theta = 0$. Thus, for large smooth spheres the scattering consists almost entirely of two types of contribution: that due to diffraction, which is independent of the nature of the surface and the composition; and the less prominent but equally important processes of specular reflection and refraction, which do depend on shape, composition, and surface quality. Owing to interference between rays scattered in the same direction it is necessary to retain the phases, of course, but the bottom line is that the ray theory provides a satisfactory description of many phenomena involving very large size parameters. For this reason it is very much worth our time to study briefly the geometric-optics description of scattering from a homogeneous dielectric sphere and the exact way in which such a description is related to the Mie theory.

4.3 Geometric optics of the transparent sphere

The above division of the physical processes involved in scattering from transparent spheres is evident in the expressions (3.140) for the partial-wave coefficients a_ℓ and b_ℓ. The terms of unity describe primarily the effects of diffraction, while S_ℓ^E and S_ℓ^M describe specular reflection, refraction, and possible internal reflections. The diffraction contributions can be uncovered immediately from (3.95) by letting $a_\ell = b_\ell = \frac{1}{2}$ for $\ell + \frac{1}{2} < \beta$, and 0 otherwise, thereby introducing a cutoff on the sums.† If $x \equiv (\ell + \frac{1}{2})\theta$ is kept fixed as

---

† The point is that $\text{Re}[\frac{1}{2}(1 - e^{2i\alpha_\ell})]$ averages to $\frac{1}{2}(1)$ for $\ell + \frac{1}{2} < \beta$, and to $\frac{1}{2}(0)$ for $\ell + \frac{1}{2} > \beta$.

$\ell \to \infty$, only very small values of θ contribute (e.g., Eq. (2.122)) and we find from Eq. (D.50) that

$$\pi_\ell \simeq \tfrac{1}{2}\ell(\ell+1)[J_0(x)+J_2(x)],$$
$$\tau_\ell \simeq \tfrac{1}{2}\ell(\ell+1)[J_0(x)-J_2(x)]. \qquad (4.19)$$

In this approximation the upper limit is the greatest integer in β and

$$S_1(\theta) = S_2(\theta) \simeq \sum_{\ell=1}^{[\beta]}(\ell+\tfrac{1}{2})J_0\left[(\ell+\tfrac{1}{2})\theta\right] \longrightarrow \int_0^\beta \lambda J_0(\lambda\theta)\,d\lambda = \beta^2\frac{J_1(\beta\theta)}{\beta\theta}, \quad (4.20)$$

where λ is the continuum extrapolation of $\ell+\tfrac{1}{2}$ as we replace the sum by an integral over all relevant impact parameters. Other than the approximation of $\sin\theta$ by θ this is just the Airy diffraction pattern of Chapter 2 depicted in Fig. 2.4, independent both of surface and of composition and leading to $Q_{\text{ext}}(\beta) \simeq 2$.

In reality, of course, diffraction and reflection cannot be completely separated, as illustrated by the Fock transition region in our study of the impenetrable sphere. We emphasize that introduction of a cutoff and direct conversion of the sum to an integral are gross approximations that can be valid only in the limit $\beta \to \infty$. The mathematical difficulty for finite β is already manifest in Eqs. (3.135): the sum over 1 by itself diverges, but in general this term is canceled out by a corresponding term of unity in the exponential.

Further investigation of this limit begins by following the fate of an incident ray with impact parameter b penetrating the sphere, as in Fig. 4.8. Although it is customary to measure the angles of incidence and refraction with respect to the normal to the surface, it is here more convenient to employ temporarily the complementary angles φ_i and φ_r, as indicated. A custom to which we shall adhere, however, is that of referring to the interior of the sphere as medium 1, and the exterior in which it is embedded as medium 2. The relative index of refraction is then $n \equiv n_1/n_2 > 1$, and in terms of these angles Snell's law becomes

$$\cos\varphi_i = n\cos\varphi_r. \qquad (4.21)$$

From Fig. 4.8 we also note that the localization principle can be expressed as

$$\cos\varphi_i = \frac{\ell+\tfrac{1}{2}}{ka} \equiv \frac{\lambda}{\beta} \leq 1; \qquad \text{or,} \quad 1+\tan^2\varphi_i = \frac{\beta^2}{\lambda^2}. \qquad (4.22)$$

Thus, a narrow range of partial waves is associated with a ray in the geometric-optics limit.

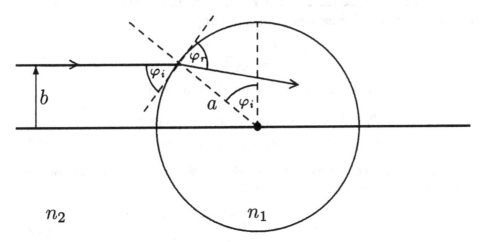

Fig. 4.8. An incident ray with impact parameter b refracted into the transparent sphere.

The immediate task here is to examine the large-β behavior of the partial-wave coefficients as written in (3.140), and toward that end we rewrite (3.141) as

$$S_\ell^E = -\frac{\zeta_\ell^{(2)'}(\beta)\psi_\ell(n\beta) - n^{-1}\zeta_\ell^{(2)}(\beta)\psi_\ell'(n\beta)}{\zeta_\ell^{(1)'}(\beta)\psi_\ell(n\beta) - n^{-1}\zeta_\ell^{(1)}(\beta)\psi_\ell'(n\beta)} \equiv -\frac{A}{B}, \qquad (4.23)$$

and similarly for $S_\ell^M = S_\ell^E(n^{-1} \to n)$. Now we appeal to the Debye asymptotic expansions, Eqs. (C.4) and (C.5), which were actually tailored to this problem (Debye 1909b), to write the leading–order expressions

$$\zeta_\ell^{(1)}(\beta) \xrightarrow[\beta > \lambda \to \infty]{} \left(\frac{\beta/\lambda}{\tan\varphi_i}\right)^{1/2} e^{i(\lambda\tan\varphi_i - \lambda\varphi_i - \pi/4)},$$

$$\zeta_\ell^{(1)'}(\beta) \xrightarrow[\beta > \lambda \to \infty]{} \left(\frac{\beta}{\lambda}\frac{\sin 2\varphi_i}{2}\right)^{1/2} i e^{i(\lambda\tan\varphi_i - \lambda\varphi_i - \pi/4)},$$

$$\psi_\ell(n\beta) \xrightarrow[\beta > \lambda \to \infty]{} \left(\frac{n\beta/\lambda}{\tan\varphi_r}\right)^{1/2} \cos(\lambda\tan\varphi_r - \lambda\varphi_r - \pi/4),$$

$$\psi_\ell'(n\beta) \xrightarrow[\beta > \lambda \to \infty]{} \left(\frac{n\beta}{\lambda}\frac{\sin 2\varphi_r}{2}\right)^{1/2} \sin(\lambda\tan\varphi_r - \lambda\varphi_r - \pi/4), \quad (4.24)$$

and $\zeta_\ell^{(2)}$ is obtained by changing the sign of i. Then,

$$B \simeq \frac{1}{2n}\left(\frac{n\sin\varphi_i}{\sin\varphi_r} + 1\right) e^{i(\lambda\tan\varphi_i - \lambda\varphi_i - \pi/4)} e^{-i(\lambda\tan\varphi_r - \lambda\varphi_r - \pi/4)}$$

$$\times \left(\frac{\sin\varphi_r}{\sin\varphi_i}\right)^{1/2}\left(1 - i\frac{n\sin\varphi_i - \sin\varphi_r}{n\sin\varphi_i + \sin\varphi_r} e^{2i(\lambda\tan\varphi_r - \lambda\varphi_r)}\right), \qquad (4.25)$$

in which we recognize the external Fresnel reflection coefficient for a plane interface in terms of these complementary angles,

$$r_{22}^{\parallel} = \frac{n \sin \varphi_i - \sin \varphi_r}{n \sin \varphi_i + \sin \varphi_r}. \tag{4.26}$$

Define, for convenience, the following phase functions:

$$\Phi_i \equiv \lambda \tan \varphi_i - \lambda \varphi_i = \beta(\sin \varphi_i - \varphi_i \cos \varphi_i),$$
$$\Phi_r \equiv \lambda \tan \varphi_r - \lambda \varphi_r = n\beta(\sin \varphi_r - \varphi_i \cos \varphi_r), \tag{4.27}$$

in which $n = m + i\kappa$ and we have introduced (4.22). With this notation, and a similar calculation for A, (4.23) yields

$$S_\ell^E = -e^{-2i\Phi_i + 2i\Phi_r} \frac{1 - ir_{22}^{\parallel} e^{-2i\Phi_r}}{1 - ir_{22}^{\parallel} e^{2i\Phi_r}}, \tag{4.28}$$

and similarly for S_ℓ^M. Although the external reflection coefficients arise naturally here, we are reminded that they are related to the internal coefficients by Eqs. (1.40) and (1.41).

As an aside we note that rays passing just outside the sphere correspond to $\lambda/\beta > 1$, suggesting the definition $\lambda/\beta = \cosh \psi$ and the alternative form (C.3) of Debye's asymptotic expansions. In (4.28) we would then have the replacement

$$e^{-2i\Phi_i} \longrightarrow e^{-2\beta(-\sinh \psi + \psi \cosh \psi)}, \tag{4.29}$$

in which the argument of the exponential is strictly negative unless $\lambda = \beta$. These rays outside the sphere are therefore exponentially damped and do not contribute to the amplitudes in geometric optics – they do, however, contribute in the general theory.

Both Debye (1909b) and van de Hulst (1946) studied these expressions for weakly absorbing spheres, $\kappa/m \ll 1$, so that n is taken to be approximately real in Snell's law and in the Fresnel coefficients;† we call this the *Debye approximation*. Debye also noted the expansion

$$-S_\ell^E = -ir_{22} e^{-2i\Phi_i} + (1 - r_{22}^2) e^{-2i\Phi_i} e^{2i\Phi_r} \sum_{p=1}^{\infty} \left(-ir_{22} e^{2i\Phi_r}\right)^{p-1}, \tag{4.30}$$

for either polarization. (Owing to the explicit factor of n in Φ_r one cannot simply replace n by n^{-1} at this point to relate the two S-functions, although that factor is the only one which is now complex.) There is an obvious interpretation for the terms in this series: the first describes direct reflection

---

† Note that both authors use the definition $\zeta_\ell = j_\ell - in_\ell$.

from the surface, the second describes direct transmission through the sphere (i.e., two successive refractions), and the pth term describes rays emerging after $p-1$ internal reflections (see, e.g., Fig. 1.8.)

Absorption

At this point we are in a position to evaluate the absorption efficiency in the limit $\beta \to \infty$, which can be rewritten from (3.106) as

$$Q_{\text{abs}}(\beta) = \frac{1}{2\beta^2} \sum_{\ell=1}^{\infty} (2\ell + 1)\left[\left(1 - |S_\ell^{\text{E}}|^2\right) + \left(1 - |S_\ell^{\text{M}}|^2\right)\right]. \tag{4.31}$$

However, to do this we should first rewrite the S-functions in terms of the usual angles of incidence and refraction: $\theta_i = \pi/2 - \varphi_i$ and $\theta_r = \pi/2 - \varphi_r$. Then, in the Debye approximation,

$$\Phi_i = \beta\left(\cos\theta_i + \theta_i \sin\theta_i - \frac{\pi}{2}\sin\theta_i\right),$$

$$\Phi_r = n\beta\left(\cos\theta_r + \theta_r \sin\theta_r - \frac{\pi}{2}\sin\theta_r\right), \tag{4.32}$$

with $n = m + i\kappa$.

Because the sphere is so large with respect to the wavelength, rays penetrating it tend to lose their coherence in traversing it, so that the phases become effectively random. Hence, the total intensity is well approximated by summing the individual intensities of each term in (4.30), and we obtain the approximation

$$1 - |S_\ell^{\text{E}}|^2 \simeq 1 - r_2^2 - \frac{(1 - r_2^2)^2 e^{-\alpha}}{1 - r_2^2 e^{-\alpha}} = (1 - e^{-\alpha})\frac{1 - r_2^2}{1 - r_2^2 e^{-\alpha}}. \tag{4.33}$$

In this expression, and that for S_ℓ^{M}, parallel and perpendicular polarization are now referred to as polarizations 2 and 1, respectively, and according to (1.40a) the reflection coefficients are either internal or external as desired. The decay parameter

$$\alpha \equiv 4\beta\kappa\cos\theta_r \tag{4.34}$$

is simply the attenuation coefficient of Eq. (2.4) multiplied by the path length of the ray through the sphere, which is the same after each successive internal reflection. Thus, the first factor after the equality sign in (4.33) represents the decrease in intensity suffered upon penetration into the sphere via the transmissivity $1 - r_2^2$, followed by traversal of the medium; the denominator then accounts for the loss of energy at each successive internal reflection.

In this limit of geometric optics we can employ the approximation used in (4.20) in (4.31), along with a further change of integration variable, to write

$$Q_{abs}^{(go)} = \int_0^{\pi/2} (1 - e^{-\alpha}) \left(\frac{1 - R_1}{1 - R_1 e^{-\alpha}} + \frac{1 - R_2}{1 - R_2 e^{-\alpha}} \right) \sin \theta_i \cos \theta_i \, d\theta_i, \quad (4.35)$$

where we now employ the reflectivities $R_j = r_j^2$, within which the angle θ_r is to be determined from Snell's law. That is, $R_1 = r_1^2 = (R_{22}^{\parallel})^2$ and $R_2 = r_2^2 = (R_{22}^{\perp})^2$. For very large spheres $\beta \to \infty$ implies that $\alpha \to \infty$, so that, for any nonzero absorption,

$$Q_{abs}^{(go)} \xrightarrow[\beta \to \infty]{} 1 - \int_0^{\pi/2} (R_1 + R_2) \sin \theta_i \cos \theta_i \, d\theta_i. \quad (4.36)$$

A similar procedure can be applied to the scattering efficiency:

$$Q_{sca}(\beta) = \frac{2}{\beta^2} \sum_{\ell=1}^{\infty} (2\ell + 1) \left(\frac{1}{2} \left| 1 - S_\ell^E \right|^2 + \frac{1}{2} \left| 1 - S_\ell^M \right|^2 \right). \quad (4.37)$$

In the Debye approximation

$$\left| 1 - S_\ell^E \right|^2 \simeq 1 + \left| S_\ell^E \right|^2 = 1 + R_1 + \frac{(1 - R_1)^2 e^{-\alpha}}{1 - R_1 e^{-\alpha}} \xrightarrow[\beta \to \infty]{} 1 + R_1,$$

so that

$$Q_{sca}^{(go)}(\beta) \xrightarrow[\beta \to \infty]{} 1 + \int_0^{\pi/2} (R_1 + R_2) \sin \theta_i \cos \theta_i \, d\theta_i \quad (4.38)$$

and once again $Q_{ext}^{(go)} = Q_{sca}^{(go)} + Q_{abs}^{(go)} \simeq 2$. The role of the terms of unity in the results (4.36) and (4.38) is notable, for they are precisely those leading to the diffraction effects in (4.20). Their intimate entanglement with the pure reflection term is evident in both results, though diffraction is still dominant. This is verified by noting that, for water, say,

$$\int_0^{\pi/2} (R_1 + R_2) \sin \theta_i \cos \theta_i \, d\theta_i \simeq 0.06641, \qquad n = 1.333. \quad (4.39)$$

As a consequence we see that our earlier expectation that the absorption efficiency might approach unity in the limit of very large β was not quite correct – its proper limit is approximately 0.9336 in this case. However, further meditation convinces us that this is the correct behavior, for there will always be *some* reflection even if everything else that enters the sphere is completely absorbed; very large α is also tantamount to very large β in the above equations.

It is somewhat interesting to plot the integrand in (4.39), as in Fig. 4.9. One sees that the dominant contribution arises from rays near grazing incidence,

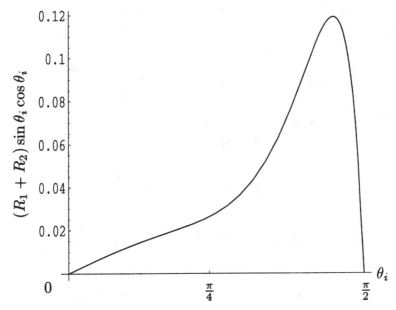

Fig. 4.9. A plot of the integrand in Eq. (4.39), illustrating the importance of rays at grazing incidence.

$\theta_i \simeq \pi/2$. This is a persistent phenomenon that will permeate much of the following work, again emphasizing the importance of the edge domain.

Radiation pressure

In his original analysis of Mie scattering Debye (1909b) was interested primarily in computing the cross section for radiation pressure, Eq. (3.128). From (3.130) and (3.140) we evaluate

$$Q_{sca}\langle\cos\theta\rangle = \frac{1}{\beta^2}\,\mathrm{Re}\left(\sum_{\ell=1}^{\infty}\frac{\ell(\ell+2)}{\ell+1}\left[\left(1-S_\ell^E\right)\left(1-S_{\ell+1}^{E\,*}\right)+\left(1-S_\ell^M\right)\right.\right.$$

$$\left.\left.\times\left(1-S_{\ell+1}^{M\,*}\right)\right]\right)+\frac{1}{\beta^2}\,\mathrm{Re}\sum_{\ell=1}\frac{2\ell+1}{\ell(\ell+1)}\left(1-S_\ell^E\right)\left(1-S_\ell^{M*}\right). \quad (4.40)$$

In the second sum, and in the Debye approximation,

$$\left(1-S_\ell^E\right)\left(1-S_\ell^{M*}\right) \simeq 1 + S_\ell^E S_\ell^{M*}$$

$$\simeq 1 + r_1 r_2 + (1-r_1^2)(1-r_2^2)\frac{e^{2i(\Phi_r - \Phi_r^*)}}{1 - r_1 r_2 e^{2i(\Phi_r - \Phi_r^*)}}$$

$$\simeq 1 + r_1 r_2 + (1-r_1^2)(1-r_2^2)\frac{e^{-\alpha}}{1 - r_1 r_2 e^{-\alpha}}$$

$$\xrightarrow[\alpha\to 0]{} \frac{2 - r_1^2 - r_2^2}{1 - r_1 r_2} \simeq 2, \quad (4.41)$$

because the reflectivities are much less than unity, except near the edge, where they approach unity. Thus, the second sum is approximately

$$\text{Re}\left(\frac{2}{\beta^2}\sum_{\ell=1}^{[\beta]}\frac{2\ell+1}{\ell(\ell+1)}\right)\longrightarrow\frac{4}{\beta^2}\text{Re}\left(\int_0^\beta\frac{\lambda\,d\lambda}{\lambda^2-\frac{1}{4}}\right)\simeq\frac{\log\beta}{\beta^2},\qquad(4.42)$$

and therefore negligible for very large β.

To evaluate the first sum in (4.40) we first note that

$$\sin\theta_i=\frac{l+\frac{1}{2}}{ka}=\frac{\lambda}{\beta},\qquad\sin\theta_i'=\frac{\lambda+\frac{1}{2}}{\beta},\qquad(4.43)$$

where primes refer to $S_{\ell+1}$. This latter quantity refers to a ray with a slightly different angle of incidence from that for S_ℓ, so we can write $\theta_i'=\theta_i+\Delta\theta_i$. Thus,

$$\sin\theta_i'=\sin\theta_i+\Delta\theta_i\cos\theta_i+\cdots\simeq\frac{\lambda}{\beta}+\Delta\theta_i\cos\theta_i,\qquad(4.44)$$

so that

$$\Delta\theta_i\cos\theta_i\simeq\frac{1}{\beta},\qquad\cos\theta_i'\simeq\cos\theta_i-\Delta\theta_i\sin\theta_i.\qquad(4.45)$$

Remembering the terms containing $\pi/2$ in (4.32), we find that

$$\Phi_i-\Phi_i'\simeq-\theta_i+\frac{\pi}{2},$$

$$\Phi_r-\Phi_r'^*=\simeq 2i\kappa\beta\cos\theta_r-\theta_r+\frac{\pi}{2},\qquad(4.46)$$

in the Debye approximation. To leading order in β we obtain

$$Q_{\text{sca}}^{(\text{go})}(\beta)\langle\cos\theta\rangle\simeq 1+\text{Re}\Bigg[\int_0^{\pi/2}e^{2i\theta_i}\Bigg(R_1-(1-R_1)^2\frac{e^{2i\theta_r}e^{-\alpha}}{1+R_1e^{2i\theta_r}e^{-\alpha}}$$

$$+\,R_2-(1-R_1)^2\frac{e^{2i\theta_r}e^{-\alpha}}{1+R_2e^{2i\theta_r}e^{-\alpha}}\Bigg)\sin\theta_i\cos\theta_i\,d\theta_i\Bigg],\quad(4.47)$$

and, as $\beta\to\infty$ with fixed $\kappa\neq 0$,

$$Q_{\text{pr}}^{(\text{go})}(\beta)\xrightarrow[\beta\to\infty]{}1-\int_0^{\pi/2}(R_1+R_2)\sin\theta_i\cos^2\theta_i\,d\theta_i.\qquad(4.48)$$

For $n=1.333+0.1i$ we find that $Q_{\text{pr}}^{(\text{go})}(\beta)\simeq 1.0369$; if $\kappa=0$ Eq. (4.47) leads to $Q_{\text{pr}}^{(\text{go})}(\beta)\simeq 0.6871$.

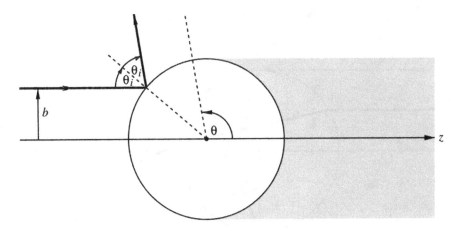

Fig. 4.10. Incident rays undergoing direct reflection define the geometric shadow cast by the sphere, thereby delineating 1-ray and 0-ray regions.

Analysis of individual rays

The standard of analysis set by Debye for this problem, not often seen since, prompts us to consider his series (4.30) and its physical interpretation by van de Hulst (1946) at greater length. In doing so we shall find that it provides an excellent guide for a treatment of high-frequency scattering from the transparent sphere.

Recall that the $p = 0$ term in the series describes a ray undergoing direct reflection from the surface, as depicted in Fig. 4.10. As is true with all rays in the geometric-optics approximation, it defines one or more *lit regions* and one or more *shadow regions*, there being one of each in this case. The shadow region here is just the expected geometric shadow cast by the sphere, which is of infinite extent in ray optics. In Fig. 4.10 $\theta_i = \pi/2 - \theta/2$ is the angle of incidence and θ is the scattering angle.

Figure 4.11 illustrates the lit and shadow regions for the directly transmitted ray, the $p = 1$ term of the series. For grazing, or tangential incidence there emerges a critical angle of refraction θ_c, leading to an angle of tangential emergence θ_t given by

$$\theta_t = \pi - 2\theta_c. \tag{4.49}$$

The critical angle is found from Snell's law: $\sin \theta_c = 1/n$. We note that in both cases of direct reflection and transmission there is both a 0-ray region and a 1-ray region, corresponding respectively to shadowed and lit regions. In geometric optics the shadow boundary is considered sharp.

The third, or $p = 2$ term of the series (4.30) refers to a ray penetrating

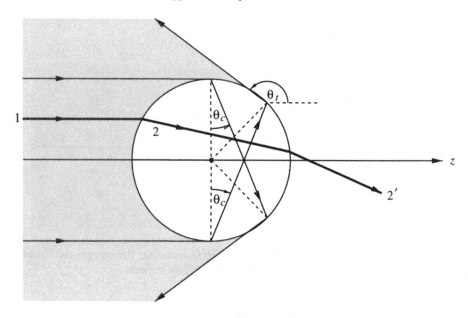

Fig. 4.11. Rays directly transmitted through the sphere also define 1-ray and 0-ray regions.

the sphere, undergoing one internal reflection, and then emerging in the backward hemisphere ($n > 1$). It is depicted in Fig. 4.12, perusal of which also indicates that this case is a bit more complicated than the previous two. The angles θ_i and θ_r range over $[0, \pi/2]$, while the scattering angle θ covers $[0, \pi]$. Some meditation reveals three possible types of trajectory for rays of this kind, classified as follows:

$$\theta = 2(2\theta_r - \theta_i) - \pi, \tag{4.50a}$$

$$\theta = \pi - 2(2\theta_r - \theta_i), \tag{4.50b}$$

$$\theta = \pi + 2(2\theta_r - \theta_i), \tag{4.50c}$$

found by simply adding up the total angle turned through by the ray.

The first of these, (4.50a), can be realized only for $n < 1$, so we omit further reference to it. Equation (4.50b) is valid for $n > 1$ if $2\theta_r - \theta_i > 0$, since θ must be $\leq \pi$. The angular relation determining this is $2\theta_{r0} - \theta_{i0} = 0$, and with Snell's law and a half-angle formula we find the condition

$$\cos \theta_{i0} = \tfrac{1}{2}(n^2 - 2), \tag{4.51}$$

which is valid only if $\sqrt{2} \leq n \leq 2$. Hence, (4.50b) holds when $0 < 2(2\theta_r - \theta_i)$ $< \pi$ and $1 \leq n \leq \sqrt{2}$; when $\sqrt{2} \leq n \leq 2$ we employ (4.50b) for $\theta_i < \theta_{i0}$ and

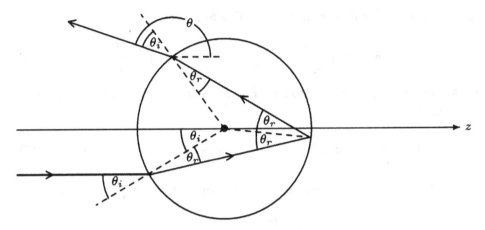

Fig. 4.12. The third term of the Debye series describes rays transmitted into the sphere and emerging after undergoing one internal reflection.

(4.50c) for $\theta_i > \theta_{i0}$. If $n > 2$ then $2\theta_r - \theta_i < 0$ and (4.50c) is the only valid relation.

For $n > 1$ the limiting ray is given by

$$\theta_i = \pi/2, \qquad \theta_r = \theta_\ell \equiv \sin^{-1}(1/n). \tag{4.52}$$

The corresponding scattering angle is $\theta = \theta_L$:

$$\theta_L = \begin{cases} 4(\pi/2 - \theta_\ell) = 4\cos^{-1}(1/n), & 1 < n < \sqrt{2}, \\ 4\theta_\ell, & n > \sqrt{2}. \end{cases} \tag{4.53}$$

To further understand the scattering angle as a function of the angle of incidence let us define

$$y \equiv \sin\left(\frac{\theta}{2}\right) = \cos(2\theta_r - \theta_i), \tag{4.54a}$$

corresponding to (4.50b). Then

$$\frac{dy}{d\theta_i} = -\frac{\sin(2\theta_r - \theta_i)}{n\cos\theta_r}(2\cos\theta_i - n\cos\theta_r). \tag{4.54b}$$

The sign of the derivative tells us whether θ is an increasing or decreasing function of θ_i, and it can change sign at only two places: $\theta_i = \theta_{i0}$, for which $\theta = \pi$ and $\sqrt{2} \le n \le 2$; and $\theta_i = \theta_{iR}$, for which $2\cos\theta_{iR} = n\cos\theta_{rR}$, and

$$\sin\theta_{iR} = s \equiv \left(\frac{4 - n^2}{3}\right)^{1/2}, \qquad \cos\theta_{iR} = c \equiv \left(\frac{n^2 - 1}{3}\right)^{1/2}. \tag{4.55}$$

The last two relations can be valid only for $1 \le n \le 2$. The corresponding scattering angle for this second case is $\theta = \theta_R$, where

$$y_R = \sin\left(\frac{\theta_R}{2}\right) = \frac{(8 + n^2)c}{3n^2}, \quad \cos\left(\frac{\theta_R}{2}\right) = \frac{s^3}{n^2}. \tag{4.56}$$

As we shall see, θ_R is the *rainbow angle*. The expression (4.53) for the limiting angle yields

$$y_L = \sin\left(\frac{\theta_L}{2}\right) = \sin(2\theta_\ell) = \frac{2N}{n^2}, \quad N \equiv (n^2 - 1)^{1/2}, \quad n > 1. \tag{4.57}$$

So, for $1 < n < \sqrt{2}$, θ_R is always smaller than θ_i; but for $\sqrt{2} \leq n \leq 2$,

$$\begin{aligned}
\theta_R < \theta_L, \quad & n < n_0, \\
\theta_R > \theta_L, \quad & n > n_0,
\end{aligned} \tag{4.58}$$

where $n_0 \equiv (6\sqrt{3} - 8)^{1/2}$.

With these results we can now describe the complete behavior of the $p = 2$ ray, beginning with $\theta = \pi$ at $\theta_i = 0$.

$1 < n < \sqrt{2}$: As θ_i increases from 0 to θ_{iR}, θ decreases from π to θ_R, the rainbow angle; θ_{iR} is a turning point and θ_R appears around the 2-ray/0-ray shadow boundary. The nomenclature means that, as θ_i continues to increase toward its limiting value of $\pi/2$, θ *decreases* to its final value of θ_L, and thus the region $\theta_R < \theta < \theta_L$ is covered twice. That is, two different incident rays can pass through every direction in this region, as illustrated in Fig. 4.13. Note that θ_R is an angle of minimum deviation, as discussed in Chapter 1.

$\sqrt{2} < n < 2$: There are now two turning points: $\theta_i = \theta_{iR}$ and $\theta_i = \theta_{i0}$; for $n < n_0$ we have $\theta_L > \theta_R$ and the rainbow occurs at a 2-ray/0-ray boundary; for $n > n_0$ we have $\theta_L < \theta_R$ and it occurs at a 3-ray/1-ray boundary – a neighborhood of the backward direction is covered by three rays.

$n \geq 2$: There are no turning points, hence only 0-ray and 1-ray regions.

Unless otherwise noted we shall restrict further discussion to the range $1 < n < \sqrt{2}$, primarily because this contains the refractive index of water – and we do live on the water planet. For $p > 1$ the total deviation of the pth ray as it undergoes $p - 1$ internal reflections is†

$$\theta = (p - 1)\pi + 2\theta_i - 2p\theta_r, \tag{4.59}$$

---

† The number π here is to be interpreted as 180°, for the angles are measured in degrees.

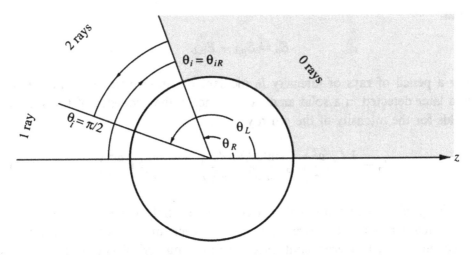

Fig. 4.13. Rays transmitted into the sphere and emerging after undergoing one internal reflection define 0-ray, 1-ray, and 2-ray regions, illustrating that the rainbow angle is an angle of minimum deviation.

and

$$\cos(\theta_{iR})_p = \left(\frac{n^2 - 1}{p^2 - 1}\right)^{1/2}, \qquad p > 1. \tag{4.60}$$

Clearly, higher-order rainbows are associated with rays incident nearer and nearer to the edge. The rainbow angle for the pth ray is found to be

$$(\theta_R)_p = (p - 1)\pi + 2\cos^{-1}\left[\left(\frac{n^2 - 1}{p^2 - 1}\right)^{1/2}\right]$$

$$- 2p\cos^{-1}\left[\left(\frac{n^2 - 1}{n^2}\frac{p^2}{p^2 - 1}\right)^{1/2}\right] \pmod{2\pi}, \tag{4.61}$$

locating an infinity of higher-order rainbows (see Chapter 6 below). For water with $n = \frac{4}{3}$, the first-order rainbow angles are $\theta_{iR} \simeq 59.39°$ and $\theta_R \simeq 137.97°$.

The intensities of emerging rays are readily computed from the Fresnel coefficients, so that the fraction of the original intensity remaining in the final transmitted ray can be written in terms of the reflectivity and transmissivity for each polarization:

$$E_{p,1} = (1 - R_1)R_1^{p-1}(1 - R_1),$$
$$E_{p,2} = (1 - R_2)R_2^{p-1}(1 - R_2), \qquad p > 0, \tag{4.62a}$$

and

$$E_p = E_{p,1} + E_{p,2}. \tag{4.62b}$$

For a pencil of rays of intensity I_0 incident on an element of surface area and later detected in a solid angle at distance r, the geometry of the sphere yields for the intensity of the pth ray

$$I_p = \frac{1}{2}\left(I_0 \frac{a^2}{r^2}\right) E_p \frac{\sin(2\theta_i)}{\sin\theta}\left(\frac{d\theta}{d\theta_i}\right)^{-1} = \frac{I_0 E_p}{r^2}\left(\frac{d\sigma}{d\Omega}\right)_p. \tag{4.63}$$

Owing to repeated internal reflection, one surely expects the intensity to decrease rapidly for increasing p – and even more so if absorption is occurring. This has been confirmed by direct numerical computation and been summarized in a marvelous table by van de Hulst (1957, p. 231).† He shows that the overwhelming portion of the emerging intensity is directly transmitted ($> 82\%$), and that more than 98.5% of the scattered energy is contained in the first three terms of the series (4.30). In addition, when all terms in p are accounted for, almost 60% of the energy can be found in the range $0°$–$30°$ of scattering angles, and another 30% in the range $30°$–$60°$. Clearly, backward phenomena do not encompass a great deal of the incident energy in scattering from large dielectric spheres. What we shall eventually see, however, is that the remaining small amount of energy can sometimes be focused into quite small angular ranges where it has an observable impact.

Despite its successes, geometric optics fails at certain special angles, as we shall see, and along focal lines such as the central axis. Short of the actual limit, there are always various diffraction corrections. Moreover, the ripple structure in the extinction efficiency discussed above requires the full Mie theory for its explanation. So one gains the impression that, for large size parameters, we may be well advised to retain the ray picture as a foundation and analyze the whole of the Mie solution for $\beta \gg 1$ along the lines suggested at the end of Chapter 2. Among many other points requiring closer scrutiny are the rigorous treatment of the several crude approximations made above, and the effects of ray interference at the boundaries in Fig. 4.13, for example. Preparatory to that adventure we first take a moment to cast the Mie theory into a form suitable for analytic continuation, and then briefly outline an extension of the results of Chapter 2 for the impenetrable sphere to its electromagnetic realization as a perfect conductor.

---

† We have reconstructed this table and find no significant differences, which is remarkable given that his calculations were carried out with a slide rule!

4.4 Perfectly conducting spheres at high frequencies

Analytic continuation into the complex angular momentum plane, as was done with the scalar amplitude of Eq. (2.100), requires some modification in the expressions (3.95) for the electromagnetic scattering amplitudes. For several reasons we introduce new angular functions $p_\ell(\cos\theta)$ and $t_\ell(\cos\theta)$, as defined in (D.52), which are more useful for extrapolation to complex indices. They contain only simple Legendre polynomials, remove the explicit appearance of numerical factors, and exclude values with $\ell = 0$. The denominator factors in (3.95) would generate kinematic poles that have nothing to do with properties of the scatterer, on the one hand. On the other, the sums in (3.95) begin with $\ell = 1$, and are therefore unsuitable for application of the Poisson sum formula (2.77). We therefore add and subtract fictitious $\ell = 0$ terms to S_1 and S_2, which play the role of 'nonsense channels' in Regge-pole theory. Then, with reference to Eqs. (3.139) and (3.140), the scattering amplitudes can be rewritten as

$$S_j(\beta,\theta) = S_j^{(+)}(\beta,\theta) + \tfrac{1}{2}[\sigma_1(\beta) - 1]\sec^2(\theta/2)$$
$$= S_j^{(-)}(\beta,\theta) - (-1)^j \tfrac{1}{2}\sigma_2(\beta)\csc^2(\theta/2), \quad j = 1,2, \quad (4.64)$$

where the first line is regular in the forward direction, the second is regular in the backward direction, and $j = 1$ and $j = 2$ refer to polarization perpendicular and parallel to the scattering plane, respectively.

The auxiliary amplitudes in (4.64) are defined as

$$S_1^{(+)}(\beta,\theta) \equiv -\frac{1}{2}\sum_{\ell=0}^{\infty}\left\{\left[S_\ell^M(\beta) - 1\right]t_\ell(\cos\theta) + \left[S_\ell^E(\beta) - 1\right]p_\ell(\cos\theta)\right\}, \quad (4.65a)$$

$$S_1^{(-)}(\beta,\theta) \equiv -\frac{1}{2}\sum_{\ell=0}^{\infty}(-1)^\ell$$
$$\times\left\{\left[S_\ell^M(\beta) - 1\right]t_\ell(-\cos\theta) - \left[S_\ell^E(\beta) - 1\right]p_\ell(-\cos\theta)\right\}, \quad (4.65b)$$

where we have employed the identities (D.53) in the second line, the phase factors S_ℓ^M and S_ℓ^E are defined in (3.141), and $S_2(\beta,\theta)$ is obtained through the interchanges $S_\ell^M \leftrightarrow S_\ell^E$;

$$\sigma_1(\beta) \equiv \frac{4ne^{2i(n-1)\beta}}{(n+1)^2 - (n-1)^2 e^{4in\beta}},$$

$$\sigma_2(\beta) \equiv \frac{n-1}{n+1}e^{-2i\beta} - \sigma_1(\beta). \quad (4.66)$$

One might suppose the functions $\sigma_i(\beta)$ to represent contributions from the central ray, given that they arise from zero angular momentum. However, reference to (2.40) reminds us that this corresponds neither to zero angular momentum nor zero impact parameter. Moreover, it is evident from (4.64) that these terms contribute for all scattering angles. Because of the reciprocity we shall consider only S_1 until it becomes necessary to state explicit results for S_2. In the exact forward and backward directions one refers to (D.54) to write

$$S_1(\beta,0) = S_2(\beta,0) = S^{(+)}(\beta,0) + \tfrac{1}{2}[\sigma_1(\beta) - 1], \qquad (4.67a)$$

$$S_1(\beta,\pi) = -S_2(\beta,\pi) = S^{(-)}(\beta,\pi) + \tfrac{1}{2}\sigma_2(\beta), \qquad (4.67b)$$

with

$$S^{(+)}(\beta,0) \equiv \frac{1}{2}\sum_{\ell=0}^{\infty}(\ell + \tfrac{1}{2})\left\{\left[S_\ell^{M}(\beta) - 1\right] + \left[S_\ell^{E}(\beta) - 1\right]\right\}, \qquad (4.68a)$$

$$S^{(-)}(\beta,\pi) \equiv \frac{1}{2}\sum_{\ell=0}^{\infty}(-1)^\ell(\ell + \tfrac{1}{2})\left\{\left[S_\ell^{M}(\beta) - 1\right] - \left[S_\ell^{E}(\beta) - 1\right]\right\}, \qquad (4.68b)$$

Perfectly conducting spheres

For our purposes perfect conductivity will be defined as the limit $n \to i\infty$, so that

$$\sigma_1(\beta) \xrightarrow[n \to i\infty]{} 0, \qquad \sigma_2(\beta) \xrightarrow[n \to i\infty]{} e^{-2i\beta}, \qquad (4.69)$$

and the S-matrix elements simplify considerably:

$$S_\ell^{E}(\beta) \xrightarrow[n \to i\infty]{} S^{E}(\lambda,\beta) \equiv -\frac{\zeta_{\lambda-1/2}^{(2)}{}'(\beta)}{\zeta_{\lambda-1/2}^{(1)}{}'(\beta)},$$

$$S_\ell^{M}(\beta) \xrightarrow[n \to i\infty]{} S^{M}(\lambda,\beta) \equiv -\frac{\zeta_{\lambda-1/2}^{(2)}(\beta)}{\zeta_{\lambda-1/2}^{(1)}(\beta)}, \qquad (4.70a)$$

where $\lambda = \ell + \tfrac{1}{2}$. With these modifications application of the Poisson sum formula, or modified Watson transformation, to (4.65) yields

$$S_1^{(+)}(\beta,\theta) = \frac{1}{2}\sum_{m=-\infty}^{\infty}(-1)^m\int_0^\infty \left\{\left[1 - S^{M}(\lambda,\beta)\right]t_{\lambda-1/2}(\cos\theta)\right.$$
$$\left. +\left[1 - S^{E}(\lambda,\beta)\right]p_{\lambda-1/2}(\cos\theta)\right\}e^{2mi\pi\lambda}\,d\lambda, \qquad (4.71a)$$

$$S_1^{(-)}(\beta,\theta) = -\frac{i}{2}\sum_{m=-\infty}^{\infty}(-1)^m\int_0^\infty \left\{\left[1 - S^{M}(\lambda,\beta)\right]t_{\lambda-1/2}(-\cos\theta)\right.$$
$$\left. -\left[1 - S^{E}(\lambda,\beta)\right]p_{\lambda-1/2}(-\cos\theta)\right\}e^{(2m+1)i\pi\lambda}\,d\lambda. \qquad (4.71b)$$

Analytic continuation into the complex λ-plane begins with the observation that the integrands are meromorphic functions of λ having only simple poles. As noted in Chapter 2, these Regge poles are located symmetrically with respect to the origin, so that only those in the right half-plane need be computed, and those close to the real axis are located near $\lambda = \beta$. For the magnetic case the poles are found at the zeros of $\zeta^{(1)}_{\lambda-1/2}(\beta)$ in the λ-plane for $\beta \gg 1$ and are given by Eq. (2.102),

$$\lambda^{M}_{n} \equiv \lambda_{n} = \beta + e^{i\pi/3}\frac{x_{n}}{\gamma} - e^{-i\pi/3}\frac{x_{n}^{2}}{60}\gamma + O(\gamma^{2}), \qquad (4.72\text{a})$$

where again $\gamma = (2/\beta)^{1/3}$ is the penumbra parameter, and $\{x_{n}\}$ are the zeros of the Airy function $\mathrm{Ai}(-x)$ numbered in order of increasing imaginary part. The poles of $S^{E}(\lambda, \beta)$ are located at the zeros of $\zeta^{(1)\prime}_{\lambda-1/2}(\beta)$ and are obtained in a similar way:

$$\lambda^{E}_{n} \equiv \mu_{n} = \beta + e^{i\pi/3}\frac{x'_{n}}{\gamma} - e^{-i\pi/3}\frac{x_{n}^{\prime 2}}{60}\gamma\left(1 - \frac{9}{x_{n}^{\prime 3}}\right) + O(\gamma^{2}), \qquad (4.72\text{b})$$

with $\{x'_{n}\}$ the zeros of $\mathrm{Ai}'(-x)$. All these poles are calculated using the Schöbe asymptotic formulas of Appendix C in the region $|\lambda - \beta| \lesssim \beta^{1/3}$, and are essentially the zeros of the cylindrical Hankel functions in the λ-plane (Streifer and Kodis 1964).

The asymptotic behavior of the angular functions is not uniform in $(0, \pi)$, which is the motivation for writing the amplitudes as we did in (4.64). Consequently we must consider three separate angular regions at first, though all will join smoothly. Unfortunately, there is not yet a uniform treatment similar to that of the scalar problem.

The wide-angle region

This angular region is defined by

$$\varepsilon_{1} \lesssim \theta \lesssim \pi - \varepsilon_{2}, \qquad \varepsilon_{1} \gg \gamma, \qquad \varepsilon_{2} \gg \beta^{-1/2}, \qquad (4.73)$$

dictated by the regions of validity of the asymptotic expansions of the angular functions. We begin with Eq. (4.71b) and focus on the magnetic contribution containing $S^{M}(\lambda, \beta)$; the electric term follows from the replacements $S^{M} \to S^{E}$ and $t \to -p$. The procedure follows very closely the treatment of $F(\beta, \theta)$ in the scalar case, Eqs. (2.104)–(2.109), so we provide a bit less detail here.

One first splits the m-sum in (4.71b) and deforms the contour off the real axis, so that symbolically

$$\sum_{m=-\infty}^{\infty} \int_0^{\infty} \lambda \longrightarrow \sum_{m=0}^{\infty} \int_0^{\infty+i\varepsilon} d\lambda + \sum_{m=1}^{\infty} \int_0^{\infty-i\varepsilon} d\lambda. \qquad (4.74)$$

We next employ contour rotation, judicious changes of variable $\lambda \to -\lambda$, and the reflection formulas

$$S^{\mathrm{M}}(-\lambda, \beta) = e^{-2\pi i \lambda} S^{\mathrm{M}}(\lambda, \beta), \qquad t_{\lambda-1/2}(\cos\theta) = -t_{-\lambda-1/2}(-\cos\theta), \quad (4.75)$$

which follow from (C.14), (D.31), and the definitions (D.52). The results of these manipulations are simplified by introduction of the auxiliary functions (D.55), in a manner reminiscent of the substitution (2.106); the reason is that their asymptotic properties and identities given in Appendix D allow us to change signs of their arguments and distort the contours as desired. We can write the amplitude (4.71b) as in Eq. (2.109), the sum of a background integral and a residue series:

$$S_1^{(-)}(\beta, \theta) = S_{1,\mathrm{geo}}^{(-)}(\beta, \theta) + S_{1,\mathrm{res}}^{(-)}, \qquad (4.76a)$$

with a geometric term

$$\begin{aligned} S_{1,\mathrm{geo}}^{(-)}(\beta, \theta) &= -\frac{1}{2} \int_{-\infty+i\varepsilon}^{\infty-i\varepsilon} \left[S^{\mathrm{M}}(\lambda, \beta) t_{\lambda-1/2}^{(1)}(\cos\theta) + S^{\mathrm{E}}(\lambda, \beta) p_{\lambda-1/2}^{(1)}(\cos\theta) \right] d\lambda \\ &= \frac{1}{2} \int_{\Gamma_1} \left[S^{\mathrm{M}}(\lambda, \beta) t_{\lambda-1/2}^{(1)}(\cos\theta) + S^{\mathrm{E}}(\lambda, \beta) p_{\lambda-1/2}^{(1)}(\cos\theta) \right] d\lambda \\ &\quad - \tfrac{1}{2} e^{-2i\beta} \mathrm{cosec}^2(\theta/2), \end{aligned} \qquad (4.76b)$$

and what will become a residue series,

$$\begin{aligned} S_{1,\mathrm{res}}^{(-)}(\beta, \theta) &= \frac{i}{2} \sum_{m=0}^{\infty} (-1)^m \int_{-\infty+i\varepsilon}^{\infty+i\varepsilon} \left[S^{\mathrm{M}}(\lambda, \beta) t_{\lambda-1/2}(-\cos\theta) \right. \\ &\quad \left. - S^{\mathrm{E}}(\lambda, \beta) p_{\lambda-1/2}(-\cos\theta) \right] e^{(2m+1)i\pi\lambda} d\lambda. \end{aligned} \qquad (4.76c)$$

In the second line of (4.76b) we have noted, as in the scalar case, that the integrand possesses a saddle point at $\bar{\lambda} = \beta \cos(\theta/2)$ and employed the asymptotic properties of the integrand to deform the contour into the same one as that in Fig. 2.10. In further deforming this contour onto the path of steepest descent through the saddle point, however, we obtain a residue contribution from the pole at $\lambda = \frac{1}{2}$ of the angular functions. (This is the pole at $\ell = 0$ which is evident in the definitions (D.52)). Marvelously, but not unexpectedly, this contribution exactly cancels out the $\ell = 0$ term in (4.64).

Evaluation of the integral in (4.76b) is performed exactly as in Chapter 2 for the scalar case (and in Appendix E), and a similar result holds for polarization 2. Hence, in this region the reflection amplitudes analogous to (2.118) are

$$S_{1,\text{geo}}(\beta,\theta) \simeq \frac{i\beta}{2} e^{-2i\beta \sin(\theta/2)} \left(1 + \frac{i}{2} \frac{1 - 2\sin^2(\theta/2)}{\beta \sin^3(\theta/2)} + O(\beta^{-2})\right), \quad (4.77a)$$

$$S_{2,\text{geo}}(\beta,\theta) \simeq -\frac{i\beta}{2} e^{-2i\beta \sin(\theta/2)} \left(1 - \frac{i}{2\beta \sin^3(\theta/2)} + O(\beta^{-2})\right). \quad (4.77b)$$

For a refractive index $n > 1$ there is generally a phase shift of π for polarization 1 upon reflection, which accounts for the sign difference here.

In Eq. (4.76c) the exponential function guarantees enough damping to close the path in the upper half-plane (see Table C.1) and we obtain a residue series that is again interpreted in terms of surface waves:

$$S_{1,\text{res}}^{(-)}(\beta,\theta) = -\pi \sum_{m=0}^{\infty} (-1)^m \left(\sum_n r_n^M t_{\lambda-1/2}(-\cos\theta)e^{(2m+1)i\pi\lambda_n}\right.$$

$$\left. - \sum_n r_n^E p_{\lambda-1/2}(-\cos\theta)e^{(2m+1)i\pi\mu_n}\right), \quad (4.78)$$

with a similar expression for $S_{2,\text{res}}^{(-)}(\beta,\theta)$ obtained by interchanging the poles and residues. The residues are calculated just as in (2.119), and near $\lambda = \beta$ are

$$r_n^M \simeq \frac{e^{-i\pi/6}}{2\pi\gamma a_n'^2}, \qquad r_n^E \simeq -\frac{e^{-i\pi/6}}{2\pi\gamma x_n' a_n^2}, \qquad (4.79)$$

with the same notation as there and in Eq. (B.11). These series are again rapidly convergent owing to exponential damping, so only the first few poles lying close to the real axis need be included. Moreover, according to the discussion following (D.51), $t_{\lambda-1/2}(-\cos\theta)$ is generally $O(\beta)$ greater than $p_{\lambda-1/2}(-\cos\theta)$ in this region, so that the magnetic contribution to $S_{2,\text{res}}^{(-)}$ can be neglected – this is not so for $S_{1,\text{res}}^{(-)}$. The dominant contributions from the residue series are then

$$S_{1,\text{res}}^{(-)}(\beta,\theta) \simeq -\left(\frac{2\pi}{\sin\theta}\right)^{1/2} e^{i\pi/4} \left(\sum_n \sqrt{\lambda_n} r_n^M e^{i\lambda_n\theta} - \frac{i}{\sin\theta} \sum_n \frac{r_n^E}{\sqrt{\mu_n}} e^{i\mu_n\theta}\right),$$

$$S_{2,\text{res}}^{(-)}(\beta,\theta) \simeq -\left(\frac{2\pi}{\sin\theta}\right)^{1/2} e^{i\pi/4} \sum_n \sqrt{\mu_n} r_n^E e^{i\mu_n\theta}. \qquad (4.80a)$$

The results, (4.77) and (4.80), are valid for $\theta \gg \gamma$ and $\pi - \theta \gg \beta^{-1/2}$.

The near-forward region

In this region, defined by $0 < \theta \lesssim \gamma$, we employ $S_1^{(+)}(\beta, \theta)$ and proceed in a manner very similar to that above. Focusing upon the magnetic contribution explicitly, we manipulate (4.71a) into the form

$$
S_1^{(+)}(\beta, \theta) = \frac{1}{2} \sum_{m=0}^{\infty} (-1)^m \left\{ \int_0^{\infty+i\epsilon} \left[1 - S^M(\lambda, \beta) \right] t_{\lambda-1/2}(\cos\theta) e^{2mi\pi\lambda} \, d\lambda \right.
$$

$$
\left. + \int_{i\infty}^0 \left[e^{2i\pi\lambda} - S^M(\lambda, \beta) \right] t_{\lambda-1/2}(\cos\theta) e^{2mi\pi\lambda} \, d\lambda \right\}
$$

$$
+ \text{electric contribution.} \tag{4.81}
$$

Except for the electric contribution, this expression is virtually identical to that of Eq. (2.111), although here we have moved the contours off the real axis. In particular, in the second line of (4.81) it has been noted that the integrand vanishes as $|\lambda| \to \infty$ in the second quadrant, so that the contour can be moved onto the imaginary axis.

The structure of (4.81) is just that of (2.110): a diffraction term, a surface-wave term, and a Fock-like transition term arising from edge effects. Analysis of the first two terms follows almost exactly that of the scalar problem, Eqs. (2.113) and (2.115), whereas in the transition region it is similar but more extensive. We find that

$$
S_1^{(+)}(\beta, \theta) = S_{1,\text{geo}}^{(+)}(\beta, \theta) + S_{1,\text{res}}^{(+)}(\beta, \theta), \tag{4.82a}
$$

where

$$
S_{1,\text{geo}}^{(+)}(\beta, \theta) \simeq \frac{\beta^2}{2} \left(\frac{\theta}{\sin\theta} \right)^{1/2} \left(\frac{2J_1(\beta\theta)}{\beta\theta} + F_1(\beta, \theta) + O(\gamma^4) \right), \tag{4.82b}
$$

$$
S_{1,\text{res}}^{(+)}(\beta, \theta) \simeq -2\pi i \left(\frac{\theta}{\sin\theta} \right)^{1/2} \sum_{m=1}^{\infty} (-1)^m
$$

$$
\times \left(\sum_n r_n^M \lambda_n J_1'(\lambda_n\theta) e^{2mi\pi\lambda_n} + \sum_n r_n^E \frac{J_1(\mu_n\theta)}{\theta} e^{2mi\pi\mu_n} \right), \tag{4.82c}
$$

and the Fock transition term $F_1(\beta, \theta)$ is a sum of integrals of products of Airy functions and Bessel-function derivatives. In obtaining these results we have employed the uniform asymptotic expansions (D.63) of the angular functions. The contributions from polarization 2 are almost identical, with F_2 obtained from F_1 by means of the interchanges $J_1' \leftrightarrow J_1$, and the residue series $S_{2,\text{res}}^{(+)}$ from $S_{1,\text{res}}^{(+)}$ by $r_n^E \leftrightarrow r_n^M$ and $\lambda_n \leftrightarrow \mu_n$.

Khare (1975) has made a careful study of the transition functions and demonstrated that they provide a smooth connection both with the near-forward direction and with the wide-angle region. Physically these functions interpolate between saddle-point behavior in the illuminated region and the exponentially-damped surface waves in the shadow region. As in the scalar case, our eventual primary interest is with the domain $\theta \ll \gamma$, in which

$$F_1(\beta, \theta) \simeq \gamma^2 \left(J_1'(\beta\theta)M_0 + \frac{J_1(\beta\theta)}{\beta\theta}M_0' \right), \tag{4.83}$$

where M_n is given by (2.134), and

$$M_n' \equiv e^{-i\pi/3} \int_0^\infty \frac{Ai'(z)}{Ai'(ze^{2i\pi/3})} z^n \, dz - e^{2in\pi/3} \int_0^\infty \frac{Ai'(z)}{Ai'(ze^{-2i\pi/3})} z^n \, dz. \tag{4.84}$$

Thus, to leading orders neither the surface waves nor the $\ell = 0$ term contribute and we find that

$$S_1(\beta, \theta) \simeq \frac{\beta^2}{2} \left(\frac{\theta}{\sin\theta} \right)^{1/2} \left[\frac{2J_1(\beta\theta)}{\beta\theta} + \gamma^2 \left(J_1'(\beta\theta)M_0 + \frac{J_1(\beta\theta)}{\beta\theta}M_0' \right) + O(\beta^{-1}) \right], \tag{4.85a}$$

$$S_2(\beta, \theta) \simeq \frac{\beta^2}{2} \left(\frac{\theta}{\sin\theta} \right)^{1/2} \left[\frac{2J_1(\beta\theta)}{\beta\theta} + \gamma^2 \left(J_1'(\beta\theta)M_0' + \frac{J_1(\beta\theta)}{\beta\theta}M_0 \right) + O(\beta^{-1}) \right], \tag{4.85b}$$

where

$$M_0 = 1.255e^{i\pi/3}, \qquad M_0' = -1.089e^{i\pi/3}. \tag{4.85c}$$

In the exactly forward direction, where $J_\nu(x) \to (x/2)^\nu$ as $x \to 0$,

$$S_1(\beta, 0) = S_2(\beta, 0) \simeq \frac{\beta^2}{2} \left[1 + \tfrac{1}{2}(M_0 + M_0') + O(\beta^{-1}) \right], \tag{4.86}$$

in addition to the much stronger statement (3.98). Here the extinction efficiency

$$Q_{\text{ext}}(\beta) \simeq 2 + 0.209\beta^{-2/3} + O(\beta^{-1}) \tag{4.87}$$

differs slightly from the scalar result (2.136) owing to the vector nature of the process. Nevertheless, this expression supplies a rigorous proof of the asymptotic value of the extinction efficiency as $\beta \to \infty$ for the perfectly conducting sphere.

The near-backward region

It remains to consider the near-backward region, defined as $\pi - \theta \lesssim \beta^{-1/2}$. All the results for the wide-angle region remain finite as $\theta \to \pi$, but the asymptotic expansions for the angular functions employed there are no longer valid in this limit. Rather, they should be replaced by the uniform expansions (D.63). Upon carrying out the explicit algebra (Khare 1975) one finds that Eqs. (4.77) remain valid up to and including $\theta = \pi$, but that the residue sums take the forms

$$S_{1,\text{res}}(\beta, \theta) \simeq -2\pi \left(\frac{\pi - \theta}{\sin \theta}\right)^{1/2} \sum_{m=0}^{\infty} (-1)^m$$

$$\times \left(\sum_n r_n^M \lambda_n J_1'[\lambda_n(\pi - \theta)] e^{(2m+1)i\pi\lambda_n}\right.$$

$$\left. - \sum_n r_n^E \frac{J_1[\mu_n(\pi - \theta)]}{\pi - \theta} e^{(2m+1)i\pi\mu_n}\right), \tag{4.88}$$

and $S_{2,\text{res}}(\beta, \theta)$ is obtained by the interchanges $r_n^M \leftrightarrow r_n^E$ and $\lambda_n \leftrightarrow \mu_n$. These series connect smoothly with (4.80), so that we have a complete asymptotic description of the short-wavelength scattering of a plane wave from a perfectly conducting sphere.

As is evident from the preceding comments, this physical system has been investigated in considerable detail by Khare (1975) in his doctoral dissertation. It had also been studied earlier in much the same manner by Senior and Goodrich (1964), with much the same results. They obtained only the leading-order term for the forward amplitudes, however, and failed to describe the necessary continuity through the Fock transition region. The techniques developed earlier by Nussenzveig (1965, 1969a, b) and extended by Khare (1975) will be seen to be precisely the right tools for studying the dielectric sphere in the following chapter.

5

Short-wavelength scattering from transparent spheres

With the framework provided by geometric optics, along with the tools developed in studying models of impenetrable spheres, we are now in a position to analyze scattering from large inhomogeneous dielectric spheres. We shall focus primarily on real refractive indices in the range $1 < n < \sqrt{2}$, though most of the results are certainly valid for complex n. The parameter values of principal interest are then

$$\beta^{1/3} \gg 1, \qquad |n-1|^{1/2}\beta^{1/3} \gg 1. \tag{5.1}$$

The second condition, to be discussed further below, is related to the damping of surface waves owing to refraction into the sphere, which will be seen to be small if n is not too close to unity; it also excludes Rayleigh–Gans scattering and some of the anomalous diffraction region.

We have already reformulated the scattering amplitudes in Eqs. (4.64)–(4.68), and application of the Poisson sum formula in preparation for continuation into the complex λ-plane results in the amplitude functions of Eqs. (4.71). The S-functions S_ℓ^{M} and S_ℓ^{E} are now given by Eq. (3.141), though at this point we find it convenient to combine both types of multipole into the single expression

$$S_\ell = -\frac{\zeta_\ell^{(2)}(\beta)}{\zeta_\ell^{(1)}(\beta)}\left(\frac{\ln'\zeta_\ell^{(2)}(\beta) - nv_j\ln'\psi_\ell(n\beta)}{\ln'\zeta_\ell^{(1)}(\beta) - nv_j\ln'\psi_\ell(n\beta)}\right), \qquad j = 1,2, \tag{5.2a}$$

where

$$\ln' f(x) \equiv \frac{d}{dx}\ln f(x). \tag{5.2b}$$

The magnetic and electric contributions are distinguished by

$$v_1 = 1, \qquad v_2 = n^{-2}, \tag{5.3}$$

respectively.

141

Continuation into the complex plane requires knowledge of the poles of the integrands in (4.71), and from (5.2) these are determined by the condition

$$\psi_{\lambda-1/2}(n\beta)\zeta^{(1)'}_{\lambda-1/2}(\beta) - nv_j\psi'_{\lambda-1/2}(n\beta)\zeta^{(1)}_{\lambda-1/2}(\beta) = 0, \quad j = 1,2, \qquad (5.4)$$

with $\lambda = \ell + \frac{1}{2}$. This equation has no solutions in λ on the real axis, but infinitely many elsewhere, and we are interested in solutions corresponding to large β and λ. Thus, we solve (5.4) by first replacing the cylindrical functions by their asymptotic expansions, which requires some serious attention to Appendix C. In particular, owing to the Stokes phenomenon one must consider the various sectors of the λ-plane as indicated in Fig. C.1. As noted there, the regions j and j' contain zeros of the Bessel functions $J_\lambda(\beta)$, as does a neighborhood of the real axis in region A; the zeros of the Hankel functions are to be found in the four off-axis shaded regions. One readily shows from the expansions in Appendix C that, outside the shaded regions, Eq. (5.4) becomes

$$(\lambda^2 - \beta^2)^{1/2} = \pm v_j(\lambda^2 - n^2\beta^2)^{1/2}, \qquad (5.5)$$

which has no solutions in λ. We thus focus on the shaded regions.

For reasons that will become very clear presently, we shall not study this pole distribution in any detail at this time. In fact, it has been so studied in considerable depth by Streifer and Kodis (1965), and by Nussenzveig (1969a), who identifies the two classes of poles indicated in Fig. 5.1. In region 1 there is a series of poles located very close to the real axis that correspond to the optical resonances, or free modes of vibration of the dielectric sphere mentioned by Debye (1909b). There is another series of poles in region 2 located a bit further from the real axis that constitute a set of broader resonances, whereas in region 3 there is an infinite number of poles approaching the negative integers very rapidly as their real parts become large. The resonances described by these Regge poles near the real axis will be studied in depth in Chapter 7.

There is a second class of poles shown in Fig. 5.1 that is associated with the zeros of the Hankel functions. These are quite similar to those found for the perfectly conducting sphere, Eqs. (4.72), and again have an interpretation in terms of surface waves. At finite energy there is an infinite number of these poles in the first quadrant, in contrast with those of the first class that, for given β, are finite in number in the right half-plane.

It is just this large but finite set of poles lying close to the real axis, however, that now presents us with a serious difficulty. By converting the partial-wave

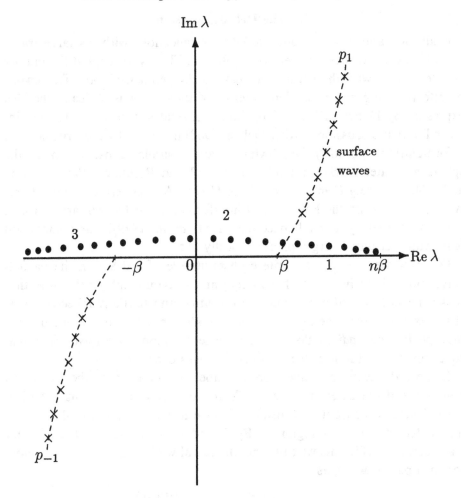

Fig. 5.1. The distribution of Regge poles for the homogeneous dielectric sphere. Those near the real axis (•) correspond to resonances and reflectivities near unity, whereas those along the curves p_1 and p_{-1} (×) are associated with damped surface waves.

series to a contour integral plus a residue sum we are still faced with a slowly-convergent series. All we have done, really, is reformulate that sum into one over a similar number of terms each of which now contains many partial waves in resonance! At high frequencies the approach to geometric optics implies that each incident ray produces an infinite series of internally-reflected rays, and hence a large number of saddle points; this presents no significant improvement on the partial-wave series. Clearly we need another approach to development of expressions like (4.71) before continuing into the complex plane.

5.1 The Debye expansion

The allusion above to the multiple internal reflections within a large transparent sphere provides the seeds of a solution. That is, we exploit the infinity of interactions with the surface intrinsic to the geometric-optics framework by reformulating our expansion in terms of them. This is basically the idea expressed by Debye in Eq. (4.30), though several approximations were involved in that discussion. Earlier Debye (1908) had actually constructed such a formulation rigorously for a cylinder, and a similar construction for the sphere was later carried out by van der Pol and Bremmer (1937), as well as by Nussenzweig (1969a) and Khare (1975). A degenerate model of this type is provided by the Fabry–Pérot interferometer, or by two large parallel plates separated by a small space (e.g., Reitz *et al.* (1979)); the comparison is relevant because a large sphere is locally flat.

Our new viewpoint is to write each multipole of index ℓ in the partial-wave sums (4.65) in terms of incoming and outgoing spherical waves that strike the surface and are partially transmitted and partially reflected there. If $r \to \infty$ we recover the above thin-film models, but for finite r one obtains multiple internal *radial* reflections. To utilize this model one solves the radial equation in a fictitious space in which r ranges over $(-\infty, \infty)$.

In accordance with convention we label the interior of the sphere as region 1 and the exterior as region 2, and envision an incoming spherical wave. There is then only a transmitted wave in region 1, and both incident and reflected waves in region 2. By introducing spherical reflection and transmission coefficients we can write the radial wavefunction in each region, for each partial wave, as

$$\phi_{2,\ell}(r) = A\left(\frac{h_\ell^{(2)}(kr)}{h_\ell^{(2)}(\beta)} + R_{22}\frac{h_\ell^{(1)}(kr)}{h_\ell^{(1)}(\beta)}\right),$$

$$\phi_{1,\ell}(r) = A\left(T_{21}\frac{h_\ell^{(2)}(nkr)}{h_\ell^{(2)}(n\beta)}\right), \tag{5.6}$$

where A is a normalization constant. Requiring continuity of ϕ and ϕ' at $r = a$ results in the relation $1 + R_{22} = T_{21}$, and just enough further information to determine R_{22} and T_{21} explicitly. Similarly, an outgoing spherical wave that interacts with the surface is described by

$$\phi_{2,\ell}(r) = A\left(T_{12}\frac{h_\ell^{(1)}(kr)}{h_\ell^{(1)}(\beta)}\right),$$

$$\phi_{1,\ell}(r) = A\left(\frac{h_\ell^{(1)}(nkr)}{h_\ell^{(1)}(n\beta)} + R_{11}\frac{h_\ell^{(2)}(nkr)}{h_\ell^{(2)}(n\beta)}\right), \tag{5.7}$$

and matching now yields $T_{12} = 1 + R_{11}$ and the explicit determination of T_{12} and R_{11}. Identification of these coefficients is straightforward and one finds expressions corresponding to the magnetic case $j = 1$ in Eqs. (5.18) below. However, complementary to this physical formulation, it is also of value to provide an algebraic derivation germane both to electric and to magnetic multipoles.

Toward this end we introduce an additional notation exploited by Nussenzweig (1969a) and occasionally found useful here:

$$[z] \equiv \ln' \psi_\ell(z) = \psi'_\ell(z)/\psi_\ell(z),$$

$$[jz] \equiv \zeta_\ell^{(j)'}(z)/\zeta_\ell^{(j)}(z), \qquad j = 1, 2, \tag{5.8}$$

with $\zeta_\ell^{(1)} = \zeta_\ell$ and $\zeta_\ell^{(2)} = \zeta_\ell^*$ only if n is real. Thus, one can rewrite (5.2a) as

$$S_\ell(\beta) = -\frac{\zeta_\ell^{(2)}(\beta)}{\zeta_\ell^{(1)}(\beta)} \frac{[2\beta] - nv_j[n\beta]}{[1\beta] - nv_j[n\beta]}, \tag{5.9}$$

which is still a phase factor, of course.

The second factor in (5.9) can now be rewritten as

$$\frac{[2\beta] - nv_j[n\beta]}{[1\beta] - nv_j[n\beta]} = 1 - \frac{[1\beta] - [2\beta]}{[1\beta] - nv_j[2n\beta]} \frac{[1\beta] - nv_j[2n\beta]}{[1\beta] - nv_j[n\beta]}. \tag{5.10}$$

With the following identity that employs the Wronskians of the Ricatti–Bessel functions, (A.20),

$$[n\beta] - [2n\beta] = \frac{W\{\psi_\ell(n\beta), \zeta_\ell^{(2)}(n\beta)\}}{\psi_\ell(n\beta)\zeta_\ell^{(2)}(n\beta)} = \frac{\frac{1}{2}W\{\zeta_\ell^{(1)}(n\beta), \zeta_\ell^{(2)}(n\beta)\}}{\psi_\ell(n\beta)\zeta_\ell^{(2)}(n\beta)}$$

$$= \frac{\zeta_\ell^{(1)}(n\beta)}{2\psi_\ell(n\beta)}([1n\beta] - 2[n\beta]), \tag{5.11}$$

the last factor in (5.10) becomes

$$\frac{[1\beta] - nv_j[2n\beta]}{[1\beta] - nv_j[n\beta]} = 1 + nv_j\frac{[n\beta] - [2n\beta]}{[1\beta] - nv_j[n\beta]}$$

$$= 1 + nv_j\frac{\zeta_\ell^{(1)}(n\beta)}{2\psi_\ell(n\beta)} \frac{[1n\beta] - [2n\beta]}{[1\beta] - nv_j[n\beta]}. \tag{5.12}$$

Further algebra converts the denominator in this expression to

$$[1\beta] - nv_j[n\beta] = \frac{\zeta_\ell^{(2)}(n\beta)}{2\psi_\ell(n\beta)}([1\beta] - nv_j[2n\beta])[1 - \rho(\ell, \beta)], \tag{5.13}$$

where

$$\rho(\ell, \beta) \equiv -\frac{\zeta_\ell^{(1)}(n\beta)}{\zeta_\ell^{(2)}(n\beta)} \frac{[1\beta] - nv_j[1n\beta]}{[1\beta] - nv_j[2n\beta]}. \tag{5.14}$$

One now verifies that (5.10) can be rewritten as

$$\frac{[2\beta] - nv_j[n\beta]}{[1\beta] - nv_j[n\beta]} = -\left(R_{22}(\ell,\beta) + \frac{T_{21}(\ell,\beta)T_{12}(\ell,\beta)}{1-\rho(\ell,\beta)}\frac{\zeta_\ell^{(1)}(n\beta)}{\zeta_\ell^{(2)}(n\beta)}\right), \qquad (5.15)$$

which introduces new functions

$$R_{22}(\ell,\beta) \equiv -\frac{[2\beta] - nv_j[2n\beta]}{[1\beta] - nv_j[2n\beta]},$$

$$T_{21}(\ell,\beta) \equiv \frac{[1\beta] - [2\beta]}{[1\beta] - nv_j[2n\beta]} = 1 + R_{22},$$

$$R_{11}(\ell,\beta) \equiv \frac{\zeta_\ell^{(2)}(n\beta)}{\zeta_\ell^{(1)}(n\beta)}\rho(\ell,\beta),$$

$$T_{12}(\ell,\beta) \equiv \frac{[1n\beta] - [2n\beta]}{[1\beta] - nv_j[2n\beta]} = 1 + R_{11}. \qquad (5.16)$$

Finally, Eq. (5.9) for the S-functions becomes

$$S_\ell(\beta) = \frac{\zeta_\ell^{(2)}(\beta)}{\zeta_\ell^{(1)}(\beta)}\left(R_{22}(\ell,\beta) + \frac{\zeta_\ell^{(1)}(n\beta)}{\zeta_\ell^{(2)}(n\beta)}\frac{T_{21}(\ell,\beta)T_{12}(\ell,\beta)}{1-\rho(\ell,\beta)}\right), \qquad (5.17)$$

and the physical interpretation of the quantities on the right-hand side is made clear by noting that Eqs. (5.16) are precisely those obtained through solution of (5.6) and (5.7). That is, in terms of Ricatti–Bessel functions the spherical reflection and transmission coefficients are

$$R_{11}(\ell,\beta) = -\frac{\zeta_\ell^{(2)}(n\beta)}{\zeta_\ell^{(1)}(n\beta)}\frac{\zeta_\ell^{(1)'}(\beta)\zeta_\ell^{(1)}(n\beta) - nv_j\zeta_\ell^{(1)}(\beta)\zeta_\ell^{(1)'}(n\beta)}{\zeta_\ell^{(1)'}(\beta)\zeta_\ell^{(2)}(n\beta) - nv_j\zeta_\ell^{(1)}(\beta)\zeta_\ell^{(2)'}(n\beta)}, \qquad (5.18a)$$

$$R_{22}(\ell,\beta) = -\frac{\zeta_\ell^{(1)}(\beta)}{\zeta_\ell^{(2)}(\beta)}\frac{\zeta_\ell^{(2)'}(\beta)\zeta_\ell^{(2)}(n\beta) - nv_j\zeta_\ell^{(2)}(\beta)\zeta_\ell^{(2)'}(n\beta)}{\zeta_\ell^{(1)'}(\beta)\zeta_\ell^{(2)}(n\beta) - nv_j\zeta_\ell^{(1)}(\beta)\zeta_\ell^{(2)'}(n\beta)}, \qquad (5.18b)$$

$$T_{21}(\ell,\beta) = \frac{\zeta_\ell^{(2)}(n\beta)}{\zeta_\ell^{(2)}(\beta)}\frac{2i}{\zeta_\ell^{(1)'}(\beta)\zeta_\ell^{(2)}(n\beta) - nv_j\zeta_\ell^{(1)}(\beta)\zeta_\ell^{(2)'}(n\beta)}, \qquad (5.18c)$$

$$T_{12}(\ell,\beta) = \frac{\zeta_\ell^{(1)}(\beta)}{\zeta_\ell^{(1)}(n\beta)}\frac{2inv_j}{\zeta_\ell^{(1)'}(\beta)\zeta_\ell^{(2)}(n\beta) - nv_j\zeta_\ell^{(1)}(\beta)\zeta_\ell^{(2)'}(n\beta)}, \qquad (5.18d)$$

where we have again used the Wronskian of Eq. (A.20). Note that all coefficients have a common denominator, and that the multipole type is determined completely by v_j.

With the expansion (5.17) we have yet another representation of the physical model in terms of elementary waveforms, the progression being from plane waves → partial waves → spherical waves, and the latter provide one more tool to help in analyzing the physical system. As a check on the integrity of this representation we observe that, as the radius of the sphere approaches infinity, the asymptotic forms (A.17) lead to the following reduction of Eqs. (5.18):

$$R_{11}(\ell,\beta) \xrightarrow[\beta\to\infty]{} \frac{nv_j - 1}{nv_j + 1},$$

$$R_{22}(\ell,\beta) \xrightarrow[\beta\to\infty]{} -\frac{nv_j - 1}{nv_j + 1},$$

$$T_{21}(\ell,\beta) \xrightarrow[\beta\to\infty]{} (-1)^\ell \frac{2}{nv_j + 1},$$

$$T_{12}(\ell,\beta) \xrightarrow[\beta\to\infty]{} (-1)^\ell \frac{2nv_j}{nv_j + 1}. \tag{5.19}$$

For $\ell = 0$ these are, in fact, exact for all values of β, and in that event are precisely the Fresnel coefficients for a plane interface at normal incidence (Eqs. (1.39)).

Before going any further we find it convenient to clean up the preceding expressions somewhat. Notice that the ratios of Hankel functions in (5.16) and (5.18) cancel out everywhere in (5.17). This suggests that it may be useful to introduce a new set of coefficients as follows:

$$R_\ell^{11}(\beta) \equiv \frac{\zeta_\ell^{(1)}(n\beta)}{\zeta_\ell^{(2)}(n\beta)} R_{11}(\ell,\beta), \qquad R_\ell^{22}(\beta) \equiv \frac{\zeta_\ell^{(2)}(\beta)}{\zeta_\ell^{(1)}(\beta)} R_{22}(\ell,\beta),$$

$$T_\ell^{12}(\beta) \equiv \frac{\zeta_\ell^{(1)}(n\beta)}{\zeta_\ell^{(1)}(\beta)} T_{12}(\ell,\beta), \qquad T_\ell^{21}(\beta) \equiv \frac{\zeta_\ell^{(2)}(\beta)}{\zeta_\ell^{(2)}(n\beta)} T_{21}(\ell,\beta). \tag{5.20}$$

Equation (5.17) now reads

$$S_\ell(\beta) = R_\ell^{22} + \frac{T_\ell^{21} T_\ell^{12}}{1 - R_\ell^{11}}, \tag{5.21}$$

and the partial-wave coefficients can be written

$$\left\{ \begin{matrix} a_\ell \\ b_\ell \end{matrix} \right\} = \frac{1}{2}\left(1 - R_\ell^{22} - \frac{T_\ell^{21} T_\ell^{12}}{1 - R_\ell^{11}} \right), \tag{5.22}$$

for electric and magnetic multipoles, respectively, depending on the value of v_j. The internal coefficients also simplify:

$$c_\ell = \frac{\zeta_\ell^{(2)}(\beta)}{\zeta_\ell^{(2)}(n\beta)} \frac{T_{21}^e(\ell,\beta)}{1 - \rho^e(\ell,\beta)} = \frac{T_\ell^{21}}{1 - R_\ell^{11}},$$

$$d_\ell = \frac{\zeta_\ell^{(1)}(n\beta)}{\zeta_\ell^{(1)}(\beta)} \frac{T_{12}^m(\ell,\beta)}{1 - \rho^m(\ell,\beta)} = \frac{T_\ell^{12}}{1 - R_\ell^{11}}. \tag{5.23}$$

As always, conservation of energy is expressed in terms of the squared magnitudes of the amplitudes. By direct calculation one verifies that, for real n,

$$|R_\ell^{11}(\beta)|^2 + \frac{1}{nv_j}|T_\ell^{12}(\beta)|^2 = 1,$$

$$|R_\ell^{22}(\beta)|^2 + nv_j|T_\ell^{21}(\beta)|^2 = 1, \qquad (5.24)$$

similar to the relations among Fresnel coefficients in Chapter 1.

Now let us return to (5.21) and expand the denominator:

$$S_\ell(\beta) = R_\ell^{22} + T_\ell^{21}\sum_{p=1}^{\infty}(R_\ell^{11})^{p-1}T_\ell^{12}, \qquad (5.25)$$

which we call the *Debye expansion* of the amplitude functions S_ℓ^{M} and S_ℓ^{E}. The terms in this series have the same physical interpretation as those in Debye's original approximate expression (4.30): the $p = 0$ term R_ℓ^{22} represents direct reflection from the spherical surface; the pth term for $p \geq 1$ corresponds to transmission into the sphere (T_ℓ^{21}) followed by propagation back and forth between $r = a$ and $r = 0$ p times while undergoing $p - 1$ internal reflections at the surface $((R_\ell^{11})^{p-1})$, culminating in a final transmission to the outside (T_ℓ^{12}). The origin here acts as a perfect reflector, and internal propagation is described by the original phase factor $\zeta_\ell^{(1)}(n\beta)/\zeta_\ell^{(2)}(n\beta)$ in (5.14). There is, however, a major difference between this expansion and Debye's approximation of 1909: although the geometric description of these terms provides a nice physical framework for large spheres, it is also exact and all the corrections to geometric optics are contained in these expressions. That is basically the difference between the spherical reflection and transmission coefficients and the Fresnel coefficients. Equation (5.25) is valid for all β, but it is only for $\beta \gg 1$ that the physical interpretation of the terms in the Debye expansion becomes cogent.

With (5.25) we have accomplished our goal of re-expressing the amplitudes as an infinite series of surface interactions. Moreover, in Eqs. (5.22) and (5.23) we find two alternative ways of presenting the scattering amplitudes: one can carry out the summation over p, as illustrated there, and study the conventional partial-wave series for each multipole of index ℓ; or, we can generate the Debye expansion and sum over ℓ for each spherical wave labeled by p. It is this second alternative that provides the desired reformulation in terms of (one hopes) a few Debye terms; the ℓ-summation for each p is then carried out by means of the modified Watson transformation. Although one does not expect to study the individual terms experimentally, the Debye decomposition does allow us to develop a deeper understanding of the fundamental mechanisms contributing to the total intensity, thereby elucidating further the entire scattering process.

A glance at (5.21) reveals that the series (5.25) has a finite radius of convergence, so it behooves us to ask under what conditions such an expansion makes sense. First of all, for any finite real $\lambda = \ell + \frac{1}{2}$ the series converges, since it is easy to show directly that

$$|R^{11}(\lambda, \beta)| < 1, \qquad \lambda \text{ real.} \qquad (5.26)$$

As $|\lambda| \to \infty$, however, $|R^{11}| \to 1$ and the series diverges. In view of this it is necessary that all λ-integrals extending to infinity be treated carefully as proper limits of finite integrals. A second point for consideration is, given convergence of the series, how fast does each individual term itself converge? This can really only be addressed as we investigate each term, but we shall see that the residue series provide the stiffest test of the method.

One might be tempted, in light of the last paragraph, to dismiss the evident singularity in (5.21) as merely a mathematical fine point. In fact, it has major physical significance, for the internal reflection coefficient can approach unity for intermediate values of ℓ (or λ) as well. Figure 5.2 illustrates the behavior of $|R_\ell^{11}|$ over a broad range of size parameters; indeed, the curves remain virtually unchanged for all values of β. Since $|R_\ell^{22}| = |R_\ell^{11}|$, we see that both reflection coefficients are at first relatively small, less than 0.2, and then increase rather abruptly to unity near $\ell + \frac{1}{2} \lesssim \beta$. This is another manifestation of the localization principle, emphasizing once again the importance of the edge domain.

If n is not exceedingly large the reflection coefficients remain quite small over nearly the full range of angles of incidence, and most rays are therefore rapidly damped by successive internal reflections – there is a correspondingly strong tendency toward transmission to the outside. In this case only a few terms in the Debye series will be significant, and any absorption will only serve to reinforce this rapid convergence.

Rays incident near the edge, however, have reflectivities very close to unity and, after penetration, will be almost totally internally reflected. As a consequence, many terms contribute to the Debye sum and the utility of the expansion comes into question. We have seen earlier that near-grazing rays also launch surface waves, which will further slow the rate of convergence for residue series arising in evaluation of these terms. Thus, although the Debye series converges tolerably fast for the most part, there are edge effects that require careful attention.

It should be noted that the denominator in (5.21) does not represent a

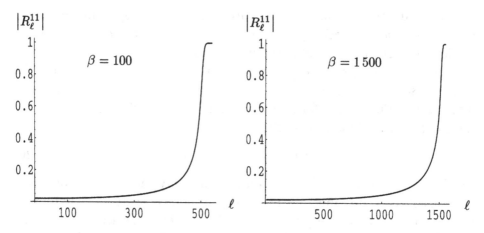

Fig. 5.2. The magnitude of the reflection coefficient, $|R_\ell^{11}|$, for $\beta = 500$ and 1500, illustrating how the dominant contributions come from the edge domain.

true singularity on the (physical) real line. Rather, it gives rise to a series of resonances arising from complex poles with imaginary parts close to the real axis. These are, in fact, just the Regge poles extracted from (5.4) and correspond to resonances in the scattering amplitudes that will be studied further in Chapter 7.

In view of these collective observations it seems of value to return to the expression (5.21) and, rather than carry out the complete expansion (5.25), rewrite $S_\ell(\beta)$ in the more physically revealing form

$$S_\ell(\beta) = R_\ell^{22} + T_\ell^{21} T_\ell^{12} + T_\ell^{21} R_\ell^{11} T_\ell^{12} + \Delta S_\ell(\beta). \tag{5.27a}$$

where the remainder term is

$$\Delta S_\ell(\beta) \equiv \frac{T_\ell^{21} (R_\ell^{11})^2 T_\ell^{12}}{1 - R_\ell^{11}}. \tag{5.27b}$$

As stated following (4.63), the first three terms contain something more than 98.5% of the scattered intensity, so it is well worth our while to investigate these separately and in some detail. Subsequent study of the remainder term will be relevant to eventual understanding of backscattering and the glory (Chapter 6), and will also lead to an explanation of the ripple structure exhibited in Figs. 3.4 and 3.5. Another global phenomenon noted in those figures, and only discussed briefly in connection with Eq. (3.100), is the slowly varying background oscillation with β in the extinction efficiency. This can now be understood completely in terms of the expressions (5.22) for the partial-wave coefficients, which govern the behavior of $Q_{ext}(\beta)$.

Let us rewrite the spherical reflection and transmission coefficients in terms of magnitude and phase:

$$R_\ell^{11} = r_\ell^{11} e^{i\varphi_\ell^{11}}, \qquad R_\ell^{22} = r_\ell^{22} e^{i\varphi_\ell^{22}},$$
$$T_\ell^{12} = (nv_j)T_\ell^{21} = t_\ell^{12} e^{i\varphi_\ell^{12}}. \tag{5.28}$$

An immediate benefit of this decomposition is that only three of these six new quantities are independent; namely, r_ℓ^{12}, φ_ℓ^{11}, and φ_ℓ^{12}. Some algebraic effort leads to the relations

$$r_\ell^{22} = r_\ell^{11}, \qquad \varphi_\ell^{21} = \varphi_\ell^{12}, \qquad t_\ell^{21} = \{(nv_j)^{-1}[1 - (r_\ell^{11})^2]\}^{1/2},$$
$$t_\ell^{21} = (nv_j)^{-1}t_\ell^{12}, \qquad \varphi_\ell^{22} = 2\varphi_\ell^{12} - \varphi_\ell^{11} - \pi. \tag{5.29}$$

The most efficient way to prove these relations is to express all quantities first in terms of the functions P_ℓ and Q_ℓ defined in (3.132), along with two other auxiliary functions,

$$M_\ell \equiv nv_j\xi_\ell(\beta)\xi_\ell'(n\beta) - \xi_\ell'(\beta)\xi_\ell(n\beta),$$
$$N_\ell \equiv nv_j\psi_\ell(\beta)\xi_\ell'(n\beta) - \psi_\ell'(\beta)\xi_\ell(n\beta). \tag{5.30}$$

For example, we find that

$$r_\ell^{11} = \left(\frac{(P_\ell + M_\ell)^2 + (Q_\ell - N_\ell)^2}{(P_\ell - M_\ell)^2 + (Q_\ell + N_\ell)^2}\right)^{1/2}, \tag{5.31a}$$

$$\varphi_\ell^{11} = \tan^{-1}\left(\frac{2(P_\ell + N_\ell + Q_\ell M_\ell)}{P_\ell^2 + Q_\ell^2 - M_\ell^2 - N_\ell^2}\right), \tag{5.31b}$$

$$\varphi_\ell^{12} = \tan^{-1}\left(\frac{P_\ell - M_\ell}{Q_\ell + N_\ell}\right), \tag{5.31c}$$

where we include both types of multipole by combining the functions in (3.132) using the multipole factor v_j.

Now we can return to the slow background oscillations in $Q_{\text{ext}}(\beta)$ and recall that we argued earlier that they must be caused by interference between diffracted and transmitted rays in the forward direction; no such oscillations occur for the impenetrable sphere. Retaining just the diffraction and transmission terms in this direction, we find from (5.22) in this approximation

$$\begin{Bmatrix} a_\ell \\ b_\ell \end{Bmatrix} \approx \frac{1}{2}\left(1 - T_\ell^{21}T_\ell^{12}\right) = \frac{1}{2}\left(1 - \frac{(t_\ell^{12})^2}{nv_j}e^{2i\varphi_\ell^{12}}\right). \tag{5.32}$$

For real n we recall that $Q_{\text{ext}} = Q_{\text{sca}}$, and use of the latter allows us to exhibit

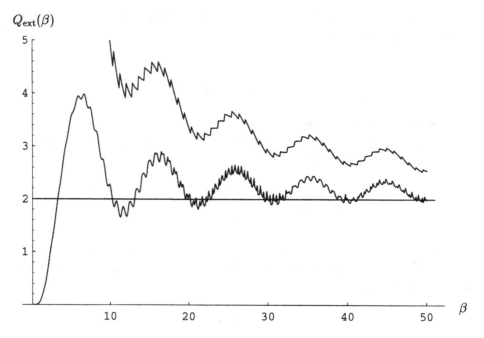

Fig. 5.3. A comparison of the extinction efficiency $Q_{ext}(\beta)$ computed with the approximation (5.33) (upper curve) with the exact expression (3.105) (lower curve), illustrating the origin of the slowly varying background.

both intensities as well as the explicit interference between the two:

$$\begin{Bmatrix} |a_\ell|^2 \\ |b_\ell|^2 \end{Bmatrix} \approx \frac{1}{4}\left(1 + \frac{(t_\ell^{12})^4}{(nv_j)^2} - \frac{2}{nv_j}(t_\ell^{12})^2 \cos(2\varphi_\ell^{12})\right). \tag{5.33}$$

Strongest transmission occurs for $\ell \ll \beta$, so that the asymptotic behavior of the Ricatti–Bessel functions given in Appendix A permits reduction of (5.31c) in this case to

$$\varphi_\ell^{12} \simeq (n-1)\beta, \tag{5.34}$$

independent of ℓ for both types of multipole. This verifies the result (3.100), and Fig. 5.3 illustrates how well this approximation to Q_{ext}, through Eq. (3.105), reproduces the slowly varying background; the larger the sphere, of course, the better the fit. The reader may wish to contemplate how this approximation overestimates $Q_{ext}(\beta)$, while the anomalous diffraction model in Fig. 4.6 underestimates it.

The scattering amplitudes

In the remainder of this chapter we shall evaluate the contributions to the scattering amplitudes from the first three terms of the Debye expansion, as identified in (5.27). Although the analysis is very much akin to what has gone before there is still some tedious, though interesting, mathematical effort required. The reader eager to assess the results immediately – perhaps, at least, in a first pass – may wish to proceed directly to Section 5.5, where the leading approximations to the amplitudes are summarized.

Let us begin by returning to (4.64)–(4.68) and rewriting them formally in terms of the Debye expansion. In doing so one must also perform the expansion explicitly for the $\ell = 0$ terms and then identify the separate contributions to each term in the general expansion; this is tantamount to making a Debye expansion of $\sigma_1(\beta)$ and $\sigma_2(\beta)$ defined in (4.66). Symbolically,

$$S_j(\beta, \theta) = \sum_{p=0}^{\infty} S_{jp}(\beta, \theta), \qquad (5.35)$$

for polarizations $j = 1$ and 2. Once again we shall require different representations in the forward and backward directions, so that, analogously to (4.64) we write

$$S_{j0}(\beta, \theta) = S_{j0}^{(+)}(\beta, \theta) - \tfrac{1}{2} \sec^2(\theta/2)$$

$$= S_{j0}^{(-)}(\beta, \theta) - (-1)^j \frac{n-1}{2(n+1)} \operatorname{cosec}^2(\theta/2)\, e^{-2i\beta}, \quad p = 0; \quad (5.36a)$$

$$S_{jp}(\beta, \theta) = S_{jp}^{(+)}(\beta, \theta) + [1 - (-1)^p] \frac{n}{(n+1)^2} \left(\frac{n-1}{n+1}\right)^{p-1} \sec^2(\theta/2)$$

$$\times e^{2i(n-1)\beta} e^{2i(p-1)n\beta},$$

$$= S_{jp}^{(-)}(\beta, \theta) + (-1)^j [1 + (-1)^p] \frac{n}{(n+1)^2} \left(\frac{n-1}{n+1}\right)^{p-1} \operatorname{cosec}^2(\theta/2)$$

$$\times e^{2i(n-1)\beta} e^{2i(p-1)n\beta}, \qquad p \geq 1. \qquad (5.36b)$$

The auxiliary functions analogous to (4.65) now consist of two sets:

$$S_{10}^{(+)}(\beta, \theta) = -\frac{1}{2} \sum_{\ell=0}^{\infty} \left\{ [R_M^{22}(\lambda, \beta) - 1] t_\ell(\cos\theta) \right.$$

$$\left. + [R_E^{22}(\lambda, \beta) - 1] p_\ell(\cos\theta) \right\}, \quad p = 0, \qquad (5.37a)$$

$$S_{1p}^{(+)}(\beta, \theta) = -\frac{1}{2} \sum_{\ell=0}^{\infty} \left\{ T_M^{21}(\lambda, \beta) T_M^{12}(\lambda, \beta) [R_M^{11}(\lambda, \beta)]^{p-1} t_\ell(\cos\theta) \right.$$

$$\left. + T_E^{21}(\lambda, \beta) T_E^{12}(\lambda, \beta) [R_E^{11}(\lambda, \beta)]^{p-1} p_\ell(\cos\theta) \right\}, \quad p \geq 1; \quad (5.37b)$$

along with the set $\{S_{10}^{(-)}, S_{1p}^{(-)}\}$ obtained via the replacement $\cos\theta \rightarrow -\cos\theta$ and inclusion of a factor $(-1)^{\ell}$ in the sums. As usual, the corresponding expressions for polarization 2 are obtained through the interchange $M \leftrightarrow E$. Note that we have introduced an obvious notation here to denote electric and magnetic contributions to the coefficients, and also have started to use the λ-notation in preparation for applying the modified Watson transformation.

Toward that end we must locate the poles in the summands, or soon-to-be integrands, of Eqs. (5.37). In line with our main aim in carrying out the Debye expansion, these are *not* the Regge poles at positions satisfying (5.4). Rather, from (5.18) we now encounter the *Regge–Debye poles*, whose locations are found as roots of the equation

$$\zeta_{\lambda-1/2}^{(1)\prime}(\beta)\zeta_{\lambda-1/2}^{(2)}(n\beta) = nv_j\zeta_{\lambda-1/2}^{(1)}(\beta)\zeta_{\lambda-1/2}^{(2)\prime}(n\beta). \tag{5.38}$$

These roots are located principally near the zeros of the Hankel functions $H_\lambda^{(1)}(\beta)$ and $H_\lambda^{(2)}(n\beta)$, and thus once more can be obtained asymptotically from the work of Streifer and Kodis (1964). The pole distribution is depicted qualitatively in Fig. 5.4, revealing no resonance structure near the real axis; indeed, the original Regge poles are all contained in the remainder term of Eq. (5.27). It should be clear from the structure of Eqs. (5.37) that all Debye terms have the same poles, and that in the pth term these are poles of order $p + 1$.

Each type of multipole has its own set of poles, and these are necessarily very similar to those of Eqs. (4.72) for the perfectly conducting sphere. As earlier, with the same notation, application of the Schöbe asymptotic expansions to the cylinder functions in (5.38) leads to the following leading-order expressions for magnetic and electric multipoles, respectively:

$$\lambda_k \simeq \beta + e^{i\pi/3}\frac{x_k}{\gamma} - \frac{i}{N} + O(\gamma), \tag{5.39a}$$

$$\mu_k \simeq \beta + e^{i\pi/3}\frac{x_k}{\gamma} - \frac{in^2}{N} + O(\gamma), \tag{5.39b}$$

where $N \equiv \sqrt{n^2 - 1}$ and $\gamma \equiv (2/\beta)^{1/3}$. These poles are found near the zeros of $H_\lambda^{(1)}(\beta)$, and there is a similar set near those of $H_\lambda^{(2)}(n\beta)$:

$$\lambda_k' \simeq n\beta + e^{-i\pi/3}\frac{x_k}{\gamma}n^{1/3} + \frac{n}{N} + O(\gamma), \tag{5.40a}$$

$$\mu_k' \simeq n\beta + e^{-i\pi/3}\frac{x_k}{\gamma}n^{1/3} + \frac{1}{nN} + O(\gamma). \tag{5.40b}$$

Recall that $\{x_k\}$ are the zeros of $\mathrm{Ai}(-x)$, and that the poles are numbered in

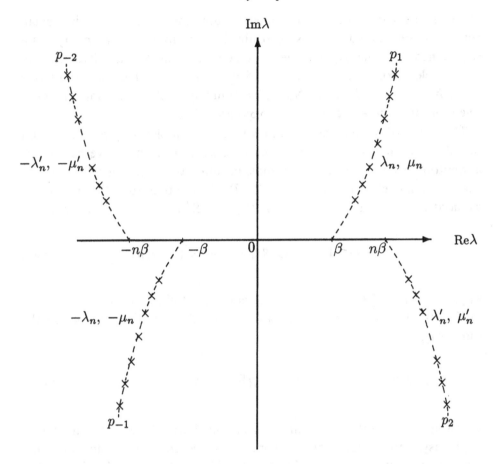

Fig. 5.4. The distribution of Regge–Debye poles $\times$ for the homogeneous dielectric sphere for $\beta \gg 1$. These poles are located near the zeros of $H_\lambda^{(1)}(\beta)$ on the curves p_1 and p_{-1}, and near the zeros of $H_\lambda^{(2)}(n\beta)$ on the curves p_2 and p_{-2}, and satisfy Eq. (5.38).

order of increasing imaginary part. Since these poles are located symmetrically with respect to the origin, we need only the positive values. Moreover, in obtaining roots asymptotically for the electric multipoles we have made the additional presumption that $|n^2\gamma/N| \ll 1$.

As with the impenetrable and perfectly conducting spheres we expect these poles to be associated with surface waves, and their imaginary parts to describe the angular damping. In (5.39) the contributions to damping from the second terms are independent of n and depend only on geometry; they represent the radiative damping arising from propagation along a curved surface. The third terms represent additional damping owing to refraction of these waves into the sphere; for n not too close to unity this is a small correction and the damping is determined primarily by the geometry. As a requirement on n this explains the second condition in (5.1), and it arises from the complete expressions underlying (5.39) and (5.40), which are asymptotic series *both* in γ and in $|n\gamma/N|$. Thus, bringing n closer to unity requires retention of higher-order terms in (5.39) and (5.40).

The utility of analytic continuation into the complex λ-plane is based on our ability to deform integration contours, which in turn is governed by the asymptotic behavior of the integrands, meaning the spherical reflection and transmission coefficients. Equations (5.37) indicate that, for either multipole, we need to understand the limits of $R^{11}(\lambda, \beta)$, $R^{22}(\lambda, \beta)$, and the product

$$U(\lambda, \beta) \equiv T^{12}(\lambda, \beta) T^{21}(\lambda, \beta) = U(-\lambda, \beta) \qquad (5.41)$$

as $|\lambda| \to \infty$. That $U(\lambda, \beta)$ is an even function of λ follows from the definitions (5.20) and the reflection properties of the Hankel functions, (C.14). In the same way,

$$R^{11}(-\lambda, \beta) = e^{2i\pi\lambda} R^{11}(\lambda, \beta), \qquad R^{22}(-\lambda, \beta) = e^{-2i\pi\lambda} R^{22}(\lambda, \beta), \qquad (5.42)$$

which are important in contour deformation. Through careful examination of the asymptotic expansions in Appendix C we are able to summarize the relevant asymptotics in Fig. 5.5, which will be of great value in all that follows. With these results we can now proceed to a study of the first three Debye terms in the short-wavelength, or high-frequency, limit.

5.2 The first Debye term – direct reflection

Analysis of the first term in the Debye expansion (5.27a), which in geometric optics corresponds to direct reflection, as indicated in Fig. 4.10, proceeds in very much the same way as that for the perfectly conducting sphere; in neither case do rays penetrate the surface. The major differences are that we now have more poles, and $\{S^M, S^E\}$ are replaced by $\{R_M^{22}, R_E^{22}\}$. As in Section 4.4, the nonuniform asymptotic behavior of the angular functions requires separate consideration of the same three angular regions, and the analysis in each is almost identical to the earlier calculations.

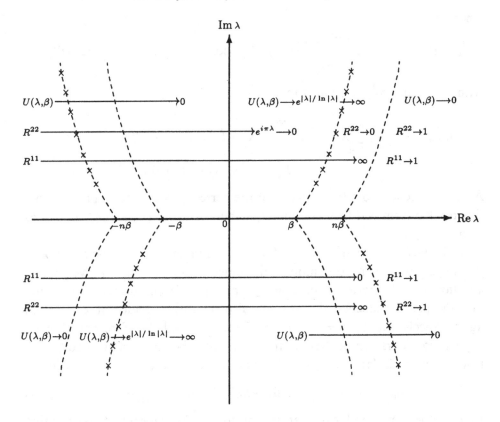

Fig. 5.5. Asymptotic behaviors of the spherical reflection and transmission coefficients in the complex λ-plane.

The wide-angle region

This is defined exactly as in (4.73), $\gamma \ll \theta \ll \pi - \beta^{-1/2}$, and from (5.36) and (5.37) the relevant expressions for the amplitudes are

$$S_{10}(\beta, \theta) = S_{10}^{(-)}(\beta, \theta) - (-1)^j \frac{n-1}{2(n+1)} \operatorname{cosec}^2(\theta/2) e^{-2i\beta}, \quad (5.43a)$$

$$S_{10}^{(-)}(\beta, \theta) = -\frac{1}{2} \sum_{\ell=0}^{\infty} \left\{ [R_M^{22}(\lambda, \beta) - 1] t_\ell(-\cos\theta) \right.$$
$$\left. + [R_E^{22}(\lambda, \beta) - 1] p_\ell(-\cos\theta) \right\}, \quad (5.43b)$$

with $S_{20}^{(-)}$ obtained from the interchange $M \leftrightarrow E$.

We now follow slavishly the same procedure as with (4.71b), which leads to a background integral and, in this model, more extensive residue series:

$$S_{10}^{(-)}(\beta, \theta) = S_{10,\text{geo}}^{(-)} + S_{10,\text{res}}^{(-)}(\beta, \theta), \quad (5.44)$$

where

$$S_{10,\text{geo}}^{(-)}(\beta,\theta) = -\frac{1}{2}\int_{-\infty+i\varepsilon}^{\infty-i\varepsilon}[R_M^{22}(\lambda,\beta)t_{\lambda-1/2}^{(1)}(\cos\theta) + R_E^{22}(\lambda,\beta)p_{\lambda-1/2}^{(1)}(\cos\theta)]\,d\lambda$$

(5.45)

with $t_{\lambda-1/2}^{(1)}$ and $p_{\lambda-1/2}^{(1)}$ defined in (D.55); and

$$S_{10,\text{res}}^{(-)}(\beta,\theta) = \frac{i}{2}\sum_{m=0}^{\infty}(-1)^m\int_{-\infty+i\varepsilon}^{\infty+i\varepsilon}[R_M^{22}(\lambda,\beta)t_{\lambda-1/2}(-\cos\theta)$$

$$-R_E^{22}(\lambda,\beta)p_{\lambda-1/2}(-\cos\theta)]e^{(2m+1)i\pi\lambda}\,d\lambda. \quad (5.46)$$

A first difference from the scenario of the perfectly conducting sphere is that this residue term will here contain contributions from $\{-\lambda_k', -\mu_k'\}$ as well as from $\{\lambda_k, \mu_k\}$.

As in (4.76b), the integrand of (5.45) contains a saddle point at $\bar\lambda = \beta\cos(\theta/2)$, and when the contour is deformed into Γ_1 in Fig. 2.10 we pick up additional residue series at the poles $\pm\lambda_k'$ and $\pm\mu_k'$. Upon further deformation into the path of steepest descent one also obtains a residue at $\lambda = \frac{1}{2}$ that again cancels out the $\ell = 0$ term in (5.43); we shall therefore omit further explicit consideration of these terms. It will also be convenient to collect all residues at the poles $\{\pm\lambda_k', \pm\mu_k'\}$ into one function, so that we now write

$$S_{10}^{(-)}(\beta,\theta) = S_{10,\text{geo}}^{(-)}(\beta,\theta) + S_{10,\text{res}}^{(-)}(\beta,\theta) + S_{10,\text{res}'}^{(-)}(\beta,\theta). \quad (5.47)$$

In a manner very similar to that leading to (2.117), the explicit contribution from the saddle point is found to be

$$S_{10,\text{geo}}^{(-)}(\beta,\theta) = \frac{1}{2}\int_{\Gamma_1}\left[R_M^{22}(\lambda,\beta)t_{\lambda-1/2}^{(1)}(\cos\theta) + R_E^{22}(\lambda,\beta)p_{\lambda-1/2}^{(1)}(\cos\theta)\right]d\lambda$$

$$\simeq \frac{i\beta}{2}\frac{\sqrt{n^2-\tilde{c}^2}-\tilde{s}}{\sqrt{n^2-\tilde{c}^2}+\tilde{s}}e^{-2i\beta\tilde{s}}\left\{1 + \frac{i}{2\beta}\left[\frac{1-2\tilde{s}^2}{\tilde{s}^3}\right.\right.$$

$$\left.\left. + \frac{2/\tilde{s}^2}{\tilde{c}^2+\tilde{s}\sqrt{n^2-\tilde{c}^2}} - \frac{2n^2-\tilde{c}^2}{(n^2-\tilde{c}^2)^{3/2}}\right] + O(\beta^{-2})\right\}, \quad (5.48)$$

where $\tilde{s} \equiv \sin(\theta/2)$ and $\tilde{c} \equiv \cos(\theta/2)$. For polarization 2 there is the similar contribution

$$S_{20,\text{geo}}^{(-)}(\beta,\theta) \simeq \frac{i\beta}{2}\frac{\sqrt{n^2-\tilde{c}^2}-n^2\tilde{s}}{\sqrt{n^2-\tilde{c}^2}+n^2\tilde{s}}e^{-2i\beta\tilde{s}}\left\{1 + \frac{i}{2\beta}\left[-\frac{1}{\tilde{s}^3} + \frac{4/\tilde{s}}{\tilde{c}^2+\tilde{s}\sqrt{n^2-\tilde{c}^2}}\right.\right.$$

$$\left. - 4\tilde{c}^2\frac{2n^4\tilde{s}^2 - n^2\tilde{c}^2(1+\tilde{s}^2-\tilde{s}^4)+\tilde{c}^6}{\tilde{s}^2(n^2-\tilde{c}^2)(n^2\tilde{s}^2-\tilde{c}^2)^2}\right.$$

$$\left.\left. + \frac{2n^2\tilde{s}}{(n^2-\tilde{c}^2)^{3/2}}\frac{2n^4-n^2\tilde{c}^2(1+\tilde{c}^2)-\tilde{c}^4}{(n^2\tilde{s}^2-\tilde{c}^2)^2}\right] + O(\beta^{-2})\right\}. \quad (5.49)$$

Note that, in both expressions, the leading factors contain the appropriate Fresnel coefficients, Eqs. (1.37) and (1.38), and as $n \to i\infty$ they reduce *almost* to the results (4.77) for the perfectly conducting sphere; 'almost' because here both polarizations suffer a phase shift of π under reflection. This appearance of the Fresnel reflection coefficients in (5.48) and (5.49) is a result of expanding the spherical reflection coefficients in a neighborhood of the real axis by means of (C.16) to obtain

$$R^{22}(\lambda, \beta) \approx \frac{\sqrt{\beta^2 - \lambda^2} - v_j\sqrt{n^2\beta^2 - \lambda^2}}{\sqrt{\beta^2 - \lambda^2} + v_j\sqrt{n^2\beta^2 - \lambda^2}}, \qquad (5.50)$$

and then evaluating these at the saddle point $\lambda = \bar{\lambda}$. That is, for a very large sphere the effects of curvature drop out and it looks like its tangent plane, yielding the exact connection between saddle points and geometric-optics rays.

The first residue term in (5.47) is identical to that of (4.78), with only a slight difference in the electric residues:

$$r_{0k}^{E} \simeq r_{0k}^{M} = \frac{e^{-i\pi/6}}{2\pi\gamma a_k'^2}, \qquad (5.51)$$

where a_k' is defined in Appendix B. The dominant forms are also those of (4.80):

$$S_{1,\text{res}}^{(-)}(\beta, \theta) \simeq \left(\frac{2\pi}{\sin\theta}\right)^{1/2} e^{i\pi/4} \left(\sum_k \sqrt{\lambda_k}\, r_{0k}^{M} e^{i\lambda_k\theta} - \frac{i}{\sin\theta}\sum_k r_{0k}^{E} e^{i\mu_k\theta}\right), \qquad (5.52a)$$

$$S_{2,\text{res}}^{(-)}(\beta, \theta) \simeq -\left(\frac{2\pi}{\sin\theta}\right)^{1/2} e^{i\pi/4} \sum_k \sqrt{\mu_k}\, r_{0k}^{E} e^{i\mu_k\theta}. \qquad (5.52b)$$

again valid for $\theta \gg \gamma$ and $\pi - \theta \gg \beta^{-1/2}$. These contributions have the same physical interpretation as before: incident rays tangential to the surface excite strongly damped surface waves, and hence rapidly convergent residue series. In principle, these waves can travel around the surface many times, shedding radiation as they go. As we shall see in higher-order terms, they can also refract at the critical angle, take shortcuts through the sphere, refract again at the critical angle and launch another surface wave. In practice, however, the strong damping indicated in (2.121), say, places limits on this continued propagation.

There remains the second residue term in (5.47), $S_{10,\text{res}'}$, which has a form very much like that in Eqs. (5.52). But the residues at the poles (5.40) are found to be approximately

$$r_{0k}^{M'} \simeq 2i\frac{n}{N}e^{2N\beta - 2\lambda_k' \cosh^{-1} n},$$

$$r_{0k}^{E'} \simeq 2i\frac{n^3}{N}e^{2N\beta - 2\mu_k' \cosh^{-1} n} \tag{5.53}$$

and one employs the reflection properties (5.42) to obtain the corresponding residues at $\{-\lambda_k', -\mu_k'\}$. With the leading-order terms from (5.40) we see that the arguments of the exponentials are negative for $n > 1$, and hence these terms are very strongly damped in comparison with those of (5.52). For $n = 1.33$, say, the residues are proportional to $e^{-0.4\beta}$. Thus, it is quite safe to neglect these poles in the remainder of this section, though they are not at all negligible if $n < 1$.

The near-forward region

In the region $0 < \theta \lesssim \gamma$ we employ the representation (5.37a) and the first line of (5.36a); as usual, polarization 2 is found from the interchange M $\leftrightarrow$ E. Completely analogous to (4.82) we obtain

$$S_{10}^{(+)}(\beta, \theta) = S_{10,\text{geo}}^{(+)}(\beta, \theta) + S_{10,\text{res}}(\beta, \theta), \tag{5.54a}$$

and sums over the residues (5.53) have been omitted. Then,

$$S_{10,\text{geo}}(\beta, \theta) \simeq \int_\beta^\infty [1 - R_M^{22}(\lambda, \theta)] P_{\lambda - 1/2}(\cos \theta) \lambda \, d\lambda$$

$$- \int_{i\infty}^\beta R_M^{22}(\lambda, \beta) P_{\lambda - 1/2}(\cos \theta) \lambda \, d\lambda$$

$$+ \frac{1}{2}\int_\beta^\infty \left\{ [1 - R_E^{22}(\lambda, \beta)] - \cos\theta\, [1 - R_M^{22}(\lambda, \beta)] P_{\lambda - 1/2}(\cos\theta) d\lambda \right.$$

$$- \frac{1}{2}\int_{i\infty}^0 [R_E^{22}(\lambda, \beta) - \cos\theta R_M^{22}(\lambda, \beta)] P_{\lambda - 1/2}(\cos\theta)\, d\lambda$$

$$+ \frac{1}{2}\int_0^\beta [t_{\lambda - 1/2}(\cos\theta) + p_{\lambda - 1/2}(\cos\theta)]\, d\lambda + O(\beta^{-1/2}), \tag{5.54b}$$

$$S_{10,\text{res}}^{(+)}(\beta, \theta) = -2\pi i \left(\frac{\theta}{\sin\theta}\right)^{1/2} \sum_{m=1}^\infty (-1)^m$$

$$\times \left(\sum_k r_{0k}^M \lambda_k J_1'(\lambda_k\theta) e^{2mi\pi\lambda_k} + \sum_k r_{0k}^E \frac{J_1(\mu_k\theta)}{\theta} e^{2mi\pi\mu_k} \right). \tag{5.54c}$$

Having exhibited this stage of the calculation explicitly, we shall anticipate the results below and assert that the residue contributions, as well as the $\ell = 0$ term in (5.36a), are negligible compared with the contribution from

(5.54b), and hence will receive no further consideration; readers are invited to substantiate this assertion at their leisure.

By including the similar terms for $S_{20}^{(+)}(\beta,\theta)$, and employing the form of (4.82b), we find that

$$S_{j0}(\beta,\theta) \simeq S_{j0,\text{geo}}^{(+)}(\beta,\theta) \simeq \frac{\beta^2}{2}\left(\frac{\theta}{\sin\theta}\right)^{1/2}$$
$$\times\left[2\frac{J_1(\beta\theta)}{\beta\theta}\left(1-\frac{iN}{\beta}\right) + F_j(\beta,\theta) + O(\gamma^4)\right], \quad 0 < \theta \lesssim \gamma. \quad (5.55)$$

The Fock terms $F_j(\beta,\theta)$ are once more integrals of products of Airy functions and derivatives of Bessel functions that provide a smooth connection between the near-forward direction and the wide-angle region; they have been studied in detail by Khare (1975). Within the region of the diffraction peak, $\theta \ll \gamma$, the Fock functions have the approximations

$$F_j(\beta,\theta) \simeq \frac{2}{\beta}\left(\frac{M_0}{\gamma} - \frac{i}{Nv_j} + \frac{8}{15}M_1\gamma - i\frac{M_0}{6N^3}(4n^2v_j^2 - 2 - v_j^2)\gamma^2 + O(\gamma^3)\right)J_0(\beta\theta)$$
$$- i\frac{NM_0}{\gamma}\frac{2}{\beta^2}\left(J_1'(\beta\theta) - \frac{n^2(2n^2+1)}{3N^2}\frac{J_1(\beta\theta)}{\beta\theta} + O(\theta/\gamma)\right), \quad (5.56)$$

whereas for θ near $\gamma = (2/\beta)^{1/3}$ they must be evaluated numerically. The functions M_m are defined in (2.134) and the first few approximated in (2.135).

In the exactly forward direction, where $J_v(x) \to (x/2)^v$ as $x \to 0$,

$$S_{10}(\beta,0) = S_{20}(\beta,0) \simeq \frac{\beta^2}{2}\left(1 + M_0\gamma^2 - i\frac{n^2+1}{2N}\gamma^3 + \frac{8}{15}M_1\gamma^4\right.$$
$$\left. + i\frac{M_0}{12N^3}(2n^6 - 4n^4 - 3n^2 + 3)\gamma^5 + O(\beta^{-2})\right). \quad (5.57)$$

By employing the uniform asymptotic expansions (D.63) for the angular functions we again find that the wide-angle region remains valid up to and including $\theta = \pi$ when we re-examine the residue sums. Consequently, the first Debye term is described uniformly throughout $\gamma \ll \theta \leq \pi$ by Eqs. (5.48), (5.49), and the explicit residue series of (4.78) when both polarizations are included and we make the replacements $r_k \to r_{0k}$. With these and (5.54) we now have a complete description of the first Debye term.

5.3 The second Debye term – direct transmission

The second term in the expansion (5.27a) corresponds to direct transmission of a ray in geometric optics, as illustrated in Fig. 4.11. For $n > 1$ there is again a 0-ray region and a 1-ray region, as well as a transition region of angular width $\Delta\theta \simeq \gamma$ connecting them smoothly. That is, the sharp shadow boundary of geometric optics is smoothed here by the physical waves. We thus consider three separate angular regions: the shadow, the lit region, and the penumbra, and we begin with the first.

The shadow region

The shadow boundary is determined by the angle of tangential emergence of rays at tangential incidence,

$$\theta_t = \pi - 2\theta_c, \qquad \theta_c = \sin^{-1}(1/n), \tag{5.58}$$

where θ_c is the critical angle of refraction. Since the width of the transition region between light and shadow is about γ, this region is defined by $\theta - \theta_t \gg \gamma$, and from (5.36) and (5.37) we see that, for $p = 1$, the amplitudes are given by

$$S_{j1}(\beta, \theta) = S_{j1}^{(-)}(\beta, \theta), \tag{5.59a}$$

$$S_{11}^{(-)}(\beta, \theta) = -\frac{1}{2}\sum_{\ell=0}^{\infty}(-1)^\ell [U_M(\lambda, \beta)t_\ell(-\cos\theta) - U_E(\lambda, \beta)p_\ell(-\cos\theta)], \tag{5.59b}$$

where $U(\lambda, \beta)$ is defined in (5.41) and $S_{22}^{(-)}$ is obtained by the interchange $M \leftrightarrow E$. Application of the Poisson sum formula and use of the reflection properties of the integrand reduces (5.59b) to

$$S_{11}^{(-)}(\beta, \theta) = \frac{i}{2}\sum_{m=0}^{\infty}(-1)^m \int_{-\infty}^{\infty} [U_M(\lambda, \beta)t_{\lambda-1/2}(-\cos\theta)$$

$$-U_E(\lambda, \beta)p_{\lambda-1/2}(-\cos\theta)]e^{(2m+1)i\pi\lambda}\, d\lambda, \tag{5.60}$$

which is uniformly regular up to $\theta = \pi$ in this region.

From Fig. 5.5 we see that $U(\lambda, \beta) \to 0$ as $|\lambda| \to \infty$ everywhere in the upper half-plane except in the region indicated, where it grows like $e^{|\lambda|/\ln|\lambda|}$. Some study of Appendix D, however, indicates that the angular functions behave like $e^{i\lambda\theta}$ in this limit, so that, for any $\theta > 0$, the path of integration in (5.60) can be closed at infinity and the integrals reduced to residue series from the *double* poles at $\{\lambda_k, \mu_k\}$ and $\{-\lambda'_k, -\mu'_k\}$. Once more the contributions from the latter set are negligible compared with those from the former and can be omitted. Hence,

$$S_{11}^{(-)}(\beta,\theta) \simeq S_{11,\text{res}}^{(-)}(\beta,\theta)$$

$$= -\pi \sum_{m=0}^{\infty}(-1)^m \left(\sum_k \text{Res}\{U_M(\lambda,\beta)t_{\lambda-1/2}(-\cos\theta)e^{(2m+1)i\pi\lambda}\}_{\lambda_k} \right.$$

$$\left. - \sum_k \text{Res}\{U_E(\lambda,\beta)p_{\lambda-1/2}(-\cos\theta)e^{(2m+1)i\pi\lambda}\}_{\mu_k} \right),$$

$$(5.61)$$

where $\text{Res}\{f(\lambda)\}_{\lambda_k}$ is the residue of $f(\lambda)$ at the pole λ_k.

The residues are computed by first employing the appropriate asymptotic expansions, such as (C.17) for $U(\lambda,\beta)$ and (D.63) for the angular functions. Owing to the structure of U the poles are double, and for this and higher-order terms it is useful to state here the general result. If $g(z)$ is an analytic function of z with simple zeros at z_i, and $f(z)$ is also analytic in z but with no zeros at z_i, then

$$F(z) \equiv \frac{f(z)}{[g(z)]^m}, \qquad m = 1, 2, \ldots \qquad (5.62a)$$

has mth-order poles at $\{z_i\}$ with residues

$$r_i \equiv \text{Res}\{F(z)\}_{z_i} = \frac{1}{(m-1)!} \frac{\partial^{m-1}}{\partial z^{m-1}}\left(\frac{f(z)(z-z_i)^m}{[g(z)]^m} \right)_{z=z_i} \qquad (5.62b)$$

Explicit expressions for several values of m are provided by Chen (1964). The dominant contribution in (5.61) is found to be

$$S_{11,\text{res}}(\beta,\theta) \simeq 2e^{i\pi/3}\frac{\beta}{\gamma N}\left(\frac{\pi-\theta}{\sin\theta}\right)^{1/2}e^{2iN\beta}\sum_{m=0}^{\infty}\sum_k (a_k')^{-2}$$

$$\times \left[e^{i\lambda_k[(2m+1)\pi-\theta_t]}\{[(2m+1)\pi - \theta_t]J_1'[(\lambda_k(\pi-\theta)] - (\pi\theta)J_1''[\lambda_k(\pi-\theta)]\} \right.$$

$$\left. -n^2 e^{i\mu_k[(2m+1)\pi-\theta_t]}\left([(2m+1)\pi - \theta_t]\frac{J_1[\mu_k(\pi-\theta)]}{\mu_k(\pi-\theta)} - i\frac{J_1'[\mu_k(\pi-\theta)]}{\mu_k}\right) \right],$$

$$\theta - \theta_t \gg \gamma, \quad \theta \leq \pi, \qquad (5.63)$$

and $a_k' = \text{Ai}'(-x_k)$. Although the electric (μ_k) and magnetic (λ_k) contributions are comparable for $\theta \simeq \pi$, the latter dominate when $\pi - \theta \gg \beta^{-1}$.

A similar result is found for $S_{21,\text{res}}^{(-)}(\beta,\theta)$, and hence for $S_{21}^{(-)}(\beta,\theta)$, and an appeal to (5.39) leads us to conclude that the *least* damped terms behave like $\exp[-(\sqrt{3}/2)x_k(\theta - \theta_t)/\gamma]$. Thus, the residue series are rapidly convergent for $\theta - \theta_t \gg \gamma$, as one expects in the shadow. Further description of these surface waves will be provided in the next section.

The illuminated region

In the opposite domain, $\theta_t - \theta \gg \gamma$, the above series are not useful; rather, we should expect a saddle-point contribution corresponding to geometric optics. To pursue this we return to Eqs. (5.59a) and (5.60), moving the integration path off the real axis to $(-\infty + i\varepsilon, \infty + i\varepsilon)$. The identities (2.99) and (D.57) then allow us to write

$$S_{11}^{(-)}(\beta, \theta) = S_{11,\text{geo}}^{(-)}(\beta, \theta) + S_{11,\text{res}}(\beta, \theta), \tag{5.64a}$$

where

$$S_{11,\text{geo}}^{(-)}(\beta, \theta) = -\frac{1}{2} \int_{-\infty+i\varepsilon}^{\infty+i\varepsilon} [U_{\text{M}}(\lambda, \beta) t_{\lambda-1/2}^{(2)}(\cos \theta) + U_{\text{E}}(\lambda, \beta) p_{\lambda-1/2}^{(2)}(\cos \theta)] \, d\lambda \tag{5.64b}$$

$$S_{11,\text{res}}(\beta, \theta) = \frac{1}{2} \sum_{m=0}^{\infty} (-1)^m \int_{-\infty+i\varepsilon}^{\infty+i\varepsilon} [U_{\text{M}}(\lambda, \beta) t_{\lambda-1/2}(\cos \theta)$$
$$+ U_{\text{E}}(\lambda, \beta) p_{\lambda-1/2}(\cos \theta)] e^{2(m+1)i\pi\lambda} \, d\lambda. \tag{5.64c}$$

The integrand in (5.64b) has a saddle point $\bar{\lambda} = kb$ related to the impact parameter b of the incident ray. From Fig. 4.11 the angles of incidence and refraction are related to each other by Snell's law and to the scattering angle by $\theta_i - \theta_r = \theta/2$, and, by following a procedure quite similar to that leading to (2.117), we find that

$$\bar{\lambda} = \beta \sin \theta_i. \tag{5.65}$$

Although it is a straightforward application of the method outlined in Appendix E and adopted in (2.117) for the impenetrable sphere, the saddle-point evaluation here is a bit tedious; it is described in further detail by Nussenzveig (1969a). The result, however, has a simple form:

$$S_{11,\text{geo}}(\beta, \theta) \simeq i\beta \left(\frac{n \sin \theta_i \cos \theta_i \cos \theta_r}{2 \sin \theta (n \cos \theta_r - \cos \theta_i)} \right)^{1/2}$$
$$\times \frac{4nv_j \cos \theta_i \cos \theta_r}{(nv_j \cos \theta_r + \cos \theta_i)^2} e^{2i\beta(n \cos \theta_r - \cos \theta_i)}$$
$$\times \left(1 - i\frac{F_1(\theta)}{16\beta \cos \theta_i} + O(\beta^{-2}) \right), \quad \theta_t - \theta \gg \gamma, \tag{5.66a}$$

where

$$F_1(\theta) = -2 \cot \theta_i \left(7 \cot \theta + \frac{\cot \theta_i}{2(1 - \chi)} \right) - \frac{9}{1 - \chi} + 15\chi - 6$$
$$+ (\chi - 1)(8\chi^2 + 5\chi + 8) \tan^2 \theta_i + \frac{16 \cot \theta_i}{\sin \theta} \left(\frac{n \cos \theta_r + \cos \theta_i}{n \cos \theta_i + \cos \theta_r} \right)^2,$$
$$\chi \equiv \frac{\cos \theta_i}{n \cos \theta_r}, \tag{5.66b}$$

and the exponential factor in (5.66a) is that appropriate to the path of the ray in Fig. 4.11. The quantity in the square-root in (5.66a), and called the *beam divergence* by van de Hulst (1957, p. 205), describes the angular spreading of the initially uniform intensity distribution of the incident beam as it traverses the scatterer, whereas the second factor is just the product $t_{12}t_{21}$ of Fresnel transmission coefficients for refraction into and out of the sphere.

We note that the condition in (5.66a) is actually mandated by the Debye asymptotic expansions of the cylinder functions. Also, if one wishes to have an explicit expression in terms of θ alone it is necessary to employ the relations mentioned prior to (5.65); we shall use these implicitly below. Finally, in deforming the path of integration in (5.64b) into the path of steepest descent one picks up a number of residue contributions from the first few poles at $\{\pm\lambda'_k, \pm\mu'_k\}$; as usual, these are negligible relative to those of (5.64c), as are similar residues to be obtained from that expression. We shall omit further mention of these contributions.

By closing the integration path in (5.64c) at infinity, as in (5.60), we obtain a residue series very similar to (5.61). Mimicking the procedure used there we readily find that

$$
S_{11,\text{res}}(\beta, \theta) \simeq 2e^{-i\pi/6} \frac{\beta}{\gamma N} \left(\frac{\theta}{\sin\theta}\right)^{1/2} e^{2iN\beta} \sum_{m=0}^{\infty} (-1)^m \sum_k (a'_k)^{-2}
$$

$$
\times \left[e^{i\lambda_k[2(m+1)\pi-\theta_t]} \{[2(m+1)\pi - \theta_t] J'_1(\lambda_k\theta) - i\theta J''_1(\lambda_k\theta)\} \right.
$$

$$
\left. + n^2 e^{i\mu_k[2(m+1)\pi-\theta_t]} \left([2(m+1)\pi - \theta_t] \frac{J_1(\mu_k\theta)}{\mu_k\theta} - i\frac{J'_1(\mu_k\theta)}{\mu_k} \right) \right],
$$

$$
\theta_t - \theta \gg \gamma, \quad \theta > 0, \tag{5.67}
$$

with the same physical interpretation.

An analogous calculation for polarization 2 yields the components of $S_{21}^{(-)}(\beta, \theta)$. The structure of $S_{21,\text{geo}}^{(-)}(\beta, \theta)$ is identical to (5.66a) with the replacements $F_1(\theta) \to F_2(\theta)$,

$$
F_2(\theta) = -2\cot\theta_i \left(7\cot\theta + \frac{\cot\theta_i}{2(1-\chi)} \right) - \frac{9}{1-\chi} + 7\chi + 2 - 8\chi''
$$

$$
+ \frac{16\cot\theta_i}{\sin\theta} \left(\frac{n\cos\theta_i + \cos\theta_r}{n\cos\theta_r + \cos\theta_i} \right)^2 + 16(\chi' - \chi)\frac{1 - \cos^2\theta_i \cos^2\theta_r}{(1+\chi')\cos^2\theta_i \cos^2\theta_r}
$$

$$
+ \tan^2\theta_i \left[4\chi'' \left((1-\chi)(\chi'' - \chi - 2) + 1 - 2\chi^2 - 2\chi'\frac{1-\chi^2}{1-\chi'^2} \right) \right.
$$

$$
\left. - (8\chi^2 + 9\chi + 8)(1-\chi) \right], \tag{5.68a}
$$

$$\chi' \equiv n^2\chi, \qquad \chi'' \equiv \frac{(1-\chi')(1+\chi)}{1+\chi'}, \qquad (5.68b)$$

and $(\cos\theta_i + n\cos\theta_r)^2 \rightarrow (n\cos\theta_i + \cos\theta_r)^2$. The residue series is obtained from (5.67) by the interchange $\mu_k \leftrightarrow \lambda_k$ and in the sum over k the factor n^2 must be moved from one exponential to the other.

Presently we shall be much interested in these quantities in the forward direction, so it is useful to record them here in that limit. We have carefully avoided the $\lambda = \frac{1}{2}$ poles of the angular functions in constructing the steepest-descent integral, so the second Debye term in the forward direction consists entirely of a geometric-type contribution and a residue series. In the former one first substitutes $\theta_r = \theta_i - \theta/2$ into Snell's law to obtain

$$\sin\theta_i = \frac{n\sin(\theta/2)}{\sqrt{n^2 - 2n\cos(\theta/2) + 1}}, \qquad (5.69)$$

which also yields $\sin\theta_r = n^{-1}\sin\theta_i$ and other trigonometric functions. Then, as $\theta \rightarrow 0$, $\chi \rightarrow n^{-1}$ and

$$S_{11,\text{geo}}(\beta,0) = S_{21,\text{geo}}(\beta,0) \approx 2i\frac{n^2\beta}{(n-1)(n+1)^2}$$

$$\times e^{2i\beta(n-1)}\left[1 + \frac{i}{16\beta}\left(\frac{15}{2n} + \frac{9}{n-1} - 8\right) + O(\beta^{-2})\right], \quad (5.70a)$$

$$S_{11,\text{res}}(\beta,0) = S_{21,\text{res}}(\beta,0)$$

$$\approx e^{-i\pi/6}\frac{\beta}{\gamma N}e^{2iN\beta}\sum_{m=0}^{\infty}(-1)^m\sum_k \frac{\eta_m}{a_k'^2}\left(e^{i\lambda_k\eta_m} + n^2 e^{i\mu_k\eta_m}\right), \quad (5.70b)$$

where $\eta_m \equiv 2(m+1)\pi - \theta_t$. Although these limits are finite and correct, it is not a valid procedure to take them directly as we have done, because our derivation is not valid for $\theta \sim \beta^{-1/2}$. One should properly apply the transformation (D.60) in (5.64b) and proceed as we did in obtaining the diffraction amplitude in Eq. (2.113). The result is identical to (5.70a); because it contains no reflection coefficients, this second Debye term will constitute the dominant contribution to the intensity in the forward and near-forward directions.†

The penumbra region

In the transition region $|\theta - \theta_t| \lesssim \gamma$ the saddle point of (5.65) approaches β, and the Debye asymptotic expansions for the Hankel functions fail. While (5.67) remains valid near the penumbra, we turn to the Schöbe expansions

† The $O(\beta^{-1})$ term in square brackets in (5.70a) differs somewhat from that of Khare (1975).

(C.17) to study $S_{j1,\text{geo}}(\beta,\theta)$ in this region. Once again the smooth transition between lit and shadow regions is described in terms of a Fock function $f(s)$ and its derivatives (Khare 1975),

$$\frac{d^m}{ds^m} f(s) = i^m \frac{e^{i\pi/6}}{2\pi} \int_\Gamma \frac{e^{isz}}{\text{Ai}^2(-z)} s^m \, dz, \tag{5.71}$$

where $s = (\theta - \theta_t)/\gamma$, and the contour Γ runs from $\infty e^{2i\pi/3}$ to 0, and to ∞ along the real line. In the shadow $f(s)$ has a residue-series expansion, and in the lit region it is susceptible to a saddle-point evaluation. These two representations provide a smooth interpolation between the two regions.

5.4 The third Debye term – internal reflection

Figure 4.12 illustrates the geometric-optics interpretation of the third term in the expansion (5.27a): a ray is transmitted into the sphere, undergoes one internal reflection, and emerges after a further transmission to the outside. In connection with that figure we learned that the ray description is rather sensitive to the value of the refractive index, so that here we shall focus on the region $1 < n < \sqrt{2}$ containing the value for water at optical frequencies. According to Fig. 4.13 there are three clearly defined angular regions characterized by the numbers of rays emerging within them in geometric optics, and delineated by the angles θ_L of (4.53) and θ_R of (4.56). Together with the three transition regions there are six in all that require careful study, and two of the transition regions contain striking new physical phenomena.

From (5.36) and (5.37) the amplitudes are given by

$$S_{j2}(\beta,\theta) = S_{j2}^{(+)}(\beta,\theta), \tag{5.72a}$$

$$S_{12}^{(+)}(\beta,\theta) = -\frac{1}{2} \sum_{\ell=0}^{\infty} \Big[T_M^{21}(\lambda,\beta) T_M^{12}(\lambda,\beta) R_M^{11}(\lambda,\beta) t_\ell(\cos\theta)$$
$$+ T_E^{21}(\lambda,\beta) T_E^{12}(\lambda,\beta) R_E^{11}(\lambda,\beta) p_\ell(\cos\theta) \Big], \tag{5.72b}$$

and the expression for polarization 2 is obtained in the usual way. Now apply the modified Watson transformation and rewrite the result using (5.41) and (5.42):

$$S_{12}^{(+)}(\beta,\theta) = -\frac{1}{2} \sum_{m=0}^{\infty} (-1)^m \int_{-\infty-i\varepsilon}^{\infty-i\varepsilon} [U_M(\lambda,\beta) R_M^{11}(\lambda,\beta) t_{\lambda-1/2}(\cos\theta)$$
$$+ U_E(\lambda,\beta) R_E^{11}(\lambda,\beta) p_{\lambda-1/2}(\cos\theta)] e^{-2mi\pi\lambda} \, d\lambda. \tag{5.73}$$

This differs from the procedure leading to (5.64) in that here the integration path has been moved completely *below* the real axis, but otherwise the next step is quite similar in the use of the identities mentioned there.

In the present case we find that

$$S_{12}^{(+)}(\beta, \theta) = S_{12,\text{geo}}^{(+)}(\beta, \theta) + S_{12,\text{res}}^{(+)}(\beta, \theta), \tag{5.74a}$$

with

$$S_{12,\text{geo}}^{(+)}(\beta, \theta) = -\frac{1}{2} \int_{-\infty-i\varepsilon}^{\infty-i\varepsilon} \left[U_{\text{M}}(\lambda, \beta) R_{\text{M}}^{11}(\lambda, \beta) t_{\lambda-1/2}^{(2)}(\cos \theta) \right. $$
$$\left. + U_{\text{E}}(\lambda, \beta) R_{\text{E}}^{11}(\lambda, \beta) p_{\lambda-12}^{(2)}(\cos \theta) \right] d\lambda, \tag{5.74b}$$

$$S_{12,\text{res}}^{(+)}(\beta, \theta) = -\frac{i}{2} \sum_{m=0}^{\infty} (-1)^m \int_{-\infty-i\varepsilon}^{\infty-i\varepsilon} \left[U_{\text{M}}(\lambda, \beta) R_{\text{M}}^{11}(\lambda, \beta) t_{\lambda-1/2}(-\cos \theta) \right.$$
$$\left. - U_{\text{E}}(\lambda, \beta) R_{\text{E}}^{11}(\lambda, \beta) p_{\lambda-1/2}(-\cos \theta) \right] e^{-(2m+1)i\pi\lambda} \, d\lambda. \tag{5.74c}$$

When the path of integration in (5.74c) is closed in the lower half-plane this term is reduced to a sum of residues from the poles at $\{-\lambda_k, -\mu_k\}$ and $\{\lambda_k', \mu_k'\}$.

As was carried out explicitly in (2.117), and implicitly in the preceding section, we develop a saddle-point integration by first making the change of variables

$$\lambda = \beta \sin \psi_1 = n\beta \sin \psi_2 \tag{5.75}$$

in (5.74b). By employing the Debye asymptotic expansions for the Hankel functions and (D.59) for the angular functions we then obtain a contribution

$$S_{12,\text{geo}}(\beta, \theta) = 2e^{i\pi/4} n\beta \left(\frac{2\beta}{\pi \sin \theta} \right)^{1/2} \int_{\Gamma} A(\psi_1, \beta, \theta) e^{i\beta\delta(\psi_1, \theta)} \, d\psi_1, \tag{5.76a}$$

where the integration path will be determined presently and

$$\delta(\psi_1, \theta) = 2 \left[2n \cos \psi_2 - \cos \psi_1 + \left(2\psi_2 - \psi_1 - \frac{\pi - \theta}{2} \right) \sin \psi_1 \right], \tag{5.76b}$$

$$A(\psi_1, \beta, \theta) = \sqrt{\sin \psi_1} \cos^2 \psi_1 \cos \psi_2 \, \frac{n \cos \psi_2 - \cos \psi_1}{(n \cos \psi_2 + \cos \psi_1)^2}$$
$$\times \left\{ 1 + \frac{i}{\beta} \left[\frac{1}{4 \cos \psi_1} \left(1 + \frac{5}{3} \tan^2 \psi_1 \right) - \frac{1}{2n \cos \psi_2} \left(1 + \frac{5}{3} \tan^2 \psi_2 \right) \right.\right.$$
$$+ \frac{7}{8} \frac{\cot \theta}{\sin \psi_1} - \frac{\tan^2 \psi_2}{\cos^2 \psi_1} (n \cos \psi_2 - \cos \psi_1) - \frac{\tan^2 \psi_1}{n \cos \psi_2}$$
$$\left.\left. - \frac{1}{\sin \theta \sin \psi_1} \frac{(n \cos \psi_2 + \psi_1)^3}{(n \cos \psi_1 + \cos \psi_2)^3} \frac{n \cos \psi_1 - \cos \psi_2}{n \cos \psi_2 - \cos \psi_1} \right] \right\}. \tag{5.76c}$$

The saddle points are determined from

$$\frac{\partial \delta}{\partial \psi_1} = 2[2\psi_2 - \psi_1 - \tfrac{1}{2}(\pi - \theta)] \cos \psi_1 = 0, \tag{5.77}$$

and are therefore described by

$$\overline{\psi}_1 = \theta_i, \qquad \overline{\psi}_2 = \theta_r,$$
$$\overline{\lambda} = \beta \sin \theta_i = n\beta \sin \theta_r. \tag{5.78a}$$

The angles θ_i and θ_r are related to θ by Eq. (4.50),

$$\theta = \pi - 2(2\theta_r - \theta_i), \tag{5.78b}$$

and have the same meaning as in that discussion; they are related to each other by Snell's law, $\sin \theta_i = n \sin \theta_r$. Some algebra produces the explicit relation

$$\cos\left(\frac{\theta}{2}\right) = \frac{2 \sin \theta_i}{n^2}\left(\sqrt{(1 - \sin^2 \theta_i)(n^2 - \sin^2 \theta_i)} - \frac{n^2}{2} + \sin^2 \theta_i\right), \tag{5.79}$$

and the constraints $0 \leq \theta_i$ and $\theta_r \leq \pi/2$ imply that the positive square-root is to be taken.

We are being a bit more explicit here than with previous calculations because this proves to be a somewhat different scenario from others, and its details need broader elaboration. From (5.78a) it is clear that the saddle points must lie in $[0, \beta]$ on the real axis, and (5–79) can be converted into a second-order algebraic equation in $z \equiv \sin \theta_i = \overline{\lambda}/\beta$. With $u \equiv 2\cos(\theta/2)$ and $v \equiv n^2/4$, this is

$$z\{[(1 - z^2)(4v - z^2)]^{1/2} + z^2 - 2v^2\} - uv = 0. \tag{5.80}$$

A solution is effected by elimination of the radical through squaring to obtain a fourth-order equation, one whose physical solutions must also satisfy (5.80). The resulting roots and their properties have been studied at some length by Nussenzveig (1969b, Appendix A), the results of which we merely summarize here. Recall the definitions of s and c in Eq. (4.55), as well as those of θ_L in (4.53) and the rainbow angle θ_R in (4.56). Explicitly, these angles are

$$\theta_{iR} = \sin^{-1}\left(\frac{4 - n^2}{3}\right)^{1/2}, \quad \theta_L = 4\cos^{-1}(1/n), \quad \theta_R = 2\cos^{-1}\left(\frac{4 - n^2}{3n^{4/3}}\right)^{3/2}. \tag{5.81}$$

We further define

$$\epsilon \equiv \theta - \theta_R, \quad p \equiv \left(\frac{4c}{3s}\right)^{1/2}, \quad q \equiv \frac{3 + c^2}{18s^2}. \tag{5.82}$$

The behavior of the roots now follows exactly the scenario depicted in Fig. 4.13. At $\theta = \theta_R$, the 0-ray/2-ray boundary, there is a double real root at $z = z_R$; for $\theta < \theta_R$, the 0-ray region, there is a pair of complex-conjugate roots that leave the real axis at right angles as θ decreases; for $\theta > \theta_R$ the roots are real and move away from the point z_R in opposite directions, the smaller one tending to zero at $\theta = \pi$ in the 1-ray region, the larger increasing and then becoming unphysical at $\theta = \theta_L$ ($\theta_i = \pi/2$), the 1-ray/2-ray boundary. The behavior of the roots in a neighborhood of $\theta = \theta_R$ is found by expanding them in powers of ϵ:

$$\left\{ \begin{matrix} z' \\ z'' \end{matrix} \right\} = s \pm cp\epsilon^{1/2} - \left(cq + \frac{s}{2}p^2 \right)\epsilon$$
$$\pm \left(\frac{5}{2}c\frac{q^2}{p} - \frac{c(17c^2 + 3)}{216s^2p} + spq - \frac{c}{6}p^3 \right)\epsilon^{3/2} + O(\epsilon^2). \quad (5.83)$$

For $\epsilon > 0$ these constitute two real saddle points near the rainbow angle, and for $\epsilon < 0$ we make the substitution $\epsilon^{1/2} \to -i(\theta_R - \theta)^{1/2}$ and obtain two complex-conjugate saddle points.

Figure 5.6 summarizes the behavior of the saddle points in the λ-plane. In the 1-ray region, $\pi > \theta > \theta_L$, there is only one saddle point given by $\bar{\lambda}/\beta = \sin\theta_i$ that moves from the origin to the point z_L as θ decreases in this range. At $\theta = \theta_L$ another saddle point appears at $\bar{\lambda}/\beta = 1$ in a normal Fock-type transition region, and as θ decreases further from θ_L to θ_R the two real saddle points move toward one another, coalescing at $\theta = \theta_R$; this is the 2-ray region, and the rainbow occurs at the 2-ray/0-ray boundary. Mathematically, the rainbow is described by the 'collision' of two saddle points in the complex plane! For $\theta < \theta_R$ the saddle points become complex and move off the real axis at right angles in complex-conjugate trajectories, describing the 0-ray region and geometric shadow. Now we are in a position to study Eqs. (5.74) in the various angular regions.

The 1-ray region

Here we have $\theta - \theta_L \gg \gamma$ and the dominant contribution comes from a single saddle point. According to the discussion associated with Eqs. (E.21)–(E.23) the direction of the steepest-descent path is determined by the sign of $\partial^2\delta/\partial\psi_1^2$ evaluated at the saddle point, and from (5.77) this is

$$\left. \frac{\partial^2\delta}{\partial\psi_1^2} \right|_{\substack{\psi_1=\theta_i \\ \psi_2=\theta_r}} = 2\left(2\frac{d\theta_r}{d\theta_i} - 1 \right)\cos\theta_i = -\frac{d\theta}{d\theta_i}, \quad (5.84)$$

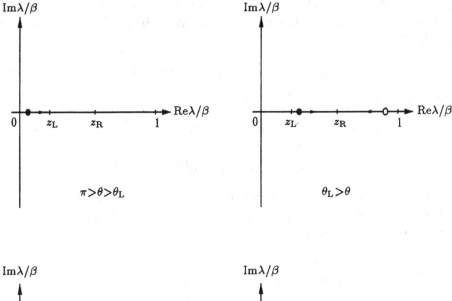

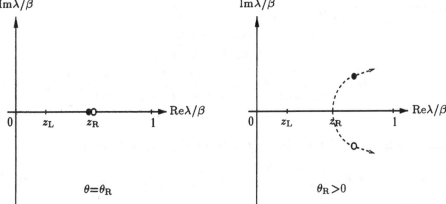

Fig. 5.6. Behaviors of the saddle points (•, ○) for the third Debye term as functions of the scattering angle θ. The arrows indicate the movement of the saddle points as θ decreases from π to θ_R, and then further into the shadow. The extent of intrusion of the trajectories into the complex plane is limited by the validity of the Debye asymptotic expansions.

where we have taken note of (5.78b). From Fig. 4.13 we know that

$$\frac{d\theta}{d\theta_i}\left\{\begin{matrix}>\\=\\<\end{matrix}\right\}0 \quad\text{for }\theta_i\left\{\begin{matrix}>\\=\\<\end{matrix}\right\}\theta_{iR}, \tag{5.85}$$

and in the 1-ray region $\theta_i < \theta_{iR}$. Hence, the right-hand side of (5.84) is positive and the path of steepest descent makes an angle of $\pi/4$ with the real axis as it crosses the saddle point. This is the contour Γ_1 in Fig. 5.7, obtained by appropriate deformation of the contour in (5.74b). In performing this deformation we sweep a number of poles at $\{\pm\lambda_k, \pm\mu_k\}$, so that we can rewrite (5.74b) as

$$S_{12,\text{geo}}^{(+)}(\beta, \theta) = S_{12,\text{geo}}(\beta, \theta) + S_{12,\text{res}}(\beta, \theta), \qquad (5.86a)$$

where $S_{12,\text{geo}}$ is given by (5.76a), and

$$S_{12,\text{res}}(\beta, \theta) = -i\pi \sum_k \left\{ \pm \text{Res}[U_\text{M}(\lambda, \beta)R_\text{M}^{11}(\lambda, \beta)t_{\lambda-1/2}^{(2)}(\cos\theta)]_{\pm\lambda_k} \right.$$

$$\left. \pm \text{Res}[U_\text{E}(\lambda, \beta)R_\text{E}^{11}(\lambda, \beta)p_{\lambda-1/2}^{(2)}(\cos\theta)]_{\pm\mu_k} \right\}, \qquad (5.86b)$$

where the sign of the Res is associated with the sign of the poles.

The saddle-point evaluation is again straightforward, though tedious, and the geometric term yields

$$S_{12,\text{geo}}(\beta, \theta) \simeq i\beta \left(\frac{\sin\theta_i}{\sin\theta} \right)^{1/2} \frac{(2n\cos\theta_i\cos\theta_r)^{3/2}}{(2\cos\theta_i - n\cos\theta_r)^{1/2}}$$

$$\times e^{2i\beta(2n\cos\theta_r - \cos\theta_i)} \frac{n\cos\theta_r - \cos\theta_i}{(n\cos\theta_r + \cos\theta_i)^3}$$

$$\times \left\{ 1 - \frac{i}{64\beta\cos\theta_i} \left[G_1 + \frac{64n^2}{\sin\theta_i\sin\theta} \frac{1-\chi'}{1-\chi} \left(\frac{1+\chi}{1+\chi'}\right)^3 \right] + O(\beta^{-2}) \right\}, \qquad (5.87a)$$

where χ and χ' are defined in (5.66) and (5.68), respectively, and

$$G_1 = 8\cot\theta_i \left(\cot\theta + \frac{\cot\theta}{2(2\chi-1)} \right) + 6(9\chi - 11) - \frac{15}{2\chi-1} - \frac{9}{(2\chi-1)^2} + \tan^2\theta_i$$

$$\times \left(56\chi^3 - 3\chi^2 + \frac{39\chi}{2} - \frac{79}{2} - \frac{33}{2(2\chi-1)} - \frac{51}{4(2\chi-1)^2} - \frac{15}{4(2\chi-1)^3} \right). \qquad (5.87b)$$

In (5.87a) we can again extract the beam divergence factor for the path *in this scenario*, as well as the product $t_{12}r_{22}t_{21}$ of Fresnel coefficients describing two refractions and an internal reflection. The expression for polarization 2 has very much the same form:

$$S_{22,\text{geo}}(\beta, \theta) \simeq i\beta \left(\frac{\sin\theta_i}{\sin\theta} \right)^{1/2} \frac{(2n\cos\theta_i\cos\theta_r)^{3/2}}{(2\cos\theta_i - n\cos\theta_r)^{1/2}}$$

$$\times e^{2i\beta(2n\cos\theta_r - \cos\theta_i)} \frac{\cos\theta_r - n\cos\theta_i}{(\cos\theta_r + n\cos\theta_i)^3}$$

$$\times \left\{ 1 - \frac{i}{64\beta\cos\theta_i} \left[G_2 + \frac{64n}{n^2\sin\theta_i\sin\theta} \frac{1-\chi}{1-\chi'} \left(\frac{1+\chi'}{1+\chi}\right)^3 \right] + O(\beta^{-2}) \right\}, \qquad (5.88a)$$

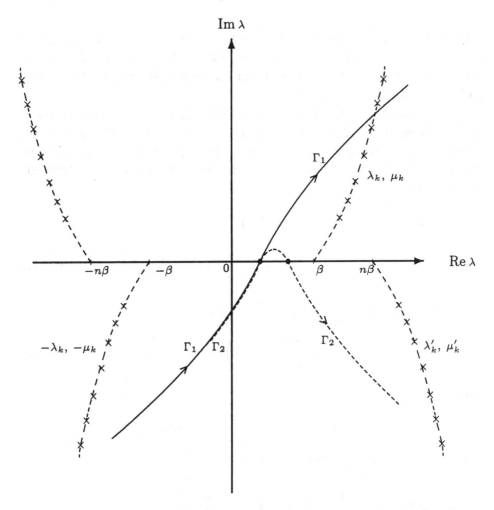

Fig. 5.7. Saddle points (•) and steepest-descent paths for evaluation of the integral in Eq (5.74b) The contour is deformed into the path Γ_1 in the 1-ray region, and into the path Γ_2 in the 2-ray region. The two paths coincide well into the third quadrant.

and

$$G_2 \equiv G_1 + 64\frac{(\chi - \chi')(2 - \chi + \chi')}{(2\chi - 1)(1 + \chi')} + 64\tan^2\theta_i$$
$$\times \left[\chi^2 - 2\chi + (\chi^2 + \sec^2\theta_r)\left(\frac{n^2 - 1}{n^2}\frac{1}{1 + \chi'} + \frac{n^2\chi\cos^2\theta_i}{n^2\chi^2 - \cos^2\theta_i}\right)\right]$$
$$- \tan^2\theta_i\,(\chi - \chi')\left(16(2\chi + 1) - \frac{48}{2\chi - 1} + \frac{8(7\chi - 3)}{1 + \chi'} + \frac{212}{(1 + \chi')(2\chi - 1)}\right.$$
$$\left. + \frac{12(4\chi^2 - 2\chi - 5)}{(1 + \chi')^2} - \frac{36}{(1 + \chi')^2(2\chi - 1)} + \frac{1}{2}(2\chi - 1)^2(1 + \chi')\right). \quad (5.88b)$$

Both these expressions are quite similar to that of (5.66a) for the direct transmission term, except for the additional factor here corresponding to the internal Fresnel reflection coefficient and the proper phase factor for the ray path in Fig. 4.12.

Residue series are provided by Eqs. (5.74c) and (5.86b) and, as usual, we can eliminate forthwith those arising from the poles $\{\lambda'_k, \mu'_k\}$ as being negligible. Evaluation of the residues at the triple poles is a bit tedious, though the leading-order contributions can be expressed simply in terms of two functions

$$\tau_1(m, \lambda, z, \varphi) \equiv e^{i\lambda\chi_m}\left[\frac{\chi_m}{z}\left(1 + \frac{\chi_m}{z}\right)J_1'(\lambda\varphi)\right.$$
$$\left. - i\frac{\varphi}{z}\left(1 + 2\frac{\chi_m}{z}\right)J_1''(\lambda\varphi) - \frac{\varphi^2}{z^2}J_1'''(\lambda\varphi)\right], \quad (5.89a)$$

$$\tau_2(m, \lambda, z, \varphi) \equiv \frac{e^{i\lambda\chi_m}}{\lambda\varphi}\left[\frac{\chi_m}{z}\left(1 + \frac{\chi_m}{z}\right)J_1(\lambda\varphi)\right.$$
$$\left. - i\frac{\varphi}{z}\left(1 + 2\frac{\chi_m}{z}\right)J_1'(\lambda\varphi) - \frac{\varphi^2}{z^2}J_1''(\lambda\varphi)\right], \quad (5.89b)$$

with

$$\chi_m \equiv (2m + 1)\pi - 4\cos^{-1}(1/n). \quad (5.89c)$$

The total residue contribution is then given by

$$S_{12,\text{res}}(\beta, \theta) \simeq -\frac{2i\beta}{\gamma}e^{i\pi/3}\left(\frac{\pi - \theta}{\sin\theta}\right)^{1/2}e^{i4N\beta}\sum_{m=0}^{\infty}(-1)^m$$
$$\times \sum_k \frac{\tau_1(m, \lambda_k, N, \pi - \theta) - \tau_2(m, \mu_k, N/n^2, \pi - \theta)}{a_k'^2}, \quad (5.90a)$$

$$S_{22,\text{res}}(\beta, \theta) \simeq -\frac{2i\beta}{\gamma}e^{i\pi/3}\left(\frac{\pi - \theta}{\sin\theta}\right)^{1/2}e^{i4N\beta}\sum_{m=0}^{\infty}(-1)^m$$
$$\times \sum_k \frac{\tau_1(m, \mu_k, N/n^2, \pi - \theta) - \tau_2(m, \lambda_k, N, \pi - \theta)}{a_k'^2}, \quad (5.90b)$$

with $a_k' \equiv \text{Ai}'(-x_k)$, and $\{x_k\}$ are the positive zeros of the Airy function.

These residue series are regular at $\theta = \pi$, but the exact backward direction turns out to be quite complicated and will subsequently be treated separately. Thus, in the 1-ray region we can require $\pi - \theta \gg \beta^{-1}$ and use the Debye asymptotic formulas to rewrite (5.90a), say, as

$$
S_{12,\text{res}}(\beta, \theta) \approx -\frac{i\beta}{N^2} e^{i\pi/12} \left(\frac{\gamma}{\pi \sin \theta}\right)^{1/2} e^{4iN\beta} \left(i \sum_k \frac{(\zeta_{2,0}^+)^2 + N\zeta_{2,0}^+}{a_k'^2} e^{i\lambda_k \zeta_{2,0}^+} + \sum_{m=1}^{\infty} (-1)^m \right.
$$

$$
\left. \times \sum_k \frac{i[(\zeta_{2,m}^+)^2 + N\zeta_{2,m}^+]e^{i\lambda_k \zeta_{2,m}^+} - [(\zeta_{2,m}^-)^2 + N\zeta_{2,m}^-]e^{i\lambda_k \zeta_{2,m}^-}}{a_k'^2} \right), \quad (5.91)
$$

where

$$
\zeta_{2,m}^{\pm} \equiv 2m\pi - 4\cos^{-1}(1/n) \pm \theta. \tag{5.92}
$$

A physical interpretation of these series in terms of surface waves readily suggests itself (Chen 1964; Nussenzveig 1969b). Tangentially incident rays undergo two critical refractions and one total internal reflection, finally re-emerging tangentially to excite further surface waves that propagate from the 2-ray/1-ray boundary into the 1-ray region, the total angular deviation being described by the angles $\zeta_{2,m}^{\pm}$. That is, the diffracted rays take two 'shortcuts' across the sphere, and there are infinitely many ways they can do this. The phase factor $e^{4iN\beta}$ accounts for the optical path difference associated with the two shortcuts. There are, in fact, two types of diffracted ray: those that undergo a critical internal reflection; and those that instead are critically refracted to the surface and, after propagating as surface waves through some angle, are critically refracted again to the inside where they proceed as rays of the first type. These two types are illustrated in Fig. 5.8.

Similar interpretations can be made for the surface waves associated with the first two Debye terms, of course. The series of (5.52) for the directly reflected term also contain the fruit of tangentially incident rays that launch surface waves, although no shortcuts are taken because there is no penetration. In the $p = 1$, directly transmitted term the series (5.63) and (5.67) contain terms corresponding to a single shortcut, critical refraction, and surface waves. Indeed, all higher Debye terms will contain contributions of this general type.

The 2-ray region

The angular region between θ_R and θ_L will be delimited by

$$
\theta_L - \theta \gg \gamma, \qquad \theta - \theta_R \ll \gamma^2/N, \tag{5.93}
$$

and contains two saddle points that can be treated independently with additive contributions; this is so as long as they are not too close together. These saddle points are partially characterized by $\theta_i \gtrsim \theta_{iR}$, so that, from

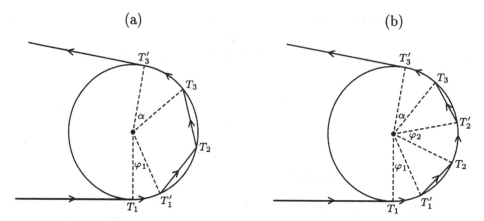

Fig. 5.8. A physical interpretation of the residue series (5.91) in terms of complex rays. (a) A type-I ray that is launched and ends with a surface wave after undergoing a total internal reflection between two refractions prior to radiating at angle θ, and $\alpha = \zeta_{2,0}^{+} - \varphi_1$. (b) The internal reflection is replaced by two successive critical refractions, so that $\alpha = \zeta_{2,0}^{+} - \varphi_1 - \varphi_2$.

(5.84) and (5.85), the saddle point with $\theta_i < \theta_{iR}$ is the one just considered above, and the second with $\theta_i > \theta_{iR}$ is traversed by the steepest-descent path at an angle of $-\pi/4$. This path is that labeled Γ_2 in Fig. 5.7, and on moving onto that contour, we pick up residue contributions from the poles at $\{-\lambda_k, -\mu_k\}$ and $\{\lambda'_k, \mu'_k\}$. For this region, then, we shall write

$$S_{12}(\beta, \theta) = \overline{S}_{12,\text{geo}}(\beta, \theta) + \overline{S}_{12,\text{res}}(\beta, \theta) + \overline{S}_{12,\text{res}'}(\beta, \theta), \qquad (5.94)$$

and again $\overline{S}_{12,\text{res}'}$ can be neglected compared with the other two terms.

Because the calculations here are virtually identical to those above for the 1-ray region, we merely summarize the results. The saddle-point contributions are

$$\overline{S}_{12,\text{geo}}(\beta, \theta) \simeq S_{12,\text{geo}}(\beta, \theta)\Big|_{\theta_i = \theta'_i} + S_{12,\text{geo}}(\beta, \theta)\Big|_{\theta_i = \theta''_i}, \qquad (5.95)$$

where $\theta'_i > \theta_{iR}$ and $\theta''_i < \theta_{iR}$ correspond respectively to z' and z'' in (5.83) and $S_{12,\text{geo}}$ is given by (5.87a). The residue contribution is

$$\overline{S}_{12,\text{res}}(\beta, \theta) \simeq -\frac{2\beta}{\gamma} e^{i\pi/3} \left(\frac{\theta}{\sin\theta}\right)^{1/2} e^{4iN\beta} \sum_{m=0}^{\infty} (-1)^m$$

$$\times \sum_k \frac{\tau_1(m + \tfrac{1}{2}, \lambda_k, N, \theta) + \tau_2(m + \tfrac{1}{2}, \mu_k, N/n^2, \theta)}{a'^2_k}. \qquad (5.96)$$

These representations are regular at $\theta = 0$, but can be simplified considerably for $\theta \gg \beta^{-1}$. The resulting expression for the residue series is identical to (5.91) without the term containing $\zeta_{2,0}^{+}$; this omission arises because, in the 2-ray region, the minimum arc required for a surface wave to emerge is $2\pi - \theta - \theta_{\mathrm{L}}$.

Corresponding results for polarization 2 are

$$\bar{S}_{22,\mathrm{geo}}(\beta, \theta) \simeq S_{22,\mathrm{geo}}(\beta, \theta)\Big|_{\theta_i = \theta_i'} + S_{22,\mathrm{geo}}(\beta, \theta)\Big|_{\theta_i = \theta_i''}, \tag{5.97}$$

where $S_{22,\mathrm{geo}}$ is given by (5.88) and

$$\bar{S}_{22,\mathrm{res}}(\beta, \theta) \simeq -\frac{2\beta}{\gamma} e^{i\pi/3} \left(\frac{\theta}{\sin\theta}\right)^{1/2} e^{4iN\beta} \sum_{m=0}^{\infty} (-1)^m$$

$$\times \sum_{k} \frac{\tau_1(m + \frac{1}{2}, \mu_k, N/n^2, \theta) + \tau_2(m + \frac{1}{2}, \lambda_k, N, \theta)}{a_k'^2}. \tag{5.98}$$

The validity criteria for these results arise generally from the need for the Debye asymptotic expansions to remain valid, and this is reflected in the first condition in (5.93). In addition, correction terms to the geometric contributions (5.95) and (5.97) will diverge near θ_{R} like $\beta[N(\theta - \theta_{\mathrm{R}})]^{3/2}$, leading to the second restriction in (5.93). In this neighborhood the saddle points also approach each other closely and can no longer be considered independent, so that qualitative modification of the saddle-point method is required.

The 0-ray region

In this region $\theta_{\mathrm{R}} - \theta \gg \gamma^2/N$, thereby violating the second condition in (5.93). The saddle points are complex conjugates here and thus one decays while the other grows. By choosing the phase convention to ensure that the trigonometric functions have positive real parts in the immediate neighborhood of $\theta = \theta_{\mathrm{R}}$, one forces the *lower* saddle point to be exponentially decreasing.† Hence, we can move the steepest-descent contour Γ_2 smoothly through the lower saddle point alone. The residue contributions remain identical to those in the 2-ray region, but the geometric contributions are those of (5.95) and (5.97) evaluated at the lower saddle point. The latter can be interpreted as arising from a 'complex ray', and it is useful to examine explicitly the angular dependence of its contribution by carrying out an expansion in $\epsilon \equiv \theta_{\mathrm{R}} - \theta$

---

† Specifically, the exponential in question is that in Eqs. (5.95) and (5.87a), say, which arises from that in the general solution (E.31). The conclusion follows from the expressions for trigonometric functions of complex argument.

(> 0 here). While (5.83) already provides an expansion for $\sin \theta_i$, we shall also need the following leading-order expansions:

$$
\left\{\begin{array}{l} \cos \theta_i' \\ \cos \theta_i'' \end{array}\right\} \simeq c \mp sp\epsilon^{1/2} + \left(sq - \frac{c}{2}p^2\right)\epsilon
$$

$$
\mp \left[\frac{s}{2p}\left(5q^2 - \frac{17c^2+3}{108s^2}\right) - cpq - \frac{s}{6}p^3\right]\epsilon^{3/2} + O(\epsilon^2), \qquad (5.99a)
$$

$$
\left\{\begin{array}{l} \cos \theta_r' \\ \cos \theta_r'' \end{array}\right\} \simeq 2c \mp \frac{sp}{2}\epsilon^{1/2} + \left(\frac{sq}{2} + \frac{p^2}{16c}(3 - 7c^2)\right)\epsilon
$$

$$
\mp \left[\frac{s}{4p}\left(5q^2 - \frac{17c^2+3}{108s^2}\right) + \frac{pq}{8c}(3 - 7c^2) - \frac{sp^3}{192c^3}(43c^2 + 9)\right]\epsilon^{3/2} + O(\epsilon^2),
$$

$$
\qquad (5.99b)
$$

$$
\left\{\begin{array}{l} n\cos \theta_r' - \frac{1}{2}\cos \theta_i' \\ \cos \theta_r'' - \frac{1}{2}\cos \theta_i'' \end{array}\right\} \simeq \frac{3c}{2} + \frac{s}{4}\epsilon \pm \frac{cp}{6}\epsilon^{3/2} + O(\epsilon^2), \qquad (5.99c)
$$

in the notation of (4.55) and (5.82). Then,

$$
S_{j2,\text{geo}}(\beta, \theta_R - \epsilon) \simeq 8\beta e^{i\pi/4} \frac{v_j(2v_j - 1)}{(1 + 2v_j)^3} \frac{n^2}{(8 + n^2)^{1/2}} \frac{c^{1/2}}{s^{3/2}}
$$

$$
\times e^{i\beta(6c - s|\epsilon|)} \exp\left[-\beta\frac{s}{2}\left(\frac{4c}{3s}|\epsilon|\right)^{3/2}\right]
$$

$$
\times [3s/(4c|\epsilon|)]^{1/2} \{1 + O[1/(\beta|\epsilon|^{3/2})]\}, \qquad (5.100)
$$

where v_j is the multipole-type indicator. It will be of importance later to note that the damping factor in the exponent is proportional to β and describes the damping as θ moves away from the 2-ray/0-ray boundary and into the shadow. Equations (5.96) and (5.98), along with (5.89) and (5.39), however, show that this factor in the residue series is proportional to $\beta^{1/3}$, so that these eventually dominate the amplitude in the deep shadow. Nearer the rainbow angle, though, damping of the amplitude is much stronger.

This completes the analysis of the third term of the Debye expansion, except for the three narrow angular regions. In the domain $|\theta - \theta_L| \lesssim \gamma$ the right-hand saddle point is very close to β, so that the Debye asymptotic expansions of the Hankel functions fail. This is again a Fock-type transition region that provides a smooth interpolation between the 1-ray and 2-ray regions, and its treatment is essentially the same as for the penumbra region for the second Debye term, to which the reader is referred. The remaining two regions – the backward direction and the 2-ray/0-ray region – pose substantially new problems in our development of scattering theory and lead to new and interesting physics. We shall complete their basic mathematical analysis here and then examine the consequences in the context of special applications in the following chapter.

The 2-ray/0-ray (rainbow) region

According to Fig. 4.13 this is the narrow domain $|\theta - \theta_R| \lesssim \gamma^2/(2N)$, with $N = \sqrt{n^2 - 1}$. The rainbow angle θ_R is an angle of minimum deviation, as discussed in connection with Fig. 1.9, and in general scattering theory is associated with an extremum in the deflection function, Eq. (1.27); in this case it is a minimum.

Our mathematical problem in evaluating the scattering amplitudes here stems from the overlapping of the two saddle points in this region, and their coalescence at $\theta = \theta_R$. However, this situation has been anticipated in discussing the extension of the method of steepest descents in Appendix E – the CFU method of (E.33)–(E.42). While the residue results (5.96) and (5.98) for the 2-ray region remain valid here, we must return to the form (5.76) for the geometric term and rewrite it for both polarizations as

$$S_{j2,\text{geo}}(\beta, \theta) = -e^{i\pi/4} \frac{n}{\sqrt{\pi \sin \theta}} \kappa^{3/2} F_j(\beta, \theta), \tag{5.101a}$$

where $\kappa \equiv 2\beta$,

$$F_j(\beta, \theta) \equiv \int g_j(\psi_1) e^{\kappa f(\psi_1, \theta)} \, d\psi_1, \tag{5.101b}$$

and

$$f(\psi_1, \theta) \equiv i \left[2n \cos \psi_2 - \cos \psi_1 + \left(2\psi_2 - \psi_1 - \frac{\pi - \theta}{2} \right) \sin \psi_1 \right]. \tag{5.101c}$$

Although we could include higher-order terms, albeit with considerable effort, it is sufficient to include here only terms up to $O(\beta^{-1})$, so that

$$g_j(\psi_1) \equiv v_j \sqrt{\sin \psi_1} \, \cos^2 \psi_1 \cos \psi_2 \frac{\cos \psi_1 - v_j n \cos \psi_2}{\cos \psi_1 + v_j n \cos \psi_2}. \tag{5.101d}$$

Following the procedure outlined in Appendix E, we change variables through the transformation

$$f(\psi_1, \theta) = \tfrac{1}{3} u^3 - \zeta(\epsilon) u + A(\epsilon), \qquad \epsilon \equiv \theta - \theta_R, \tag{5.102}$$

and then make the identifications

$$\tfrac{2}{3}\zeta^{3/2}(\epsilon) = i[n(\cos\theta'_r - \cos\theta''_r) - \tfrac{1}{2}(\cos\theta'_i - \cos\theta''_i)] \xrightarrow[|\epsilon|\ll 1]{}$$

$$i\frac{2(c\epsilon)^{3/2}}{3(3s)^{1/2}}\left(1 + \frac{875c^6 - 1257c^4 + 657c^2 + 45}{5760(cs)^3}\epsilon + O(\epsilon^2)\right), \quad (5.103)$$

$$A(\epsilon) = i[n(\cos\theta'_r + \cos\theta''_r) - \tfrac{1}{2}(\cos\theta'_i + \cos\theta''_i)]$$

$$\xrightarrow[|\epsilon|\ll 1]{} i\left(3c + \tfrac{1}{2}s\epsilon + \frac{c(11c^2 - 15)}{72s^2}\epsilon^2 + O(\epsilon^3)\right), \quad (5.104)$$

where c and s are defined in (4.55), $\{\theta'_i, \theta'_r\}$ and $\{\theta''_i, \theta''_r\}$ correspond to the two saddle points, and we have employed the expansions (5.99). Note that the right-hand sides of (5.103) and (5.104) are proportional to the difference (sum) of the optical paths through the sphere for the two rays.

To proceed further it is necessary to exercise some care in specifying the phase factors so that the regular branch of the transformation (5.102) is followed; this corresponds to the one on which Eq. (E.35) holds. Upon completing this examination we are able to solve (5.103) as

$$\zeta(\epsilon) = e^{-i\pi}\frac{c\epsilon}{(3s)^{1/3}}\left(1 + \frac{875c^6 - 1257c^4 + 657c^2 + 45}{8640(sc)^3}\epsilon + O(\epsilon^2)\right), \quad (5.105)$$

and the path of integration in (5.101b) can be chosen such that the transformed path in the u-plane runs from $\infty e^{-i\pi/3}$ through the saddle points $\{\zeta^{1/2}, -\zeta^{1/2}\}$ to $\infty e^{i\pi/3}$. One then follows the procedure outlined at the end of Appendix E, generalizing it slightly to account for both polarizations. The final result of the asymptotic expansion, up to terms of $O(\beta^{-1})$, is

$$F_j(\beta, \theta) \simeq 2\pi i(2\beta)^{-1/3}e^{2\beta A(\epsilon)}\left\{[p_{0j} - (2\beta)^{-1}(q_{1j} + 2\zeta(\epsilon)q_{2j})]\,\mathrm{Ai}[(2\beta)^{2/3}\zeta(\epsilon)]\right.$$

$$\left. -(2\beta)^{-1/3}[q_{0j} - 2(2\beta)^{-1}p_{2j}]\,\mathrm{Ai}'[(2\beta)^{2/3}\zeta(\epsilon)]\right\}, \quad (5.106)$$

where $p_{mj}(\epsilon)$ and $q_{mj}(\epsilon)$ are the CFU coefficients.

Evaluation of the coefficients is somewhat tedious, but the first few have been obtained by Nussenzveig (1969b) and Khare (1975). As indicated in Appendix E, we generalize (E.37) to account for both polarizations and find for the leading coefficients

$$p_{0j}(\epsilon) = \tfrac{1}{2}[G_j(\theta''_1, \epsilon) + G_j(\theta'_1, \epsilon)],$$

$$q_{0j}(\epsilon) = \tfrac{1}{2}\zeta^{-1/2}[G_j(\theta''_1, \epsilon) - G_j(\theta'_1, \epsilon)]. \quad (5.107)$$

For example,

$$p_{01} \simeq \frac{4i(3s)^{1/6}c}{27\sqrt{3}n}[1 + O(\epsilon)],$$

$$q_{01} \simeq -\frac{28 - 31s^2}{27\sqrt{3}n(3s)^{7/6}}[1 + O(\epsilon)]. \qquad (5.108)$$

Eventually, however, one must evaluate these numerically for a specific application and refractive index.

By evaluating the coefficients to $O(\beta^{-1})$ and employing the appropriate asymptotic expansions from (B.10), the result (5.106) can be extended smoothly into the 2-ray region where (5.95) and (5.97) are regained. Closer to the rainbow angle everything can be expanded in powers of ϵ. As an example, if we look only at polarization 1, the forms (5.108) can be used to obtain the gross simplification

$$F_1(\beta, \theta) \approx -\frac{8\pi^{1/2}}{27\beta^{1/2}}\frac{c}{n}e^{6ic\beta + is\beta\epsilon}\left(\frac{4c}{3s}\epsilon\right)^{-1/4}$$

$$\times \sin\left[\frac{s\beta}{2}\left(\frac{4c}{3s}\epsilon\right)^{3/2} + \frac{\pi}{4}\right]\left[1 + O\left(\frac{1}{\beta\epsilon^{3/2}}\right)\right], \qquad (5.109)$$

where we have employed (4.56) at $\theta = \theta_R$. The oscillations arise from interference between the two geometric-ray contributions, and are just those exhibited by the Airy functions in Fig. B.1. Further oscillations can emerge from interference between the rainbow rays and those from direct reflection. (Note, however, that this expression cannot be extended to $\epsilon = 0$, since we have employed the asymptotic formulas for the Airy functions for large β). We shall return to these points in the next chapter where these results are applied to optical phenomena in water.

The near-backward (glory) region

This angular domain is defined generally by

$$\theta = \pi - \varepsilon, \qquad 0 < \varepsilon \lesssim \beta^{-1/2}, \qquad (5.110)$$

though we shall focus principally on the exactly backward direction $\theta = \pi$. As will become evident both here and in the following, and as noted earlier with respect to Fig. 3.6, a detailed description of scattering processes in this region is rather complicated and will require continual fine tuning.

We can approach this task by appealing to the work already done regarding the 1-ray region. There it was noted that the residue series (5.90) were uniformly valid up to $\theta = \pi$, so we need only make this substitution to acquire these contributions. The geometric contributions of (5.87) and (5.88), however, are *not* valid in the backward direction, which is evident from (5.36); the term $S_{12,\text{geo}}^{(+)}$ must be replaced by $S_{12,\text{geo}}^{(-)}$, as found from (5.37b). One then carries out the saddle-point integrations anew, after noting that there is a single saddle point at $\lambda = 0$, corresponding to the central ray, for which $\theta_i = \theta_r = 0$.

If we wish to evaluate only the leading-order contributions in this region, up to $O(\beta^{-1})$, then it is not really necessary to perform these additional calculations. The leading factors in (5.87a) and (5.88a) *are*, in fact, uniformly valid up to $\theta = \pi$, and are readily evaluated by taking the appropriate limits. Although it is clear that $\cos\theta_i = \cos\theta_r = 1$ at $\theta = \pi$, a bit more care is needed in describing how the saddle point $z = \sin\theta_i$ approaches the origin. This behavior can be uncovered from (5.80), in which for small z we can neglect terms in z^2. The resulting approximate equation has the solution

$$\sin\theta_i \approx \frac{n\cos(\theta/2)}{2-n}, \qquad \theta \to \pi. \tag{5.111}$$

For both polarizations we thus have

$$S_{j2,\text{geo}}(\beta,\pi) \approx (-1)^{j+1} i\beta \frac{2n^2(n-1)}{(2-n)(n+1)^3}[1 + O(\beta^{-1})]e^{2i(2n-1)\beta}. \tag{5.112}$$

This is indeed the central-ray contribution, for it contains precisely the amplitude and phase factor for double traversal of the sphere along the central axis (e.g., compare with (5.87a)).†

In the residue series (5.90) we let $\theta = \pi$ and retain only the $m = 0$ terms, for higher-order surface waves take more than one turn around the sphere. We find that

$$S_{j2,\text{res}}(\beta,\pi) \approx (-1)^j \frac{i\beta^{4/3}}{2^{1/3}N^2}e^{i\pi/3}e^{i4N\beta}\sum_k (a_k')^{-2}$$
$$\times\chi_0\Big[(\chi_0 + N)e^{i\lambda_k\chi_0} - n^2(n^2\chi_0 + N)e^{i\mu_k\chi_0} + O(\beta^{-1})\Big], \tag{5.113}$$

with $N = \sqrt{n^2 - 1}$, and from (5.89c) $\chi_0 = \pi - 4\cos^{-1}(1/n)$. There is an immediate surprise in these last results when they are compared with (5.112): at least in the leading factors, the residue contributions are $O(\beta^{1/3})$ *greater* than the geometric terms. This 'Regge-pole dominance' is a new effect in that, while it is common in shadow regions, this is its first appearance in a lit region. To exemplify this let us take $n = \frac{4}{3}$ and keep only the leading term

---

† It's a bit farfetched to refer to these as geometric rays, for in geometric optics the intensity in the backward direction is infinite, because it is a focal line. More properly, they are the WKB approximations to the geometric rays.

in the residue series. With the help of (5.39) and Tables B.1 and B.2, we find the gross approximation

$$\left| \frac{S_{j2,\text{res}}(\beta, \pi)}{S_{j2,\text{geo}}(\beta, \pi)} \right| \simeq 4.6\beta^{1/3} e^{-0.4\beta^{1/3}}. \tag{5.114}$$

For $\beta = 10^2$ the right-hand side is about 3.3, decreasing to unity near $\beta = 10^3$.

Physically the enhancement of the surface-wave contribution is an axial-focusing effect† that brings together an entire cone of diffracted rays in the backward direction, analogous to the Poisson spot in the forward direction. This effect is responsible for the difference in leading factors between (5.91) and (5.113), resulting in an amplification factor of $\beta^{1/2}$. As we shall see in Chapter 6, the effectiveness of this virtual ring source in water is made possible by the value of the parameter χ_0, which is somewhat more than $14°$ in that medium.

5.5 A summary of the first three Debye contributions to the scattering amplitudes

In an effort to separate the significant results from the wealth of detail found in the preceding sections, we summarize here the dominant contributions to the amplitudes from the first three Debye terms throughout the full range of scattering angles, $0 \le \theta \le \pi$. The following presentation provides expressions for the scattering amplitudes in the form of geometric-optics and classical-diffraction contributions plus leading-order corrections. Omission of the latter would simply improve the geometric-optics results of Chapter 4 by including interference in the observable intensities.

For clarity some of the notation of the previous sections has been changed slightly. In that sense, the present section is self-contained, as well as restricted to real refractive indices in the range $1 < n < \sqrt{2}$.

Direct reflection ($p = 0$)

$$S_{j0,\text{geo}}(\beta, \theta) \approx i\frac{\beta}{2} r_j(n, \theta) e^{-2i\beta \sin(\theta/2)} \left(1 + \frac{i}{2\beta} F_{j0}(\theta) + O(\beta^{-2}) \right), \quad \gamma < \theta \le \pi,$$

$$\approx \frac{\beta^2}{2} \left(\frac{\theta}{\sin \theta} \right)^{1/2} \left[2\frac{J_1(\beta\theta)}{\beta\theta} \left(1 - \frac{iN}{\beta} \right) + \frac{2}{\beta} F'_{j0}(\beta, \theta) \right], \quad \theta \ll \gamma,$$

$$\xrightarrow[\theta \to 0]{} \frac{\beta^2}{2} [1 + O(\gamma^2)], \tag{5.115}$$

---

† The effect is also present in the forward direction, of course, as indicated in Eqs. (5.70), but there the diffraction peak dominates.

where F_{j0} and F'_{j0} are given in Table 5.1, and r_j are the Fresnel amplitudes:

$$r_j(n,\theta) = \frac{\sqrt{n^2 - \cos^2(\theta/2)} - \varepsilon_j n^2 \sin(\theta/2)}{\sqrt{n^2 - \cos^2(\theta/2)} + \varepsilon_j n^2 \sin(\theta/2)}, \quad \varepsilon_j \equiv \begin{cases} 1/n^2, & j = 1 \\ 1, & j = 2. \end{cases}$$

$$(5.116)$$

We have here made the replacement $\theta_i = (\pi - \theta)/2$, and recalled that $\gamma = (2/\beta)^{1/3}$ and $N = \sqrt{n^2 - 1}$. Among other things, the functions F_{j0} provided corrections to $r_j(n, \theta)$ arising from surface curvature.

Direct transmission ($p = 1$)

$$S_{j1,\text{geo}}(\beta, \theta) \approx i\beta \left(\frac{n \sin\theta_1 \cos\theta_1 \cos\theta_2}{2 \sin\theta (n \cos\theta_2 - \cos\theta_1)} \right)^{1/2}$$

$$\times e^{2i\beta(n\cos\theta_2 - \cos\theta_1)} \frac{4nv_j \cos\theta_1 \cos\theta_2}{(nv_j \cos\theta_2 + \cos\theta_1)^2}$$

$$\times \left(1 - i\frac{F_{j1}(\theta)}{16\beta \cos\theta_1} + O(\beta^{-2}) \right), \quad [\pi - 2\sin^{-1}(1/n)] - \theta \gg \gamma,$$

$$\xrightarrow[\theta \to 0]{} 2i\frac{n^2\beta}{(n-1)(n+1)^2} e^{2i\beta(n-1)} \left[1 - i\frac{F_{j1}(0)}{16\beta} + O(\beta^{-2}) \right], \quad (5.117)$$

where F_{j1} is given in Table 5.1 for $j = 1$ and 2, and θ_1 and θ_2 are defined in conjunction with that table.

One internal reflection ($p = 2$)

$$S_{j2,\text{geo}}(\beta, \theta) \approx i\beta \left(\frac{n \sin\theta_i \cos\theta_i \cos\theta_r}{2 \sin\theta (2 \cos\theta_i - n \cos\theta_r)} \right)^{1/2}$$

$$\times e^{2i\beta(2n\cos\theta_r - \cos\theta_i)} \frac{4nv_j \cos\theta_i \cos\theta_r}{(nv_j \cos\theta_r + \cos\theta_i)^2}$$

$$\times \frac{nv_j \cos\theta_r - \cos\theta_i}{nv_j \cos\theta_r + \cos\theta_i} \left(1 - i\frac{F_{j2}(\theta)}{64\beta \cos\theta_i} + O(\beta^{-2}) \right),$$

$$\theta - \theta_L \gg \gamma (\text{1-ray}), \quad (5.118)$$

with θ_i, θ_r, and F_{j2} defined in conjunction with Table 5.1; and for $\theta_L - \theta \gg \gamma$, $\theta - \theta_R \ll \gamma^2/N$ (2-ray), this is replaced by

$$\overline{S}_{j2,\text{geo}}(\beta, \theta) \approx S_{j2,\text{geo}}(\beta, \theta)\big|_{\theta_i = \theta'_i} + S_{j2,\text{geo}}(\beta, \theta)\big|_{\theta_i = \theta''_i}. \quad (5.119)$$

In this expression the two saddle-point angles are defined by

$$\left\{\begin{array}{c} \sin\theta_i' \\ \sin\theta_i'' \end{array}\right\} \simeq s \pm cp\epsilon^{1/2} - \left(cq + \frac{s}{2}p^2\right)\epsilon + O(\epsilon^{3/2}), \qquad (5.120a)$$

$$\epsilon = \theta - \theta_R, \quad p = [4c/(3s)^{1/2}], \qquad q = \frac{3+c^2}{18s^2}, \qquad (5.120b)$$

$$s = \left(\frac{4-n^2}{3}\right)^{1/2}, \qquad c = \left(\frac{n^2-1}{3}\right)^{1/2}, \qquad (5.120c)$$

$$\theta_R = 2\cos^{-1}\left(\frac{4-n^2}{3n^{4/3}}\right)^{3/2}, \qquad \theta_L = 4\cos^{-1}(1/n), \qquad (5.120d)$$

and the corresponding θ_r are found from Snell's law.

In the rainbow region,

$$S_{j2,\text{geo}}(\beta,\theta) \approx -e^{i\pi/4}\frac{n}{\sqrt{\pi\sin\theta}}(2\beta)^{3/2}F_{jR}(\beta,\theta), \quad |\theta-\theta_R| \lesssim \gamma^2/(2N), \quad (5.121)$$

where F_{jR} is given in Table 5.1 in the so-called Airy approximation, in which only leading-order terms in ϵ and β are retained in (5.106).

In the backward direction both geometric and residue contributions must be included:

$$S_{j2,\text{geo}}(\beta,\pi) \approx (-1)^{j+1}i\beta\frac{2n^2(n-1)}{(2-n)(n+1)^3}e^{2i(2n-1)\beta}[1+O(\beta^{-1})], \quad (5.122)$$

$$S_{j2,\text{res}}(\beta,\pi) \approx (-1)^j\frac{i\beta}{\gamma^2}e^{i\pi/3}e^{i4N\beta}F_G(\beta), \qquad (5.123)$$

and F_G is given in Table 5.1.

In Table 5.1 the following definitions and numerical values apply:

$$\bar{s} = \sin(\theta/2), \quad \bar{c} = \cos(\theta/2), \quad \gamma = (2/\beta)^{1/3}, \quad N = \sqrt{n^2-1}; \quad (5.124a)$$

$$\theta = 2(\theta_1-\theta_2), \quad \sin\theta_1 = n\sin\theta_2, \quad \chi = \frac{\cos\theta_1}{n\cos\theta_2}; \qquad (5.124b)$$

$$\theta = \pi - 2(2\theta_r-\theta_i), \quad \sin\theta_i = n\sin\theta_r, \quad \chi = \frac{\cos\theta_i}{n\cos\theta_r}; \qquad (5.124c)$$

$$M_0 \simeq (1.25\,513)e^{i\pi/3}, \qquad M_1 \simeq (0.532\,291)e^{i2\pi/3}; \qquad (5.124d)$$

$$\lambda_0 = \beta + e^{i\pi/3}x_0/\gamma, \quad x_0 \simeq 2.3381, \quad a_0' \simeq 0.70\,121; \qquad (5.124e)$$

$$A(\epsilon) \approx i(1.519+0.431\epsilon), \qquad \zeta(\epsilon) \approx -0.369\epsilon; \qquad (5.124f)$$

$$p_{01}(0) \approx 0.381i, \qquad p_{02}(0) \approx 0.0078i. \qquad (5.124g)$$

Table 5.1. *Functions describing the asymptotic expansions of the scattering amplitudes*

Function	Definition
$F_{10}(\theta)$	$\dfrac{1-2\bar{s}^2}{\bar{s}^3} + \dfrac{2/\bar{s}}{\bar{c}^2+\bar{s}\sqrt{n^2-\bar{c}^2}} - \dfrac{2n^2-\bar{c}^2}{(n^2-\bar{c}^2)^{3/2}}$
$F_{20}(\theta)$	$-\dfrac{1}{\bar{s}^3} + \dfrac{4/\bar{s}}{\bar{c}^3-\bar{s}\sqrt{n^2-\bar{c}^2}} - 4\bar{c}^2\dfrac{2n^4\bar{s}^2-n^2\bar{c}^2(1+\bar{s}^2-\bar{s}^4)+\bar{c}^6}{\bar{s}^2(n^2-\bar{c}^2)(n^2-\bar{s}^2-\bar{c}^2)^2}$ $+\dfrac{2n^2\bar{s}}{(n^2-\bar{c}^2)^{3/2}}\dfrac{2n^4-n^2\bar{c}^2(1+\bar{c}^2)-\bar{c}^4}{(n^2\bar{s}^2-\bar{c}^2)^2}$
$F'_{j0}(\theta)$	$\left(\dfrac{M_0}{\gamma} - \dfrac{i}{Nv_j} + \dfrac{8M_1}{15}\gamma - i\dfrac{M_0}{6N^3}(4n^2v_j^2-2-nv_j^2)\gamma^2\right)J_0(\beta\theta)$ $-i\dfrac{NM_0}{\gamma\beta}\left(J'_1(\beta\theta) - \dfrac{n^2(2n^2+1)}{3N^2}\dfrac{J_1(\beta\theta)}{\beta\theta}\right)$
$F_{11}(\theta)$	$2\cot\theta_1\left(7\cot\theta + \dfrac{\cot\theta_1}{2(1-\chi)}\right) - \dfrac{9}{1-\chi} + 15\chi - 6$ $+(\chi-1)(8\chi^2+5\chi+8)\tan^2\theta_1 + \dfrac{16\cot\theta_1}{\sin\theta}\left(\dfrac{n\cos\theta_2+\cos\theta_1}{n\cos\theta_1+\cos\theta_2}\right)^2$
$F_{21}(\theta)$	$-2\cot\theta_1\left(7\cot\theta + \dfrac{\cot\theta_1}{2(1-\chi)}\right) - \dfrac{9}{1-\chi} + 7\chi + 2 - 8\chi''$ $+\dfrac{16\cot\theta_1}{\sin\theta}\left(\dfrac{n\cos\theta_1+\cos\theta_2}{n\cos\theta_2+\cos\theta_1}\right)^2 + 16(\chi'-\chi)\dfrac{1-\cos^2\theta_1\cos^2\theta_2}{(1+\chi')\cos^2\theta_1\cos^2\theta_2}$ $+\tan^2\theta_1\left[4\chi''\left((1-\chi)(\chi''-\chi-2)+1-2\chi^2-2\chi'\dfrac{1-\chi^2}{1-\chi'^2}\right)\right.$ $\left.-(8\chi^2+9\chi+8)(1-\chi)\right]$
$F_{j1}(0)$	$\dfrac{15}{2n} + \dfrac{9}{n-1} - 8$
$F_{12}(\theta)$	$G(\theta) + \dfrac{64n^2}{\sin\theta_i\sin\theta}\dfrac{1-\chi'}{1-\chi}\left(\dfrac{1+\chi}{1+\chi'}\right)^3$
$F_{22}(\theta)$	$G(\theta) + \dfrac{64}{n^2\sin\theta_i\sin\theta}\dfrac{1-\chi}{1-\chi'}\left(\dfrac{1+\chi'}{1+\chi}\right)^3 + 64\dfrac{(\chi-\chi')(2-\chi+\chi')}{(2\chi-1)(1+\chi')}$ $+64\tan^2\theta_i\left[\chi^2-2\chi+(\chi^2+\sec^2\theta_r)\left(\dfrac{n^2-1}{n^2}\dfrac{1}{1+\chi'}+\dfrac{n^2\chi\cos^2\theta_i}{n^2\chi^2-\cos^2\theta_i}\right)\right]$ $-\tan^2\theta_i(\chi-\chi')\left(16(2\chi+1)-\dfrac{48}{2\chi-1}+\dfrac{8(7\chi-3)}{1+\chi'}+\dfrac{212}{(1+\chi')(2\chi-1)}\right.$ $\left.+\dfrac{12(4\chi^2-2\chi-5)}{(1+\chi')^2}-\dfrac{36}{(1+\chi')^2(2\chi-1)}+\dfrac{1}{2}(2\chi-1)^2(1+\chi')\right)$
$G(\theta)$	$8\cot\theta_i\left(\cot\theta + \dfrac{\cot\theta_i}{2(2\chi-1)}\right) + 6(9\chi-11) - \dfrac{15}{2\chi-1} - \dfrac{9}{(2\chi-1)^2}$ $+\tan^2\theta_i\left(56\chi^3-3\chi^2+\dfrac{39\chi}{2}-\dfrac{79}{2}-\dfrac{33}{2(2\chi-1)}-\dfrac{51}{4(2\chi-1)^2}\right.$ $\left.-\dfrac{15}{4(2\chi-1)^3}\right)$
$F_{JR}(\beta,\theta)$	$\dfrac{2\pi i}{(2\beta)^{1/3}}e^{2\beta A(\epsilon)}p_{0j}(0)\,\mathrm{Ai}\left[(2\beta)^{2/3}\zeta(\epsilon)\right]$
$F_G(\beta)$	$\dfrac{\chi_0}{(a'_0)^2}\left[(\chi_0+N)e^{i\chi_0(\lambda_0-i/N)} - n^2(n^2\chi_0+N)e^{i\chi_0(\lambda_0-in^2/N)}\right]$

6

Scattering observables for large dielectric spheres

Following the rather lengthy mathematical onslaught in the theoretical development of scattering from a large dielectric sphere in the preceding chapter, it is perhaps time to take a break from these labors and apply the results to prediction of some measurable quantities. As well as judging the large-sphere theory against the exact expressions provided by the Mie solution, some pleasure may also be found in using the results to explore various meteorological optical phenomena.

It cannot be emphasized too strongly that the preceding theory is *exact*, in the same sense that the Mie solution is exact. In principle we can calculate amplitudes and cross sections to all orders in β – it is not an approximate theory. In practice, however, one need only retain a few terms in the asymptotic expansions and in the residue series to achieve reasonable accuracy, so that we can appropriately approximate the theoretical expressions and reduce the calculational labor. However, this is also in contrast with the partial-wave expansions, which *must* include indices ℓ to at least $O(\beta)$. The difference between approximate theories and approximations to the exact theory is that in the latter we are able to control the estimates rather precisely. A separate issue relates to how many terms must be retained in the Debye expansion itself, and this will be addressed as the need arises.

6.1 The various efficiency factors

In most scattering applications one is interested in the energy balance; additionally, phenomena such as radiative transfer require attention to momentum balance. The former is described by $Q_{ext} - Q_{abs} = Q_{sca}$, and the latter in terms of the radiation pressure, or Q_{pr}. As we have seen, these cross sections contain an irregular high-frequency component that we have called the *ripple*. Although this ripple has the appearance of noise, that is by no

187

means its origin and we shall see subsequently that it is well understood and has important physical implications.

Nevertheless, quite often these rapid oscillations are of little interest, because measurements are made on polydisperse systems; that is, there is a natural size dispersion in the sample that quenches the ripple, resulting in a smoother average cross section. Thus, in some cases it is useful to average the theoretical expressions over a small interval of β so as to more readily compare the results with observation. One way to accomplish this, of course, is to treat the ripple *as if* it were noise and simply filter the high-frequency components out of the Fourier transform. For our purposes, though, there are actually good theoretical reasons for performing the averaging process directly when called for.

Extinction

In general we shall consider the refractive index to be complex, $n = m + i\kappa$, so that the effects of small amounts of absorption can be studied by varying κ. Equation (3.73) provides the extinction efficiency in terms of the forward scattering amplitudes (optical theorem), and these are obtained from (5.115)–(5.117).

There is an additional contribution here that is easily included, because it is almost effortless to also obtain the central-ray contributions for all p. That is, we consider all rays incident along the optical axis, such that those undergoing an odd number of internal reflections exit in the backward direction and those undergoing an even number exit in the forward direction. Although the reflectivities are small here, the sum of these terms can be significant. Briefly, to obtain these contributions we return to Eqs. (5.36), in which (5.37) are employed for $\theta = 0$ and even p, and their counterparts for $\{S_{10}^{(-)}, S_{1p}^{(-)}\}$ for $\theta = \pi$ and odd p. At these angles the integrals have saddle points at $\lambda = 0$ and differ from $S_{j1,\text{geo}}$ by powers of $e^{i\pi\lambda}R_{11}$. The steepest-descent paths are identical to that of the earlier calculations and the effort becomes very simple if we retain only corrections of order β^{-1}. The results are, for $p = 2j$,

$$S_{p,\text{geo}}^{C}(\beta, 0) \simeq -i\frac{\beta}{2}\frac{n^2}{(n+1)^2}\frac{e^{2i(n-1)\beta}}{j - (n-1)/2}\left(\frac{n-1}{n+1}\right)^{2j}e^{4inj\beta}, \quad p\text{-odd}, \quad (6.1)$$

$$S_{p,\text{geo}}^{C}(\beta, \pi) \simeq i\frac{\beta}{2}\frac{n^2}{(n+1)^2}\frac{e^{-2i\beta}}{j - (n-1)/2}\left(\frac{n-1}{n+1}\right)^{2j}e^{4inj\beta}, \quad p\text{-even}. \quad (6.2)$$

Hence, successive reflections of the central ray reduce its contribution by a factor $(n-1)^2/(n+1)^2$, which for water is 0.02. However, in Q_{ext} we shall sum these over all j. Incidentally, the last two factors in each of the above equations are just those that occur in the theory of the Fabry–Pérot interferometer (e.g., Reitz *et al.* (1979)).

On collecting together these several contributions we find that the extinction efficiency up to $O(\beta^{-2})$ is given by

$$
\begin{aligned}
Q_{ext}(\beta) \approx{} & 2 + 2\operatorname{Re}(M_0)\,\gamma^2 + \operatorname{Im}\left(\frac{n^2+1}{N}\right)\gamma^3 + \frac{16}{15}\operatorname{Re}(M_1)\,\gamma^4 \\
& - \frac{1}{6}\operatorname{Im}\left(M_0\frac{n^2+1}{N^2}(2n^4 - 6n^2 + 3)\right)\gamma^5 + O(\gamma^6) \\
& - \frac{8}{\beta}\operatorname{Im}\Bigg\{\frac{n^2}{(n-1)(n+1)^2}e^{2i(n-1)\beta}\left[1 - \frac{i}{16\beta}\left(\frac{51}{6} + \frac{1}{2n} - \frac{9n}{n-1}\right)\right. \\
& \left.\left. - \sum_{j\geq1}\frac{(n-1)^{2j+1}}{(n+1)^{2j}}\frac{e^{4inj\beta}}{2j-n+1} + O(\beta^{-2})\right]\Bigg\},
\end{aligned}
\tag{6.3}
$$

where $\gamma \equiv (2/\beta)^{1/3}$, $N \equiv \sqrt{n^2 - 1}$, n can be complex, and M_0 and M_1 are estimated in (5.124d).

Several features of the approximation (6.3) deserve comment, beginning with the observation that it is independent of polarization, as it must be. The terms have been separated into two groups, so that the first two lines correspond to reflection and diffraction; the first term in this group comes equally from reflection and forward diffraction (blocking), whereas the remainder arise from edge diffraction (the Fock terms). The third and fourth lines arise from axial rays. Geometric optics and classical diffraction are regained by retaining and combining only the first terms in each group.

Figure 6.1a compares the asymptotic expression (6.3) with the exact Mie series (3.99) over a β-interval $(1, 50)$ in the case of no absorption. It is evident that the fit for $\beta \gtrsim 2$ is excellent, essentially defining the average extinction curve. There is no ripple in (6.3), of course, since this is to be found in the remainder term of the Debye expansion (5.27b) and is not included here. The comparison is extended to higher values of β in Fig. 6.1b, where the fit from (6.3) remains excellent. In this plot the curve corresponding to geometric optics and classical diffraction alone has been included, and is seen to be quite deficient even for large size parameters.

Absorption has been added in Fig. 6.2, in which it is seen that the ripple in the exact result remains only for low values of β, and disappears completely as κ is increased. The magnitude of the efficiency tends to decrease with increasing absorption, and the quality of the fit tends to begin to be good at slightly higher values of β; in any event, for $\beta \gg 1$ it remains excellent.

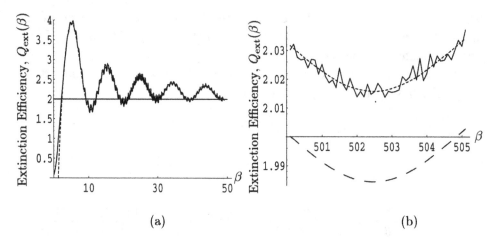

(a)   (b)

Fig. 6.1. A comparison of the extinction efficiencies $Q_{ext}(\beta)$ for $n = 1.333$ given by the exact Mie solution. (——) and by the asymptotic result (6.3) (----). (a) $0 < \beta < 50$. (b) β near 500, including the contribution from geometric optics and classical diffraction (– – –).

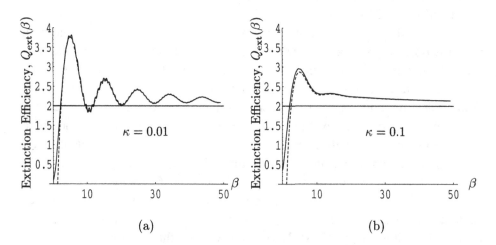

(a)   (b)

Fig. 6.2. A comparison of the extinction efficiencies $Q_{ext}(\beta)$ for $n = 1.333 + i\kappa$ given by the exact Mie solution and by the asymptotic result (6.3) with absoprtion. (a) $\kappa = 0.001$. (b) $\kappa = 0.1$.

A final point of interest to be gleaned from (6.3) is that, for real n, it provides a mathematical explanation for why the extinction curve oscillates about an axis slightly greater than the geometric-optics limit of $Q_{ext} = 2$, which it nevertheless approaches. A physical explanation can be found from Eq. (2.38) for the impenetrable sphere, remnants of which are seen in Fig. 6.1 for small β. At these values the wave nature of the process is still evident.

Absorption

A somewhat different approach will be taken with the absorption cross section, which was written explicitly in terms of the S-functions in (4.31) and a geometric-optics expression developed in (4.35). Although the Poisson sum formula, or modified Watson transformation, can be applied to the partial-wave series, continuation into the complex λ-plane is not a productive option owing to the presence of complex conjugation in (4.31). Nussenzveig and Wiscombe (1980) have shown how an asymptotic expression can be derived if one is content with an average, or 'coarse-grained', result. As mentioned earlier, this procedure has the effect of smoothing the ripple, which is often desirable in comparing with measurements, and it also considerably simplifies the resulting expressions.

Using the magnetic-multipole contribution to (4.31) as an example, we can write

$$Q_{\text{abs}}(\beta) \equiv \frac{1}{\beta^2} \sum_{\ell=1}^{\infty} (\ell + \tfrac{1}{2})(1 - |S_{\ell}^M|^2)$$

$$= \frac{1}{\beta^2} \sum_{m=-\infty}^{\infty} e^{-im\pi} \int_0^{\infty} [1 - |S^M(\lambda, \beta)|^2] e^{2\pi i m \lambda} \lambda \, d\lambda. \qquad (6.4)$$

One next replaces the Ricatti–Bessel functions in $S^M(\lambda, \beta)$ by their asymptotic forms for large β, though this must be done somewhat carefully. Then split the integration interval into $(0, \beta) + (\beta, \lambda^+) + (\lambda^+, \infty)$, where λ^+ is a cutoff defining the upper bound to the edge strip; rays with impact parameters beyond this, according to (4.29), contribute negligibly. As was seen in Chapter 2 for the impenetrable sphere, the above-edge region provides the dominant corrections to the Fresnel amplitudes arising from surface curvature. Such corrections are obtained here by applying the Schöbe expansions (C.17) in the interval (β, λ^+), which effectively means to those cylinder functions with argument $n\beta$. The Debye asymptotic expansions employed in Chapter 4 are used in $(0, \beta)$, though use of the Schöbe expansions slightly below β introduces further below-edge corrections to the geometric-optics results. Finally, the coarse-graining procedure of Nussenzveig and Wiscombe is held to eliminate all but the $m = 0$ term in (6.4).

This approach results in three distinct contributions to $Q_{\text{abs}}(\beta)$: a modification of the geometric-optics result (4.35), an above-edge integral, and a below-edge correction. The first is the most important and differs from (4.35) in that the Debye approximation discussed in connection with (4.30) is *not* made. Specifically, the angle of refraction determined by Snell's law is now complex, owing to the complex nature of $n = m + i\kappa$; in turn, the Fresnel

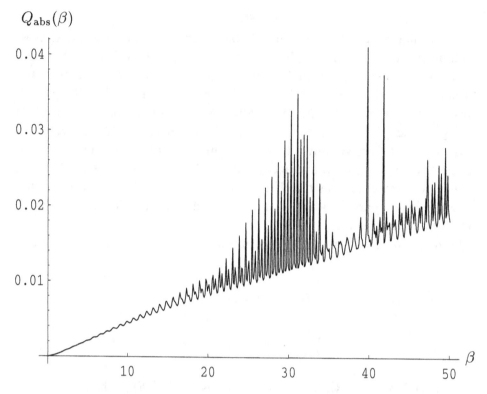

Fig. 6.3. The exact Mie solution for the absorption efficiency $Q_{abs}(\beta)$ for $n = 1.333 + i10^{-4}$ over a small range of β.

coefficients are now complex and the absorption parameter (4.34) becomes

$$\alpha = 4\beta\,\mathrm{Im}(n\cos\theta_r + \theta_r\sin\theta_i). \tag{6.5}$$

In addition, the reflectivities are now the squared magnitudes of the Fresnel amplitudes.

The two correction terms, although they are significant, are somewhat complicated and not particularly illuminating – and must be evaluated numerically in any event. Nevertheless, Nussenzveig and Wiscombe show that the asymptotic forms are in excellent agreement with the result of numerically summing the Mie series (3.106), and in doing so observe that the computing time for the former is roughly $O(\beta)$ less than that for the latter.

We shall not record these complete asymptotic expressions here because, aside from the conceptual insight that the dominant physical corrections to the geometric-optics result come from the edge strip, they seem to be of little utility. As discussed in Appendix F, computation of the spherical Bessel functions of real index becomes expensive only for complex arguments, and then only for imaginary parts greater than about 10^{-4}. Figure 6.3 exhibits the Mie result for $Q_{abs}(\beta)$ when κ is very small, illustrating that there is particularly erratic behavior when there is little absorption. There is no computational difficulty in this case, though the computing time for all the efficiencies increases with the size parameter. Figure 6.4 compares the Mie summations with both the geometric-optics result of (4.35) and its modification discussed above for four different values of κ. We see that for $\kappa = 0.01$ the erratic oscillations have settled down into a mild ripple and that the two virtually identical asymptotic results follow somewhat below the Mie curve. As absorption increases with $\kappa = 0.1$ the ripple has disappeared, obviating any need for averaging, and both the geometric-optics approximation and its modification have started to level off. At $\kappa = 0.5$ these asymptotic approximations have leveled off to almost constant values and the exact curve has begun to approach them with increasing β. Finally, at $\kappa = 1.0$ we gain the impression that the exact curve actually approaches the modified geometric-optics value of about 0.761 for large β, as we should expect with $\kappa \neq 0$. An interesting feature of the last case is the very rapid initial rise of the absorption efficiency for $\beta \gtrsim 0$, in agreement with our observation following Eq. (4.5) that absorption, when it occurs, dominates the scattering for very small size parameters. Although there are no computational difficulties for the relatively low values of β considered in Fig. 6.4, it is certainly true that the computing time increases with β, and with κ. However by the time this problem kicks in, the exact curve for $Q_{abs}(\beta)$ has approached very closely to the essentially β-independent constant value of geometric optics, which is readily computed numerically from the modified version of (4.35) and quoted above. Although the edge corrections are still evident for $\beta \sim 50$, it is simpler to employ the exact Mie series to include them than to separately program their asymptotic forms.

The radiation pressure

A similar analysis applies to evaluation of the radiation pressure – or, rather, to the quantity $Q_{sca}(\beta)\langle\cos\theta\rangle$ defined in (3.130). The asymptotic development is even more elaborate here, owing to the appearance of S-functions with different partial-wave indices. One first employs recurrence relations to convert all cylinder functions of index $\ell + 1$ into those and their derivatives of index ℓ prior to introducing the Poisson sum formula. Once again the asymptotic analysis leads to three distinct contributions to $Q_{pr}(\beta)$: a modification of the geometric-optics result (4.47), plus two integrals

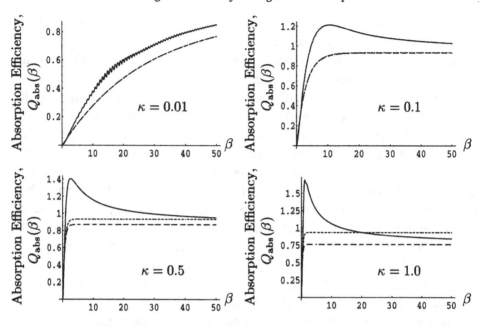

Fig. 6.4. A comparison of the exact Mie series (——) for $Q_{\text{abs}}(\beta)$ with both the geometric-optics approximation (- - - -) and its modification for complex refractive index $n = m + i\kappa$ (– – –) for several values of κ.

representing edge corrections. The modifications to $Q_{\text{pr}}^{(\text{go})}(\beta)$ are the same as those described above for $Q_{\text{abs}}(\beta)$.

Figure 6.5 compares the exact expression (3.130) for $Q_{\text{sca}}(\beta)\langle\cos\theta\rangle$ with the geometric-optics result (4.47) for several values of κ. In these plots the modified form of $Q_{\text{pr}}^{(\text{go})}$ is essentially indistinguishable numerically from (4.47) itself and so has been omitted. Although the exact curves continue to approach the geometric-optics value of about 0.964 as β becomes very large, the remaining discrepancies again reflect the edge corrections that are already contained in the Mie series.

In Fig. 6.6 we plot the radiation pressure efficiency itself from (3.128) for several values of κ, from which we garner some interesting physical observations. From Eq. (6.5) it is evident that the product $\kappa\beta$ can be employed as a control parameter for these curves, such that $Q_{\text{pr}}(\beta)$ is relatively small for $\kappa\beta \lesssim 1$ and increases toward a plateau near unity as the parameter approaches and exceeds unity. Although the ripple is clearly evident for small κ, it tends to disappear with increasing absorption. Physically, for small κ and large β forward diffraction dominates the scattering and returns most of the momentum to the field. At larger values of κ, however, the effects of diffraction and reflection are comparable and the efficiency rapidly

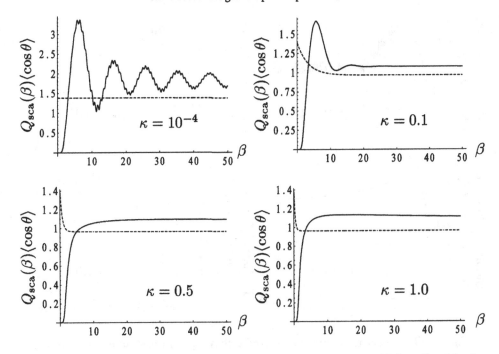

Fig. 6.5. A comparison of the exact Mie series (——) for $Q_{sca}(\beta)\langle\cos\theta\rangle$ with the geometric-optics approximation (- - - -) for several values of κ.

returns to unity following a strong surge for very small particles, for which absorption is also large. A number of these conclusions had already been reached in earlier numerical studies by Irvine (1965).

6.2 Meteorological optical phenomena

In Chapter 4 we have already explored the physical processes underlying several atmospheric phenomena, in terms of light scattering from small particles. Rayleigh scattering was seen to be at the heart of the blueness of the sky and the redness of sunsets, and the forward diffraction peak provides diffraction coronae in numerous other scenarios.

Lower in the atmosphere one observes light interacting with larger particles, such as direct backscattering from dew drops on grass. Observers note a bright halo around their heads, known as the *Heiligenschein*, which is produced from a blade of grass as primary reflector and the drop acts as a lens. The effect is also produced by animals' eyes, and exploited in highway signs using glass beads. Many other atmospheric phenomena arising from

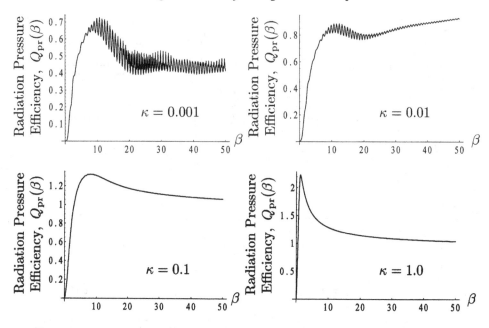

Fig. 6.6. The exact radiation-pressure efficiency $Q_{pr}(\beta)$ for several values of κ.

scattering of light are discussed in books by Minnaert (1959) and Tricker (1970).

However, it is mist and raindrops that so often capture our attention when scattering light, and we shall use these targets for our primary applications in this chapter. Raindrops are usually about 1 mm in diameter, departing from sphericity and breaking up at 5–6 mm. Thus, at the peak of the solar spectrum we are dealing with size parameters on the order of $\beta = 5000$, though much smaller in mist.

The rainbow

Almost everyone possesses a clear visual image of a rainbow, as illustrated on the front cover of this book. The primary bow exhibits a violet innermost band that blends gradually with various shades of blue, green, yellow, and orange, with red outermost. This colorful arc is usually seen after a rain shower, as well as in the spray of a waterfall or garden sprinkler, and its brilliant splash of color is surely one of the most pleasing of visual natural phenomena.

Higher in the sky one occasionally notices the secondary bow in which the colors appear in reverse order, and the region between the two is considerably darker than the surrounding sky. This dark region, which is present even when the secondary bow is not evident, is known as *Alexander's dark band*. Apparently it was first described by Alexander of Aphrodisias around 200 A.D. Less commonly observed is the series of faint bands on the inner side of the primary bow that are usually alternately pink and green. These are the so-called *supernumerary arcs* and, along with the above features, are the major focus of a theoretical description.

Perhaps the first technical contribution to an understanding of the rainbow was made by Aristotle, who suggested that it is an unusual type of reflection of sunlight from clouds. He considered the light to be reflected at a fixed angle, thereby giving rise to a circular cone of 'rainbow rays'. Thus, he correctly explained the circular shape of the rainbow and realized that it was not an object, but a set of directions along which light is scattered. The elevation from observer to primary bow is about 42°, and the secondary bow is about 8° higher. These angles were first measured in 1266 by Roger Bacon (Bridges 1964). Technically, however, it is more useful to measure the total angle through which the sun's rays are deflected, the scattering angle, yielding about 138° and 130°, respectively.

Although Aristotle and Bacon may have been the first to ponder the rainbow in a technical way, its lore is surely as old as mankind itself. For Homer it is the trail left by the goddess Iris (*The Iliad*, V, 350); indeed, in some languages the rainbow is called the 'Arc of Iris'. The bible describes the seat of God as a rainbow (*Revelations* 4:5), and the double bow in ancient China was considered to represent a male–female pair. There has probably been no culture without some interpretation of the rainbow, but only in modern science do we equate a desire to understand it with an enhancement of its beauty. Much of the earlier myth, art, and magic surrounding the rainbow has been collected by Boyer (1959) and Graham (1979), and Nussenzveig (1977) provides a succinct history of the development of its theory.

The technical story resumes with the German monk Theodoric von Freiburg (1304), who suggested that each raindrop generates its own rainbow. This major improvement on Aristotle's hypothesis did not receive much attention over the next three centuries, however, until it was rediscovered by Descartes. Both men understood that the rainbow consists of rays that enter a drop and undergo one internal reflection, and that the secondary bow arises after two internal reflections. They also noted that only one color at a time could be scattered by a water drop in a particular angular direction, and thus concluded that each color in the rainbow comes to the eye from a different set of drops. With the discovery of Snell's law around 1621 Newton was able to apply the laws of reflection and refraction to the problem and introduce the notion of dispersion in connection with the refractive index. One then understood that what we observe is essentially a collection of slightly displaced monochromatic rainbows. Huygens computed the angles for the primary bow in 1652, as did Newton later, who also verified them experimentally.† Figure 6.7 illustrates the state of understanding at this point. There are infinitely many rainbows, of course, corresponding to all possible internal reflections of a ray, and the first 19 were produced and studied in the laboratory by Billet over 100 years ago (Billet 1868); he also studied the supernumeraries.

Between them Descartes and Newton were able to explain all the more conspicuous features of the rainbow. In particular, Decartes (1637) laboriously applied the laws of geometric optics to discover that the rainbow angle is an *angle of minimum deviation* for the ray that has been once internally reflected. In an ordinary prism this minimum angle occurs at that particular angle of incidence at which the refracted ray inside the prism makes equal angles with the two prism faces, as in Fig. 1.9. However, here the rainbow angle is a caustic direction, giving an infinite light intensity according to geometric optics (e.g., Eq. (1.27)). This discovery by Descartes explains both the special character of the rainbow angle and the existence of Alexander's dark band. Nevertheless, the geometric theory of optics was completely unable to account for the supernumerary arcs. Not until Thomas Young demonstrated interference phenomena in 1803 was an explanation forthcoming, a suggestion made by Young himself (Young 1804).

Two rays emerging at the rainbow angle can arrive there by slightly different paths if they began with slightly different impact parameters. These rays interfere constructively very close to the rainbow angle, but as the scattering angle increases destructive interference occurs, and the result is a periodic variation of intensity leading to alternately light and dark bands. Because the interference depends on path lengths, the pattern of the supernumerary arcs depends on droplet size, unlike the rainbow angle itself. Hence, the arcs are rarely seen in drops with diameters larger than a millimeter. In fact, this provides one means by which to estimate droplet size (e.g., Malkus *et al.* (1948)). Moreover, because raindrops tend to grow as they fall, one is most likely to see supernumerary arcs at the peak of the bow, which is borne out by observation. Indeed, these arcs and their features are readily produced and confirmed in the laboratory (Walker 1980).

On the one hand, no rays emerge at scattering angles smaller than the rainbow angle, thereby explaining the shadow region. On the other hand,

---

† The experiments were done in 1666 and reported in his Cambridge lectures of 1699–1671.

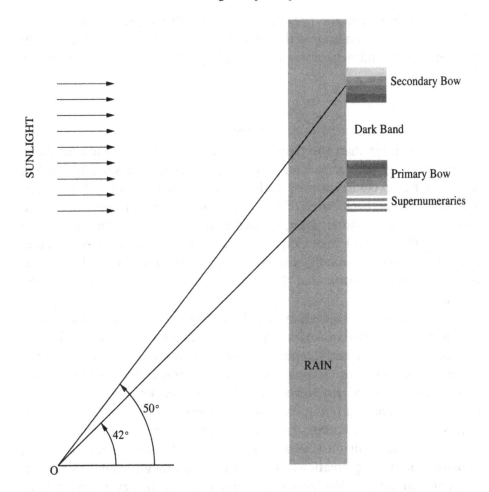

Fig. 6.7. Angles of observation for the primary (42°) and secondary (50°) rainbows in water (with corresponding scattering angles of 138° and 130°, respectively). The role of chromatic dispersion and the origin of Alexander's dark band are also illustrated. Dark to light shading denotes red to blue.

there are many rays emerging at greater angles, so that one understands the enhanced general brightness of the sky beneath the primary bow.

One final physical property of the rainbow that must be noted is that it is almost completely polarized. This phenomenon arises from a remarkable coincidence: the internal angle of incidence for the rainbow ray is very close to the Brewster angle, at which the reflected ray is completely plane polarized. It was Brewster himself, in fact, who verified the polarization of the rainbow (Brewster 1833), an experiment that today is almost trivial: if you are wearing polarized glasses while observing a rainbow, just cock your head from side to side to make it disappear and reappear at will!

The preceding remarks provide a good qualitative picture of the rainbow, but a truly quantitative theory was not forthcoming until the subtle phenomenon of diffraction was better understood, to which we turn in the following section.

The glory

Though it is rarer than the rainbow, the glory is hardly less impressive. It is seen in a very narrow backscattering angle arising from sunlight incident on fine mist or cloud droplets with diameters on the order of 5–20 μm. As shown on the back cover of this book, one observes a series of colored rings about the anti-solar point, ordered in the same way as in the rainbow with red outermost. Occasionally there are additional sets of rings of larger diameter. Glories are usually observed from mountain tops with clouds close below, or from airplanes, and the appropriate shadow serves to locate the anti-solar point.

The first recorded observation of a glory seems to have taken place atop a mountain in the Peruvian Andes, being described in some detail by a Spanish ship captain. Subsequently, reported sightings became relatively frequent in the Harz mountains of Germany, where it was called the Brocken Spectre, and on the summit of Ben Nevis in Scotland. An account of these and other early observations is given by Perntner and Exner (1910).

It is well known, of course, that Christian and other iconographies employ halos about the heads of sainted ones. Benvenuto Cellini considered himself anointed through observations of his own halo, though what he describes is almost surely the *Heiligenschein* (Cellini 1571). However, one can be equally certain that the glory has inspired such iconography throughout recorded history, owing to the very narrow backscattering angle to which it is restricted. If several persons are in more or less the same position, each nevertheless sees the glory only about his or her own head, much as with the *Heiligenschein*. Thus, as with Cellini, each person is their own saint.

One of the more amusing anecdotes in the history of the glory concerns C. T. R. Wilson, who had observed many of them. He was so delighted with the phenomenon that he set out to reproduce the effect in his laboratory in 1895, but failed. He succeeded in this effort, however, in building the first cloud chamber, for which he won the Nobel prize; such is the nature of serendipity!

The angular diameters of the rings are found to vary inversely as the size of the drops producing them, and the effect disappears completely if the drops become large enough ($\beta \gtrsim 10^3$). There are indications that the glory is strongly polarized, the first ring is rather hazy, and the intensity decreases fairly slowly from the center. General characteristics of the rings are variable, often changing significantly during a single observation. More detailed descriptions both of experimental and of numerical studies of the glory have been provided by Bryant and Jarmie (1974). These laboratory investigations reveal that the glory is a remarkably complicated phenomenon, containing intricate features not seen in its natural observation. Unfortunately, there is essentially no history of a quantitative theory of the glory matching that of the rainbow, though in Section 6.4 below we shall describe some relatively recent early attempts at physical explanations.

6.3 Theory of the rainbow

Following the work of Descartes and Newton we are able to describe most of the salient features of rainbows by means of geometric optics. After $p - 1$ internal reflections the rainbow of order $p - 1$ occurs at the angle of minimum deviation given by Eq. (4.61) and originates from a ray having an angle of incidence found from (4.60). The wavelength dependence of the refractive index of water determines the rainbow angles precisely, so that the width of a bow is predicted quite well by computing the angles in the range $n_{red}(1.331)$–$n_{blue}(1.343)$. Usually we shall employ for convenience the value $n = \frac{4}{3}$, corresponding closely to orange. Reference to Eqs. (1.37) and (1.38) permits ready evaluation of the intensities through (4.62); the high intensity at the rainbow angle follows from (4.63), but is unphysically infinite. Finally, the dark band between the first-order ($p = 2$) and second-order ($p = 3$) bows is a result of no rays from either emerging in that region. Table 6.1 exhibits the rainbow angles, widths, and intensities of the first 19 rainbows, and Fig. 6.8 illustrates Billet's 'rose', providing a useful visual glimpse of their locations with respect to the scatterer.

For several reasons, only the first two rainbows are usually seen in nature, though scattered sightings of the tertiary bow have been reported.† A first obvious reason for this is the reduction in intensity from repeated internal reflections alone, as indicated in Table 6.1. On the one hand, this is ameliorated, according to (4.60), by a decrease in damping as the incident rays move closer to glancing incidence and an increase in reflectivity; on the other hand, the bows become broader with increasing p, so that the dispersion of critical rays is greater. More debilitating, however, is the general background brightness of the sky, along with direct and transmitted glare from the first two Debye terms. Although higher-order bows are very difficult to observe

† The author is persuaded that he has observed one in front of a late-afternoon sun.

Scattering observables for large dielectric spheres

Table 6.1. *The first 19 rainbows of water*

p	θ_R(red)	θ_R(blue)	$\Delta\theta_R$	θ_{iR}(red)	θ_{iR}(blue)	$E_{p,red}$ (10^{-4})
2	137.63	139.35	1.72	59.53	58.83	914.3
3	230.37	233.48	3.11	71.91	71.52	389.9
4	317.53	321.89	4.37	76.89	76.62	215.1
5	42.76	48.35	5.58	79.67	79.46	136.4
6	127.08	133.86	6.78	81.46	81.28	94.2
7	210.90	218.87	7.96	82.72	82.57	69.0
8	294.41	303.55	9.14	83.65	83.52	52.7
9	17.71	28.02	10.32	84.36	84.25	41.6
10	100.86	112.35	11.49	84.94	84.83	33.7
11	183.92	196.57	12.66	85.40	85.31	27.8
12	266.89	280.71	13.82	85.79	85.70	23.3
13	349.81	4.79	14.98	86.11	86.03	19.9
14	72.68	88.83	16.15	86.39	86.32	17.1
15	155.51	172.82	17.32	86.64	86.57	14.9
16	238.31	256.79	18.48	86.85	86.78	13.1
17	321.09	340.73	19.64	87.03	86.97	11.6
18	43.84	64.65	20.81	87.20	87.14	10.4
19	126.58	148.55	21.97	87.35	87.29	9.3
20	209.31	232.43	23.13	87.48	87.43	8.4

naturally, they are easily studied in the laboratory, as Billet demonstrated. More recent investigations of this kind have been carried out by Walker (1976, 1980).

The Brewster angle at $\tan n \simeq 53°$ is very close to the angle of incidence of the primary bow, as indicated in Table 6.1. In consequence, we see from (1.37a) that the parallel component is strongly suppressed; this is confirmed also by (4.62a) and (3.71). Geometric optics provides no fundamental reason why this should be so, nor does it account for the supernumerary bows. For this we must turn to the wave theory and the more recondite development of Airy (1838).

The observed variability of the rainbow is the principal key to this advance, for this must reflect a sensitive dependence on drop size. For example, the bright sharp colors seen for raindrops of 1 mm diameter tend to weaken and overlap, melding into a 'white' rainbow in fog or smaller drops. This provides strong evidence for the importance of diffraction.

Considering Descartes' discovery of the rainbow caustic, or confluence, of geometric rays, Airy applied Huygens' principle to the wavefront of a droplet to find the intensity distribution in a neighborhood of that caustic. By assigning an amplitude to the initial wavefront he could then construct

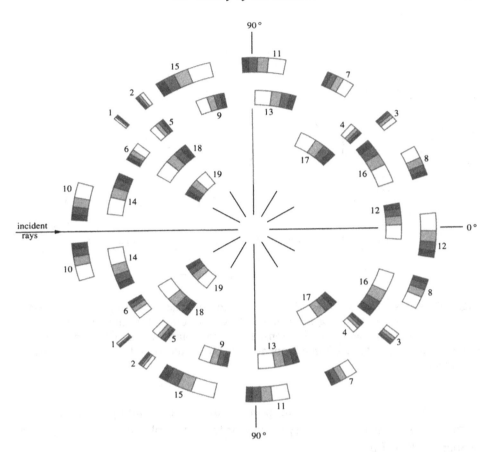

Fig. 6.8. A reconstruction of Billet's 'rose' illustrating the approximate widths, orientation, and scattering angles for the first 19 rainbows of water. Although the angular widths are accurate, the color (gray scale) gradations within each bow are only approximate. Dark to light shading denotes red to blue.

an amplitude distribution at any other point, and proceed to a rigorous description of the entire rainbow. Unfortunately, one had to guess the initial amplitude.

This notion of a minimum deviation guided Airy to choose a cubic wavefront inside the droplet such that it has an inflection point where it intersects the Descartes rainbow ray. By estimating the amplitudes along this wavefront through the theory of diffraction, Snell's law, and the Fresnel coefficients, he arrived at a description of the primary bow in terms of the *rainbow integral*, now called the Airy function $Ai(x)$. This result, although it was new to Airy, should come as no surprise here, for it parallels closely the procedure used in the CFU method of Appendix E to extend the method

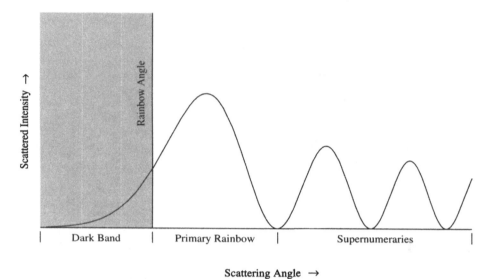

Fig. 6.9. Airy's rainbow integral as a function of the scattering angle, Ai($-x$), provides a visually satisfying mathematical description of the main rainbow peak along with supernumerary oscillations, for both polarizations. The deep shadow corresponds to the faster-than-exponential decay of the Airy function for positive arguments.

of steepest descent to two overlapping saddle points, leading to (E.42). The immediate success of this theory is evident from a re-plotting of Ai($-x$) from Appendix B in Fig. 6.9.

The first maximum in the Airy function corresponds to the principal rainbow peak, and the oscillations along the positive real axis represent the supernumerary arcs. Thus, diffraction smoothes out the intensity pattern, muting the singularity of geometric optics that would occur at $\theta = \theta_R$; indeed, the peak is somewhat displaced from this angle. Evidently the maxima can be located precisely from the properties of Ai($-x$), as will be done presently. Along the negative axis Ai($-x$) falls off faster than exponentially, like $\exp(-\frac{2}{3}|x|^{3/2})$, providing a striking realization of the shadow in the dark band. (Here $x \propto \theta - \theta_R$.) One must extend the calculations to both polarizations, of course, and there is an Airy pattern for each color determined by the refractive index of water. The purity of the colors depends on the extent to which the various monochromatic rainbows overlap, and this in turn is determined by the drop size. For droplets of diameter $\lesssim 0.01$ mm, as noted above, the overlap is so great that a white rainbow is sometimes observed.

Although the Airy theory is quite successful in its essential description of the rainbow, there remain a few shortcomings. One is the need to guess the amplitude distribution along the initial wavefront; another is that it does not provide a complete analytic picture of the scattering process in terms of the detailed interaction of waves with the target. In addition, as observed by van de Hulst (1957), the theory is generally valid for $\beta \gtrsim 5000$ and for $|\theta - \theta_R| \lesssim 0.5°$. One can extend the theory by improving on Airy's cubic guess (e.g., Mobbs (1979)), but the basic drawback remains. More recent numerical comparisons with the exact Mie series indicate that many of the theory's features remain reliable at smaller size parameters (Wang and van de Hulst 1991), so that its applicability need not be as restricted as had originally been thought. Rather than examine further properties separately, however, it is more useful to move on to the full asymptotic theory of Chapter 5 and exhibit the Airy approximation as a special case.

The asymptotic theory

From Eqs. (5.101)–(5.106) the dominant contributions to the scattering amplitudes in the rainbow region are given by

$$S_{j2,\text{geo}}(\beta, \theta) \simeq 2\pi i e^{i\pi/4} \frac{n(2\beta)^{7/6}}{\sqrt{\pi \sin \theta}} e^{2\beta A(\epsilon)}$$
$$\times \Big\{ [p_{0j} - (2\beta)^{-1}(q_{1j} + 2\zeta(\epsilon)q_{2j} + O(\beta^{-2}))] \, \text{Ai}[(2\beta)^{2/3}\zeta(\epsilon)]$$
$$- (2\beta)^{-1/3}[q_{0j} - 2(2\beta)^{-1}p_{2j} + O(\beta^{-2})] \, \text{Ai}'[(2\beta)^{2/3}\zeta(\epsilon)] \Big\}, \quad (6.6)$$

with $\epsilon = \theta - \theta_R$. The CFU coefficients $p_{mj}(\epsilon)$ and $q_{mj}(\epsilon)$ must be evaluated for a definite value of the refractive index from equations such as (5.107) and (E.38). Fortunately, not many coefficients are required usually, and for $|\epsilon| \ll (2\beta)^{-1/3}$ they can be expanded in powers of ϵ. Table 6.2 exhibits the first few for $n = 1.33$, as computed by Khare (1975), and Khare and Nussenzveig (1974). Complete expressions for $\zeta(\epsilon)$ and $A(\epsilon)$ are given in Eqs. (5.103)–(5.105), and through second order in ϵ are

$$\zeta(\epsilon) \simeq -0.369\epsilon - 0.0745\epsilon^2 + O(\epsilon^3),$$
$$A(\epsilon) \simeq i[1.519 + 0.431\epsilon - 0.115\epsilon^2 + O(\epsilon^3)]. \quad (6.7)$$

Extraction of Airy's result from these equations is straightforward: retain only terms through $O(\epsilon)$ in $A(\epsilon)$ and $\zeta(\epsilon)$, approximate $p_{0j}(\epsilon)$ by $p_{0j}(0)$, and neglect all other CFU coefficients. Among other things, this abandons Ai' completely, which has observable consequences. Let us now assess how these results describe the rainbow phenomena.

Whereas geometric-optics-type contributions to the amplitudes are typically $O(\beta)$, in the rainbow region they are $O(\beta^{7/6})$, so that the maximum

Table 6.2. *The leading CFU coefficients in Eq. (6.6)*

Coefficient	$j = 1$	$j = 2$
$p_{0j}(\epsilon)$	$i(0.0381 - 0.031\epsilon - 0.19\epsilon^2)$	$i(0.00786 + 0.046\epsilon - 0.078\epsilon^2)$
$q_{0j}(\epsilon)$	$0.0227 - 0.15\epsilon - 0.59\epsilon^2$	$0.108 - 0.015\epsilon - 0.43\epsilon^2$
$q_{1j}(\epsilon)$	$0.40 + 3.0\epsilon$	$0.42 + 2.3\epsilon$
$p_{2j}(\epsilon)$	$-1.4i$	$-0.64i$
$q_{2j}(\epsilon)$	-4.1	-3.1

intensity is enhanced by a factor of $\beta^{1/3}$. We shall see later that further enhancement arises for higher-order bows in other directions.

The argument of the Airy functions in (6.6) is $(2\beta)^{2/3}\zeta$, and the maxima of $\mathrm{Ai}(-x)$ occur at the zeros of $\mathrm{Ai}'(-x)$, which are given in Table B.1. These not only locate all the peaks of the supernumerary arcs, but also tell us that, for raindrops and visible light, the angular distance of the main peak from θ_R is $\epsilon \simeq 0.005°$. From (6.6) and Fig. 6.9 we therefore estimate the angular width of the main peak to be $O(\gamma^2)$, much narrower than the typical penumbra of $O(\gamma)$. Thus, the fraction of the total incident energy actually going into the rainbow is only $O(\gamma)$.

From Fig. B.1 we expect that Ai and Ai' are more or less of the same order within the main rainbow peak, so that the dominant contribution to the amplitudes comes from the coefficient p_{0j}. However, Table 6.2 indicates that $|p_{02}| \approx 0.2|p_{01}|$, so that, at the rainbow angle perpendicular polarization ($j = 1$) dominates and the degree of polarization is about 96%. The reasons for this can be found from Eqs. (E.25b) and (5.76c), which show that the CFU coefficients are proportional to the Fresnel amplitudes at angles of incidence θ_{iR}. As we have already noted, however, this is very close to Brewster's angle, thereby suppressing parallel polarization ($j = 2$). A survey of experimental observations of rainbow polarization can be found in Volz (1961).

All of these features to this point are shared by the Airy approximation, for the main corrections to that come from the Ai' term in (6.6). Because $\mathrm{Ai}'(z) \approx -z^{1/2}\,\mathrm{Ai}(z)$ for large z (e.g., (B.10)), the contributions become comparable outside the main rainbow peak, thereby limiting the Airy approximation to large β and small ϵ, as noted by van de Hulst. For polarization 1 the main deviations will arise for the supernumeraries and are not substantial. However, for polarization 2 the dominance of p_{01} over p_{02} means that the Ai' correction is more important. Among other things, this implies that the supernumerary peaks for the two polarizations should be out of phase, with those for $j = 1$ being near the minima for $j = 2$, and vice versa. This interchange between maxima and minima for the two polarizations was noted long ago by Bricard (1940), and its origin can be found in the 2-ray interference responsible for the supernumeraries. The reflection amplitude for polarization 2 changes sign at the Brewster angle; this occurs for one of the rays, which means that constructive interference for one will be destructive for the other.

Both (6.6) and its Airy approximation explain fully the shadow side of the rainbow, as well as the dark band between primary and secondary bows, through the asymptotic properties of the Airy function, Eqs. (B.10). Figure 6.8 indicates that the fourth- and fifth-order rainbows ($p = 5$ and 6) lurk within this shadow, though investigations by Wang and van de Hulst (1991) reveal their contributions to be insignificant. It is perhaps most useful at this point to summarize these results graphically by numerical comparison with the exact Mie series. In the angular region of the rainbow the dominant contributions come from direct reflection ($p = 0$) and the 2-ray/0-ray transition region of the single internal reflection ($p = 2$). The intensities based on these terms have been compared with the Mie series and the Airy approximation over the range $50 \le \beta \le 1500$ by Khare and Nussenzveig (1974) and Khare (1975) for $n = 1.33$, and we have re-calculated the appropriate curves for $\beta = 500$ and $1\,500$ in Figs. 6.10 and 6.11, respectively. One observes in both a rapid oscillation superimposed on the rainbow pattern, which arises from interference between the direct-reflection and rainbow terms.

At $\beta = 500$ the fit of the full asymptotic theory to the exact Mie curve is excellent for both polarizations, whereas the Airy approximation does quite well for polarization 1, but begins to deviate markedly outside the main peak. The phase shift mentioned above begins to appear here in polarization 2.

Because the rainbow peaks decrease in width as $\beta^{-2/3}$ (for fixed p), Fig. 6.11 presents a better view of the supernumeraries. Once again the comparison with the exact results is excellent, though the Airy theory slips a bit after the second supernumerary. Polarization 2 remains a good fit for the full theory, while the Airy approximation has broken down considerably, even within part of the main peak; the interchange of maxima and minima is apparent. One can attribute the small oscillations in the Mie curve to higher-order Debye terms still being felt.

In Chapter 3 we found that electromagnetic scattering from a sphere is completely determined by $|S_1|^2$, $|S_2|^2$, and their phase difference $\delta = \delta_1 - \delta_2$. To complete the analysis of the rainbow we also compute δ for the exact,

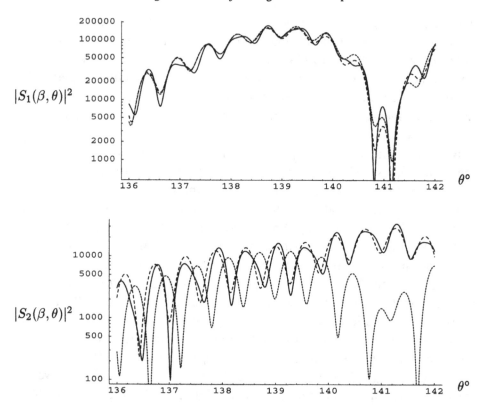

Fig. 6.10. A comparison of rainbow intensities for the asymptotic theory (- - - -) and its Airy approximation (······), for both polarizations, with the exact curve (———) from the Mie theory for $n = 1.33$ and $\beta = 500$.

asymptotic, and Airy theories and compare them in Fig. 6.12. Once again the excellent agreement between the first two is apparent (they can hardly be distinguished), but the Airy curve exhibits marked departures from the Mie result.

Finally, we should point out that, for natural incident light, the scattered intensity is given by $\frac{1}{2}(i_1 + i_2)$, so that the dominance of polarization 1 dilutes somewhat deviations occurring in the other polarization. This means that the Airy theory can still provide a useful description of many features of low-order natural rainbows, as emphasized by Wang and van de Hulst (1991).

Higher-order rainbows

Our description of the primary rainbow is readily extended to higher-order bows by the same techniques. The rainbow angle of incidence for all Debye

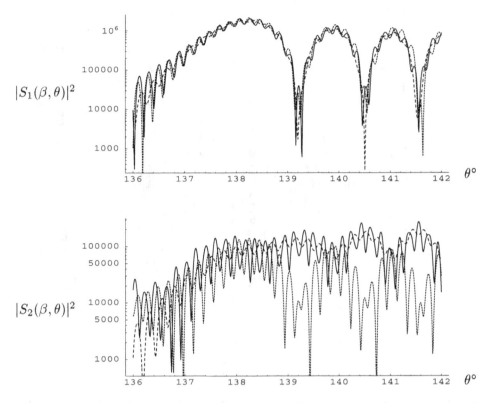

Fig. 6.11. A comparison of rainbow intensities for the asymptotic theory (- - - -) and its Airy approximation (·····), for both polarizations, with the exact curve (——) from the Mie theory for $n = 1.33$ and $\beta = 1500$.

terms was given in Eq. (4.60),

$$\cos(\theta_{iR})_p = \left(\frac{n^2 - 1}{p^2 - 1}\right)^{1/2}, \quad p > 1, \tag{6.8}$$

and the rainbow angle itself is obtained from (4.61). Numerical values for the first 19 bows are found in Table 6.1. From (6.8) we immediately see that the angle of incidence increases monotonically with increasing p, implying that higher-order bows arise from rays in the edge strip. Since the reflectivity increases near the edge as well, the damping owing to internal reflections tends itself to weaken with increasing p.

Scattering amplitudes for arbitrary p in the rainbow region can be calculated using exactly the same procedure as that leading to Eqs. (5.101)–(5.106), resulting in the same form as (6.6). Incorporating constants into the coeffi-

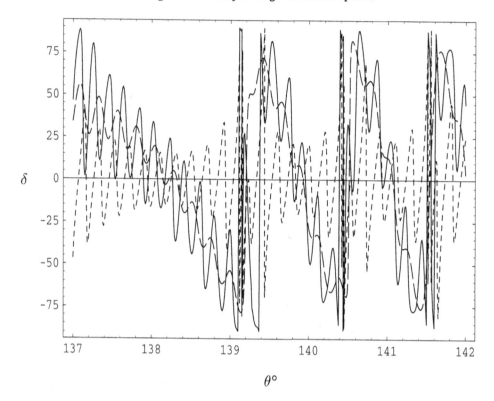

Fig. 6.12. A comparison of the amplitude phase difference $\delta = \delta_1 - \delta_2$ for the asymptotic theory (······) and its Airy approximation (---) with the exact curve (——) from the Mie theory for $n = 1.33$ and $\beta = 1500$.

cients, we express this as

$$S_{jp,\text{geo}}(\beta, \epsilon) \simeq \beta^{7/6} e^{2\beta A_p(\epsilon)} \Big\{ c_{j,p}(\beta, \epsilon) \, \text{Ai}\big[(2\beta)^{2/3} \zeta_p(\epsilon)\big]$$

$$+ d_{j,p}(\beta, \epsilon) \beta^{-1/3} \, \text{Ai}'\big[(2\beta)^{2/3} \zeta_p(\epsilon)\big] \Big\}, \qquad (6.9)$$

with

$$\begin{Bmatrix} A_p(\epsilon) \\ \tfrac{2}{3}[\zeta_p(\epsilon)]^{3/2} \end{Bmatrix} = \frac{i}{2}\big[pn(\cos\theta_r' \pm \cos\theta_r'') - (\cos\theta_i' \pm \cos\theta_i'')\big], \qquad (6.10)$$

which are generalizations of (5.103) and (5.104). For small ϵ all these functions can be expanded in powers of ϵ (Khare 1975). For example,

$$|\zeta_p(\epsilon)| \approx \frac{\sqrt{n^2 - 1}}{2^{2/3} p} \epsilon, \qquad p \gg 1, \qquad (6.11)$$

demonstrating that the main peak becomes flatter and the width increases with p, in agreement with Table 6.1 and Fig. 6.8.

The Airy theory provides a progressively poorer fit as p increases. One can see this through reference to Fig. 5.2, which indicates that the reflection coefficients approach unity very rapidly in the edge domain. Since the CFU coefficients in (6.9) are proportional to derivatives of the reflection coefficients, the higher-order coefficients are enhanced as p increases. In addition, higher-order Debye terms themselves contain higher powers of the reflection amplitudes. Detailed comparisons of rainbows in specific Debye terms $p = 2$–6 with corresponding Airy approximations have been made by Hovenac and Lock (1991), which further support these conclusions.

6.4 Theory of the glory

In Fig. 3.5 we caught a first glimpse of the complexity to be expected in the backscattered intensity. Add to this the phenomenon of the glory and one begins to anticipate a physical richness in this region matched only by a neighborhood of the rainbow angle. Indeed, we shall see that a description of scattering in the backward region is considerably more complicated than that of scattering at the rainbow angle.

Very early attempts at explaining the glory range from attributing it to secondary scattering of the forward diffraction peak (Perntner and Exner 1910), through interference between direct reflection rays and those emerging after one internal reflection (Ray, 1923), to a description in terms of 'backward diffraction' (Bucerius 1946). The reader will readily see that none of these proposals can really encompass even the qualitative features already noted, and it will become clear presently that geometric optics involving the $p = 0$ and $p = 2$ rays cannot produce anywhere near a proper description.

The first truly important theoretical constructs are due to van de Hulst (1947, 1957). To illustrate one of his major ideas, let us suppose that for some angle of incidence other than zero there is a paraxial ray that, after one internal reflection, emerges exactly in the backward direction, as in Fig. 6.13. (The central ray is just the $p = 2$ geometric-optics contribution alluded to above.) An incident plane wave emerges as a circular wavefront with virtual focus at F. Normally F is just a virtual point source, but for $\theta = \pi$ the axial symmetry implies the existence of an entire focal circle. Thus, a virtual ring source of radius b is generated, giving rise to *toroidal* wavefronts. This effect is known as *axial focusing*, and one might expect it to lead to an enhancement of the backscattering arising from constructive interference along the axis from all points on the ring source.

A more quantitative understanding of how this increase in intensity might come about is suggested by the analogous discussion for the forward direc-

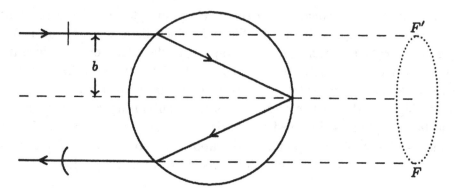

Fig. 6.13. A possible glory ray representing a portion of a plane wavefront that exits the sphere with a spherical wavefront and virtual point source F. Spherical symmetry implies that all such rays at this angle of incidence lead to a virtual ring source FF' and consequently a torroidal wavefront for scattered radiation.

tion, beginning with Eq. (4.19). We recall from (D.54b) that, in the exactly backward direction, the angular functions become

$$t_\ell(-1) = -p_\ell(-1) = (-1)^\ell (\ell + \tfrac{1}{2}). \tag{6.12}$$

In other directions the amplitude S_1 is dominated by magnetic-multipole contributions and S_2 is dominated by electric multipoles, but along the axis their contributions are comparable. This is the cross-polarization effect mentioned in Chapter 3, and simulates an interference between the two polarizations that also persists in a small neighborhood of the backward direction.

To visualize this effect we shall presume that the dominant contributions to the glory arise in the edge domain, as turns out to be the case. With reference to (3.120) and (D.63) we can approximate the scattering amplitudes in the domain $\theta = \pi - \varepsilon, 0 \le \varepsilon \ll 1$, by

$$S_1(\beta, \pi - \varepsilon) \approx 2S^M(\beta)J_1'(\lambda\varepsilon) + 2S^E(\beta)\frac{J_1(\lambda\varepsilon)}{\lambda\varepsilon}$$

$$S_2(\beta, \pi - \varepsilon) \approx -2S^E(\beta)J_1'(\lambda\varepsilon) - 2S^M(\beta)\frac{J_1(\lambda\varepsilon)}{\lambda\varepsilon}, \tag{6.13}$$

where S^M and S^E are the magnetic- and electric-multipole contributions identified in (3.120):

$$S_1(\beta, \pi) = S^M(\beta) + S^E(\beta) = -S_2(\beta, \pi). \tag{6.14}$$

A plot of the squares of the angular functions in Fig. 6.14 exhibits clearly

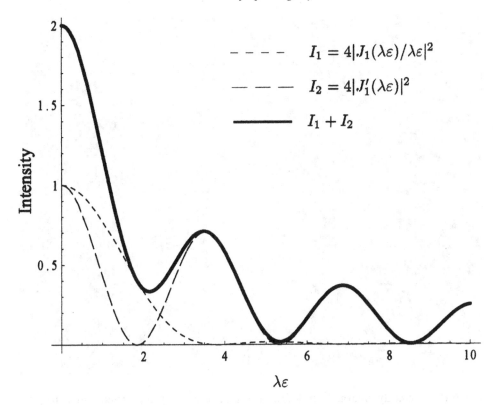

Fig. 6.14. Angular distributions of intensities in the glory, from Eq. (6.13).

the cross-polarization effect, which occurs only over the range $\lambda\varepsilon \lesssim 1$; the concomitant enhancement for $\theta \sim \pi$ is evident.

Figure 6.14 also encompasses some other glory features, such as the ring structure and the prediction that ring radii are inversely proportional to β. The haziness of the first dark ring is understandable, as is the overall dominance of polarization 2.

Unfortunately, there is no glory ray of the type indicated in Fig. 6.13 for the $p = 2$ Debye term unless $\sqrt{2} < n < 2$, which excludes water – this is a standard homework problem in geometric optics. The best scattering angle available from geometric optics is 165°, so there is a 'gap', or 'missing' angle of about 15°. In his 1947 work van de Hulst conjectured that this gap may well be spanned by continuing the geometric ray into the backward direction by means of a surface wave. That this, in fact, can be done was first demonstrated unequivocally by Nussenzveig (1969b) for the scalar model, and is clearly evident in our discussion of Eq. (5.113). We return to this presently.

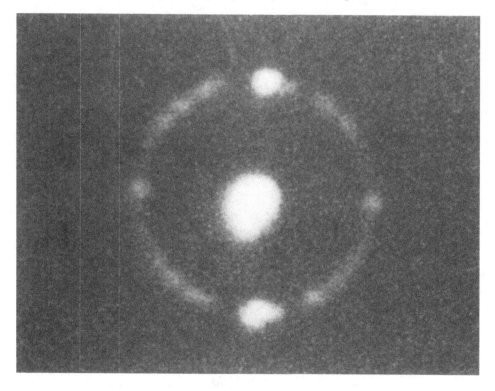

Fig. 6.15. A greatly enlarged photograph of an acoustically supported water droplet viewed at 180° with respect to the illuminating beam. Surface waves radiating from the circumference, along with the backscattered central ray, are evident. The focusing at the poles and equator is apparently caused by a slightly ellipsoidal distortion of the droplet (Fahlen and Bryant (1966); with permission).

In the mid-1960s experimental confirmation of the importance of surface-wave contributions to the glory was provided in a series of beautiful experiments carried out by Fahlen and Bryant (1966). They scattered a focused He–Ne laser beam from single water droplets suspended in an acoustic field. Figure 6.15 presents one of their results illustrating the appearance of surface waves in the backward direction, as well as the central ray. Illumination of the entire drop generates a circumferential glow that corresponds exactly to the virtual ring source! In a subsequent study (Fahlen and Bryant 1968) they also investigated the backscattered intensity as a function of the size parameter, uncovering a number of periodicities (or quasiperiodicities) that will appear theoretically below. Further evidence for the role of surface waves in backscattering can also be found in Ashkin (1980).

Numerical studies

At about the same time as the above experiments, the general availability of large computers made it feasible to sum a substantial number of terms in the Mie series numerically, and numerous computational studies of scattering functions in the glory region were forthcoming. Notable among these was the work of Dave (1969), as well as that of Bryant and Cox (1966), and Shipley and Weinman (1978). Such computations are relatively easy to carry out on a desktop computer nowadays and we can readily uncover a number of further features of the backscattered intensity distributions.

In Fig. 6.16 we show the exact total intensity $I(\theta) = |S(200, \theta)|^2$ and degree of polarization $P(\theta)$ for $n = 1.33$ and $\beta = 200$. This confirms our inferences from Fig. 6.14, and suggests that the outer rings are parallel polarized.

Figure 6.17 shows the backscattered normalized phase function as a function of β, along with that predicted by geometric optics. The latter is clearly at least an order of magnitude smaller than the exact result. Figure 6.17 also indicates that there is a strong enhancement of the ripple in this direction, although it is present in a milder form in all directions, as well as in cross sections. There is, in addition, a quasiperiodic variation in β with period $\Delta\beta \simeq 0.8$ that one can judge by eye, and another about half that length. Fourier analysis of the amplitude spectrum by Shipley and Weinman (1978) uncovered a number of other quasiperiodicities. Size distributions and even a small amount of absorption will tend to dampen any high-frequency fluctuations in observations, of course, but the major background oscillations should be observable.

The asymptotic theory

The dominant contributions from the $p = 0$ and $p = 2$ Debye terms at $\theta = \pi$ are found from (5.115), (5.122), and (5.123):

$$S_{j0,\text{geo}}(\beta, \pi) \approx (-1)^{j+1} i \frac{\beta}{2} e^{-2i\beta} \left(\frac{n-1}{n+1} \right) [1 + O(\beta^{-1})], \tag{6.15a}$$

$$S_{j2,\text{geo}}(\beta, \pi) \approx (-1)^{j+1} i\beta \frac{2n^2(n-1)}{(2-n)(n+1)^3} e^{2i(2n-1)\beta} [1 + O(\beta^{-1})], \tag{6.15b}$$

$$S_{j2,\text{res}}(\beta, \pi) \approx (-1)^j i \frac{\beta}{\gamma N^2} e^{i\pi/3} e^{i4N\beta} \sum_k (a_k')^{-2}$$

$$\times \zeta_2 [(\zeta_2 + N) e^{i\lambda_k \zeta_2} - n^2(n^2\zeta_2 + N) e^{i\mu_k \zeta_2} + O(\beta^{-1})], \tag{6.15c}$$

where the poles λ_k and μ_k are given by (5.39), the Airy-function parameters x_k and a_k' are tabulated in Appendix B, $\gamma = (2/\beta)^{1/3}$, $N = \sqrt{n^2 - 1}$, and in

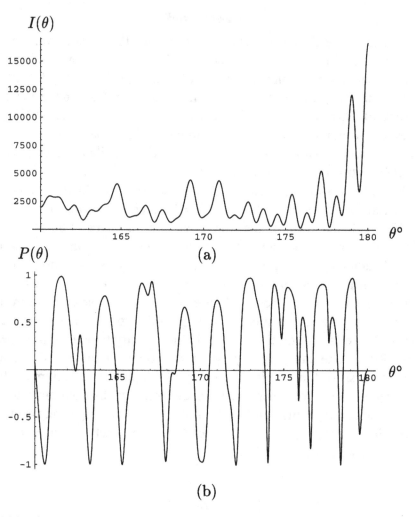

Fig. 6.16. The total intensity (a) and polarization (b) for $n = 1.33$ and $\beta = 200$ near the backward direction.

place of (5.89c) and (5.92) we shall find it useful to employ the notation

$$\zeta_2 \equiv \pi - 4\cos^{-1}(1/n). \qquad (6.16)$$

Higher-order axial rays, as noted from (6.2), contribute powers of $(n-1)^2/(n+1)^2$ and can be neglected.

At the end of Chapter 5 we saw that the residue series (6.15c) contains an enhancement factor of $\beta^{1/2}$, which is just a manifestation of van de Hulst's axial focusing effect. We also observed there that for $\beta \sim 10^2$ the surface-wave contribution dominates that from axial rays, but as $\beta \to 10^3$

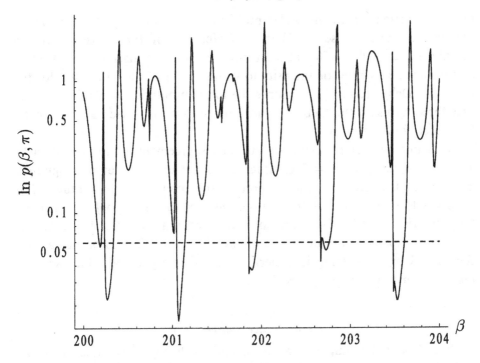

Fig. 6.17. The logarithm of the normalized backscattered phase function near $\beta = 200$. The geometric-optics approximation (----) is at least an order of magnitude smaller than the exact Mie result (solid line).

the reverse becomes true. This behavior is governed by the surface-wave damping factor

$$d_{2,k} \sim e^{-\eta_k \beta^{1/3} \zeta_2}, \qquad \eta_k \equiv x_k 2^{-4/3} - \begin{cases} 1/N, & \lambda = \lambda_k \\ n^2/N, & \lambda = \mu_k, \end{cases} \qquad (6.17)$$

where $\{x_k\}$ are tabulated in Appendix B. For $n = \frac{4}{3}$ we find that $\zeta_2 \simeq 0.25$, or $14.4°$. That is, ζ_2 is the missing angle required to bring a tangentially incident ray undergoing two shortcuts around to the backward direction, thereby validating van de Hulst's conjecture.

If the amplitudes provided by Eqs. (6.15) are compared with the corresponding terms from the Mie series for $\beta = 200$ and $n = 1.33$, say, one finds large discrepancies; the phase is wrong and the intensities differ by an order of magnitude. (Curiously enough, the intensity obtained from the asymptotic theory is within 10% of the exact value for all terms, but this is fortuitous.) The first difficulty is that the coefficients in the residue series, obtained from (5.89), are inadequate approximations and it is necessary to retain terms at least of order γ^2 in these series. This is a very tedious task that was carried out in the scalar problem by Nussenzveig (1969b), and in the electromagnetic case by Khare and Nussenzveig (unpublished). In the neighborhood of $\beta = 1500$, at least, this resolves the discrepancy in phase and brings the asymptotic intensity to within 90% of the Mie result for the first three Debye terms. In this sense the total amplitude calculated from the first three terms of the Debye expansion appears to provide a sound basis for a theory of the glory.

Unfortunately, there remains significant disagreement with the total back-scattered intensity at $\beta = 200$, which increases to an order of magnitude at $\beta = 1500$. Moreover, examination of the intensity over a range of β uncovers little evidence of the various quasiperiodicities in β revealed by the numerical studies. One can only conclude that the first few terms of the Debye expansion are incapable of explaining the glory in particular and the overall backscattering in general. It is necessary to take into account higher-order terms in the series (5.25).

Higher-order Debye terms

Without doubt the simplest way to assess which higher-order terms in the Debye expansion might yield significant contributions to the backscattering is to compute their intensities numerically from the Mie series for the first 50 values of $p \geq 2$, say. If that were all that is involved, however, one may as well just sum the original partial-wave series and be done with it, from which one would learn little about the processes involved in the structure of the glory. We shall employ this tool to verify our results later, but first it is more edifying to identify the strongest contributing terms physically.

As a first step we ask whether there are any glory rays for $p \geq 2$ – that is, noncentral rays that would be reflected directly backward after $p-1$ internal reflections. To answer the question systematically we recall Eq. (4.59) for the angle of deviation for such a ray and set $\theta = s\pi$, where s is a positive integer; s is the number of times the ray crosses the optical axis within the sphere before exiting. With the help of Snell's law this equation yields the condition for a glory ray in terms of the angle of refraction,

$$n \sin \theta_r = \cos[p\theta_r - \pi(p - s)/2], \qquad (6.18)$$

and s must be odd for a glory ray.† As the angle of incidence ranges over $(0, \pi/2)$ the angle of refraction is seen to have the bounds

$$(p - s - 1)\frac{\pi}{2p} \leq \theta_r \leq (p - s)\frac{\pi}{2p}. \qquad (6.19)$$

---

† We shall see subsequently that even values of s predict *forward* glories.

In turn, these limits place bounds on the allowed values of the refractive index for which a glory ray exists:

$$\sec[s\pi/(2p)] > n > 0, \qquad p - 1 > s,$$
$$\operatorname{cosec}[\pi/(2s)] \le n \le p, \qquad p - 1 = s, \tag{6.20}$$

following some algebra.

Quite possibly, on its way to the upper bound, θ_r may reach a value for which there is an angle of minimum deviation, in which case there can be two glory rays under the condition

$$\sec[s\pi/(2p)] \le n \le n_{\max}(p, s), \qquad p - 1 > s > 0, \tag{6.21}$$

where $n_{\max}(p, s)$ is to be determined. Indeed, the existence of such an angle is just the condition for a rainbow, which from Snell's law and (4.59) is $\tan\theta_i = p\tan\theta_r$. The condition that *both* a glory and a rainbow appear in the same direction is thus

$$p\tan\theta_r = (-1)^p[\tan(p\theta_r)]^p, \qquad \rho = (-1)^{p-s-1}. \tag{6.22}$$

This can be solved and used in Eq. (6.18) to find $n_{\max}(p, s)$.

Now back to our question. Some inspection shows that, for water, the lowest-order backward glory ray corresponds to $p = 5$ and $s = 3$, and an angle of incidence near 34°. For size parameters in the region of the glory the associated intensity is completely negligible, the internal reflection coefficients being too small at this angle. To minimize internal reflection damping we should investigate rays incident in the edge strip, where the internal spherical reflection coefficient is near unity.

Such a study begins with Appendix C and Eq. (5.24), in which we employ the Schöbe asymptotic expansions of the cylinder functions with argument β, since $|\lambda - \beta| = O(\beta^{-1/3})$ near the edge, and the Debye formulas for those with argument $n\beta$. In an obvious notation we find that

$$|R_\ell^{11}(\beta, j)|^2 = 1 - \frac{1}{nv_j}|T_\ell^{12}(\beta)|^2$$
$$\approx 1 - \frac{\gamma}{\pi Nv_j}\left|\operatorname{Ai}[-e^{-i\pi/3}\gamma(\lambda - \beta)]\right|^{-2}$$
$$\equiv 1 - c_j\beta^{-1/3}, \tag{6.23}$$

where $c_j = O(1)$, and $j = 1, 2$ denotes the polarization. After $p \gg 1$ internal reflections the total internal damping takes the form

$$|R_\ell^{11}(\beta, j)|^2 \approx e^{-pc_j\beta^{-1/3}}. \tag{6.24}$$

This result, along with the axial focusing enhancement of order $\beta^{1/2}$, implies that Debye terms with p up to several times $\beta^{1/3}$ can make appreciable contributions through tangentially incident rays. Which of these terms can we expect to be major contributors?

Possible contributions that can partially overcome the effects of internal reflection damping and benefit from axial focusing can arise from higher-order backward glory rays, rainbows, and surface waves. From Eq. (6.20) we find that, at grazing incidence, the lowest-order backward glory ray for water occurs for $p = 24$ and $s = 11$, yielding a refractive index

$$n = \sec(11\pi/48) \simeq 1.330\,07. \tag{6.25}$$

Though not essential, it will be useful to adopt this value of n as a model for further study. There exist glory rays for higher p values, but these contain far too many internal reflections, as well as slightly different values of n.

At the value of n in (6.25) the glancing ray undergoes 23 internal reflections, transmitting a small amount of energy at each point before radiating in the backward direction at the 24th; it continues doing this during the second 180° of deviation so that it arrives back at exactly the original point of incidence, to be internally reflected once more. Thus, this edge ray that is very weakly damped forms a closed orbit that is an inscribed 48-sided stellated polygon – there are 48 vertices at the surface. Moreover, there is an infinite number of such rays with $\Delta s = 22$ and $\Delta p = 48$, so that Nussenzveig (1969b) has appropriately dubbed it a *geometric resonance*. It bears some resemblance to the scattering resonances to be discussed in Chapter 7, but has no connection to the electromagnetic vibrational modes of the sphere itself.

Turning to the higher-order surface-wave contributions, we find residue series that appear very much like that for $p = 2$, Eq. (6.15c), with some modifications. First of all, ζ_2 is replaced by

$$\zeta_p \equiv \pi - 2p\cos^{-1}(1/n), \quad \text{mod } 2\pi. \tag{6.26}$$

This is not surprising, because the general difference among Debye terms for $p \geq 1$ is that the identical poles are of order $p + 1$. As with the van de Hulst conjecture for $p = 2$, then, ζ_p provides the missing angle to be covered by surface waves for values of p associated with rays within one shortcut from the backward direction. There is, of course, a corresponding surface-wave damping factor of the form

$$d_{p,k} \sim e^{-\zeta_p \alpha_k \beta^{1/3}}, \tag{6.27}$$

where α_k is a rapidly increasing function of k. Therefore, the relative importance of these contributions is determined by the damping for large p from (6.27) and that for large β from (6-17). Note that the high-p cutoff decreases with β, so that these terms become less important in the range of size parameters in which we observe the natural glory.

A second important modification of the surface-wave contributions for $p \geq 2$ concerns the geometric phase factor, which is now $\exp(2ipN\beta)$. In addition, the leading factor of i in (6.15c) is now replaced by

$$(-1)^p(-1)^{(s+1)/2}, \tag{6.28}$$

where s was defined in connection with (6.18).

The third major source of higher-order contributions is from rainbows of order $p-1$, which arise from rays nearer the edge as p increases. For example, Fig. 6.8 reveals the shadow of the tenth-order rainbow ($p = 11$) to be near the backward direction; indeed, it is almost a glory ray: $n(p = 11, s = 5) \simeq 1.327$. The bright side of the 14th-order rainbow is not too far away. Quantitatively, Eq. (4.61) can be approximated for $p \gg 1$ and we find the angular distance of the geometric rainbow angle from $\theta = \pi$ to be given approximately by

$$\epsilon_{R,p} \equiv |\pi - \theta_{R,p}| \approx |\zeta_p - 2N/p|, \qquad p \gg 1. \tag{6.29}$$

Negative values of ζ_p indicate rainbows with their dark sides encompassing $\theta = \pi$. Although the Airy function's decay here is something like $\exp(-\beta\epsilon_{R,p}/p)^{3/2}$, this damping of a complex ray penetrating the dark side of the rainbow is now countered both by axial focusing of order $\beta^{1/2}$ and by the usual rainbow enhancement of order $\beta^{1/6}$. Thus, the $p = 11$ term yields a rainbow shadow within $\epsilon_{R,p} \approx 3°$ of the backward direction and should provide a significant contribution.

Derivation of the contributions from higher-order rainbows is straightforward, and the results have precisely the same form as that of Eq. (6.6). The functions $A(\epsilon)$ and $\zeta(\epsilon)$ in (5.103)–(5.105) are modified by the replacement $n \to pn$, and the CFU coefficients depend on p. For $p = 11$ the first two for each polarization are found to be (Khare 1975)

$$p_0^M \simeq 1.12 \times 10^{-3}, \qquad q_0^M \simeq 1.46 \times 10^{-2}$$
$$p_0^E \simeq 1.31 \times 10^{-4}, \qquad q_0^E \simeq 9.95 \times 10^{-3}. \tag{6.30}$$

With this physical picture of the dominant processes contributing to the backscattered intensity we are in a position to better appreciate the predictions from the Mie series itself. Following Khare and Nussenzveig (1975), we present in Fig. 6.18 the results of numerically summing the partial-wave series for three different values of β, with $n = 1.33007$. We have used (6.29) to distinguish surface-wave contributions (– – –) from rainbow

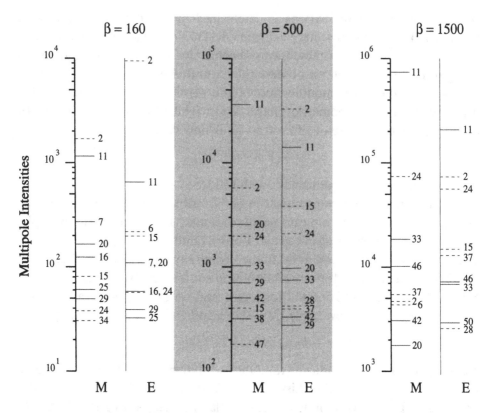

Fig. 6.18. Relative multipole strengths of the dominant Debye terms contributing to the total backscattered intensity for $n = 1.33007$ and three values of the size parameter β, computed from the exact Mie series. Rainbow (solid lines) and surface-wave (dashed lines) contributions are distinguished both for magnetic (M) and for electric (E) contributions.

and glory-ray terms (——) for the two multipole intensities at each value of β. This representation has the appearance of an 'energy-level' diagram that allows ready comparison of the relative intensities for most of the first 50 Debye terms, though we have found differences from the earlier work in the relative strengths of some higher-order terms.

For $\beta = 160$, chosen as very close to the estimated average size parameter for natural glory observations, the $p = 2$ surface wave is dominant for both multipoles, as expected. The tenth-order rainbow, however, contributes substantially to the magnetic intensity, and at $\beta = 500$ is strongly dominant. For $\beta = 1500$ the $p = 11$ term is completely dominant, and the $p = 24$ term is plainly very strong. (Interference among all these terms will also contribute to the total intensity, but need not be considered for the purposes of this diagram.)

The structure of the backscattered intensity pattern for a large dielectric sphere is indeed rather complex and involves varied physical phenomena. Further complicating it is a feature not yet considered in detail – namely, the ripple. Although this high-frequency oscillation is present in all directions, and particularly noticeable in the various cross sections, it is enhanced along with everything else in the backward direction. Figure 6.16 illustrates its distinct presence even at size parameters near those at which natural glories arise, although these fluctuations are almost always washed out by size dispersion and some absorption. The origin of the ripple is found in the remainder term (5.27b) of the Debye expansion and, as will be verified in Chapter 7, is associated with resonances described by the Regge poles of Fig. 5.1. Although it is not a dominant factor in the magnitude of the intensity here, either theoretically or experimentally, the ripple does represent an important structural feature, and will become of much greater interest in other contexts.

To summarize the role of higher-order Debye terms we present some numerical results in Table 6.3, which also reflects a changed attitude in our approach to theoretical predictions. Although analytic continuation into the complex angular-momentum plane has been essential to our understanding of the fundamental scattering processes, there comes a time when it is more useful to switch the computational paradigm to more common ground. In an era when computer speed and memory are no longer major issues in studying models based on the partial-wave series, we now find it most convenient to employ a number of well-known numerical techniques to explicate many of our physical results. Thus, Table 6.3 is based completely on numerical evaluation of individual terms in the Debye expansion through approximate summation of the appropriate partial-wave series. It refers to two representative size parameters and first compares the exact Mie results for magnetic and electric contributions to the backscattering with those from a sum of the leading Debye terms: $p = 0, 2, 11$, and 24. As expected, these apparently dominant terms in the Debye expansion constitute a substantial contribution to the intensity – in the case of $\beta = 200$ about 62%, rising to over 78% for $\beta = 1500$. In both cases there are other Debye terms that are stronger than one or the other of these nominally dominant ones, as one notes from Fig. 6.18 and sees in the remainder of Table 6.3. These entries result from including *all* Debye terms up to a maximum of $p = P_{\max}$. Increasing $P_{\max}$ brings the intensity closer to the exact results, of course, but the approach is by no means monotonic; Table 6.3 suggests that there is an oscillation about the exact value, which one presumes occurs owing to continued changes in constructive and destructive interference. Because the

Table 6.3. *Intensities and amplitudes at* $\theta = \pi$ *for* $n = 1.33007$

Amplitudes	$\beta = 200$		
	M	E	$|S_1|^2$
Exact	$92.20 - 54.35i$	$34.22 + 121.1i$	$20\,445$
Leading Debye	$109.0 - 48.37i$	$2.849 + 56.22i$	$12\,579$
$P_{max} = 25$	$96.03 - 60.34i$	$38.11 + 62.24i$	$17\,995$
$P_{max} = 50$	$86.99 - 54.80i$	$52.68 + 102.2i$	$21\,753$
$P_{max} = 200$	$92.82 - 53.20i$	$34.21 + 120.1i$	$20\,612$

	$\beta = 1500$		
	M	E	$|S_1|^2$ $(\times 10^{-6})$
Exact	$-738.2 - 795.8i$	$46.64 - 1065i$	3.939
Leading Debye	$-443.6 - 781.4i$	$144.3 - 953.2i$	3.098
$P_{max} = 25$	$-586.7 - 838.1i$	$113.0 - 1052i$	3.798
$P_{max} = 50$	$-770.0 - 903.5i$	$25.64 - 1034i$	4.306
$P_{max} = 200$	$-707.6 - 795.9i$	$30.62 - 1073i$	3.951

remainder term is not included in the Debye series, the remaining discrepancy at large P_{max} provides some measure of the contribution of the ripple to the backscattered intensity.

Figure 6.19 compares the exact backscattered intensity $|S(\beta, \pi)|^2$ with two different approximations for β near 200. The curve based on just the leading contributions from $p = 2$ and 11 is seen to fit the general pattern quite well. Inclusion of the first 25 Debye terms leads to a result that is virtually indistinguishable from that containing the leading two terms and has therefore not been included. The ripple spikes, of course, are not to be found in this approximation. What is evident, however, is that the leading terms in the Debye expansion provide an excellent description of the natural glory, and for larger size parameters inclusion of a few higher-order terms continues to give a good fit to the exact result.

Features of the glory

The physical insights provided by the asymptotic theory, combined with the partial-wave summations of individual Debye terms, allow us to describe very closely the observed features of the glory. As a beginning, the variation of features from one observation to another – and even within a single observation – reflects sensitive dependences both on n and on β. Such sensitivities are evident in the numerical calculations, that for β being obvious

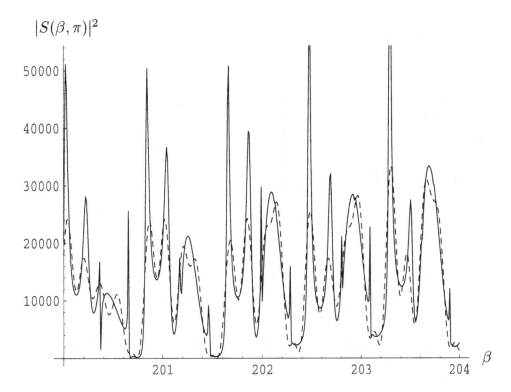

Fig. 6.19. A comparison of backscattered intensities $|S(\beta,\pi)|^2$ for the exact Mie result (——) and the sum of the Debye $p = 0, 2$, and 11 terms (- - - -) for β near 200 and $n = 1.33007$.

in Fig. 6.19, and that for n in Fig. 6.20. Dispersion implies that the two are related, of course. Sharp peaks due to the ripple effect are effectively washed out in observations of the natural glory.

The angular distribution of rings has already been accounted for in Eqs. (6.13) and Fig. 6.14, originally demonstrated by van de Hulst (1947, 1957). In those equations $\lambda\epsilon \sim \beta\epsilon$, and thus the angular width of the rings is $O(\beta^{-1})$. Note, also, that $J_1'(z) = O(z^{-1/2})$ and $J_1(z)/z = O(z^{-3/2})$, so that away from the center $S_1 \sim S^M$ and $S_2 \sim S^E$. However, in the dominant term $S_{j2,\text{res}}$ for natural glories, (6.15c) indicates that $|S^E| \sim 4|S^M|$, and therefore the outer rings are mostly parallel polarized. At larger values of β the $p = 11$ rainbow with dominant perpendicular polarization enforces the opposite result. These predictions appear to be in agreement with the meager observational evidence available on polarization.

Dominant contributions have their origin in the edge strip where incident rays penetrate and undergo almost total internal reflection, some diffraction

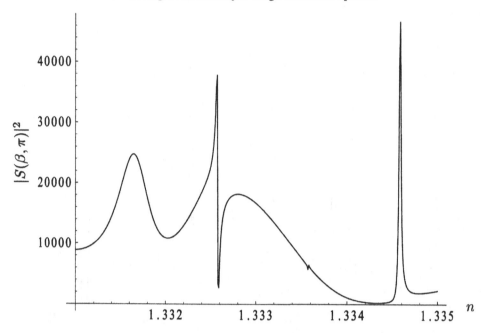

Fig. 6.20. The variation of the backscattered intensity $|S(\beta, \pi)|^2$ with the refractive index for β near 200.

losses occurring owing to surface curvature. Such losses decrease with increasing size parameter, so that higher-order Debye terms can be important in that event. These origins validate the insights both of van de Hulst (1957) and of Bryant and his collaborators (Bryant and Jarmie 1974). In addition, leading correction terms also originate near the edge, and all relevant contributions are enhanced by axial focusing and/or the rainbow factor.

Most of the slowly varying background intensity in the region of naturally occurring glories, $150 < \beta < 300$, can be accounted for by Debye terms $p = 0, 2$, and 11, as seen in Fig. 6.19. This description is provided by the approximation

$$S_j(\beta, \pi) \approx S_{j2,\text{res}} + S_{j11,\text{R}}(\beta, \pi) + S_{j2,\text{geo}}(\beta, \pi) + S_{j0,\text{geo}}(\beta, \pi), \qquad (6.31)$$

in order of decreasing importance, and $S_{j11,\text{R}}$ is the contribution from the tenth-order rainbow. The source of the variations in this background is the interference among these amplitudes when the intensity is computed, and lends strong support to the physical picture provided by the Debye expansion.

The major oscillatory structure – not quite periodic, but *quasiperiodic* – arises from the contribution to the phase from the path of each shortcut through the sphere. For p shortcuts this is

$$2pn\beta \cos \theta_c = 2pN\beta, \tag{6.32}$$

where $N = \sqrt{n^2 - 1}$, and the critical angle is defined in (1.44). The corresponding phase factor for $p = 2$ surface waves is evident in (6.15c), and the overall numerical factor aside from this was given in (6.28). Similar factors arise in higher-order rainbow expressions from the function $A_p(\epsilon)$, Eq. (6.10). Thus, for our model with $n = 1.33007$ a change in β by

$$\Delta\beta = \frac{5\pi}{22N} \simeq 0.814 \tag{6.33}$$

will change the phases of the dominant Debye terms by something close to π (mod 2π), thereby explaining the major quasiperiod in the background.

The superposition (6.31) produces further quasiperiods through mutual interference. For example, within a quasiperiod the phase difference between the $p = 11$ rainbow and $p = 2$ surface-wave terms is governed by $\Delta p = 9$, yielding a difference of $45\pi/11 \approx 4\pi$. This describes a slowly varying oscillation of two peaks within a quasiperiod (6.33), and hence a quasiperiod $\Delta\beta \simeq 0.41$.

A similar phase difference of $65\pi/11 \approx 6\pi$ between the $p = 11$ and $p = 24$ terms, relevant for larger values of β, provides a quasiperiod of $\Delta\beta \simeq 0.27$ describing three complete cycles within the main quasiperiod (6.33). The two quasiperiods $\Delta\beta = 0.27$ and 0.814 have been observed in experiments by Fahlen and Bryant (1968). (See, also, Saunders (1970)).

Returning to (6.31), there are two further quasiperiodicities that arise from the interference between $S_{j2,res}$ and $S_{j0,geo}$ ($\Delta\beta \simeq 1.1$) and between $S_{j2,res}$ and $S_{j2,geo}$ ($\Delta\beta \simeq 14$), any others being negligible. One verifies these by isolating the β-dependent phases in (6.15) and setting differences equal to 2π. One needs a very good eye, of course, to pick out all of these quasiperiods in Fig. 6.19!

Forward glories

It was noted earlier that Eq. (6.18) predicts glory rays in the forward direction for even values of s. This is also to be expected from the general discussion in Chapter 1, in which Eq. (1.18) predicts a focal singularity both in the forward direction and in the backward direction. The major observational difficulty here is the strong dominance of the forward diffraction peak, with amplitude of order β^2. Despite the axial focusing enhancement of order $\beta^{1/2}$ for a glory ray, and possible additional rainbow enhancements of order $\beta^{1/6}$, the Airy pattern still swamps everything else.

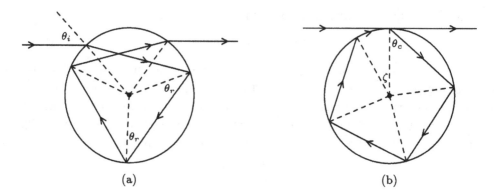

Fig. 6.21. Forward glory paths for the $p = 4$ Debye term and $n = 1.33$. (a) A real glory ray with $\theta_i \simeq 47.65°$. (b) An edge ray plus a surface wave bridging the gap of $\zeta \simeq 30°$.

To maximize the prospects for observation one should look for glory rays with a minimal number of internal reflections, the aim being to minimize damping from these as well as from absorption over an extended optical path length. This also suggests that rays incident near the edge will be optimal, since they have maximum internal reflectivity.

For water with refractive index $n = 1.33$ we find from (6.20) that the lowest-order forward glory ray corresponds to $p = 4$, $s = 2$, with an angle of incidence of $\theta_i \simeq 47.65°$. There is also a contribution from a $p = 4$ 'almost' glory ray, similar to the van de Hulst $p = 2$ term, which originates from tangential incidence. Figure 6.21 illustrates these two forward rays for $n = 1.33$, and one readily sees that the angle ζ to be made up by surface-wave propagation is given by $\zeta = 8\theta_c - 2\pi \simeq 30°$, where $\theta_c = \sin^{-1}(1/n)$ is the critical angle.

The glory-ray contribution here is certainly not great, given the relatively low angle of incidence, and a 30° gap leads to a good deal of radiation damping. Nevertheless, when the background contribution to Q_{ext} from the $p = 0$ and $p = 1$ terms is compared with that including the $p = 4$ term, as in Fig. 6.22, we indeed discern a glory contribution that is not insignificant. Thus, the $p = 4$ term should be included in (6.3) for maximum accuracy, though its detection experimentally is problematic.

An improved scenario for detection of a forward glory would be to reduce the number of internal reflections, and the smallest number for producing a glory ray is 2 ($p = 3$) when $n \geq 2$. If n is reduced to something just less than 2 so as to remain in the region of refractive indices we have been considering, then there is again a small gap to be made up by surface waves. In this

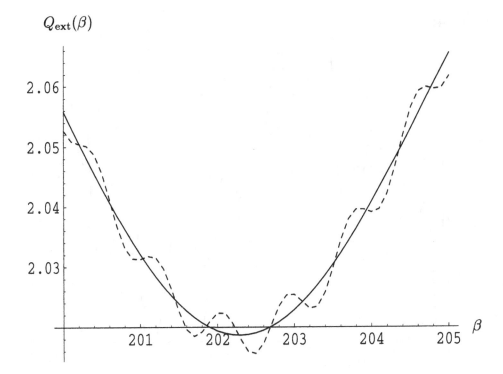

Fig. 6.22. A comparison for β near 200 of the slowly varying background contribution to $Q_{ext}(\beta)$ of the $p=0$ and $p=1$ Debye terms (——) with the similar curve that includes the $p=4$ term (- - -). There is clearly a contribution from the forward glory.

case $\zeta = 6\theta_c - \pi$ and, for example, $n = 1.9$ yields $10.5°$. Addition of a small amount of absorption dampens out most of the extraneous oscillations, so that comparison of the Mie result with the asymptotic form (6.3) in Fig. 6.23 reveals a nearly sinusoidal oscillation about the latter, rather than the irregular ripple. This, in fact, is a manifestation of the glory contribution, arising from interference between the $p=3$ term and the dominant direct reflection and transmission terms. One can expect similar contributions to the other cross sections.

Several possibilities for experimental detection of the forward glory have been proposed by Nussenzveig and Wiscombe (1980), employing liquid droplets or glass spheres with $n \lesssim 2$. Perhaps the most promising of these utilizes the strong polarization of the forward glory.

Forward glory scattering from glass spheres has been observed by Langley and Morrell (1991). In these experiments the refractive index was varied to reveal the glory oscillations in the intensity similar to those in Fig. 6.22.

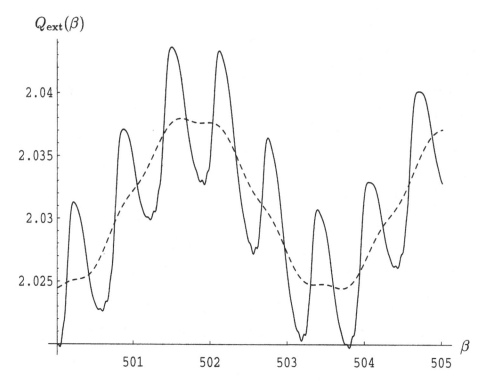

Fig. 6.23. A comparison of the exact Mie result (——) for Q_{ext} with the asymptotic form (6.3) (- - -) for $n = 1.9 + 10^{-4}i$ and β near 500. With the ripple suppressed, the forward glory oscillations in the exact curve are manifest.

These effects have also been observed in forward scattering from bubbles by Langley and Marston (1991), a subject to which we shall return in Chapter 8.

6.5 General rainbow and glory scattering

In their pioneering work on semiclassical scattering Ford and Wheeler (1959a, b) noted that rainbow and glory scattering were features to be expected in general scattering scenarios, and we discussed these singular effects briefly in Chapter 1. To emphasize the breadth of the phenomena we shall take a moment here to provide a very brief survey of their appearance in physical systems other than water droplets. This section is by no means meant to be a thorough review, but only an introduction.

Rainbow scattering

Berry (1966) first utilized the CFU method to obtain a uniform approximation for the scattering amplitude in the rainbow region. He applied the technique to a Lennard-Jones potential, and it has been applied to several other models in particle scattering; indeed, it provides the basis for the treatment in Chapter 5.

An atomic rainbow was first observed by Beck (1962) in the scattering of K from Kr and HBr. Similar results for the scattering of Na from Hg were obtained by Buck and Pauly (1971). The data exhibit a clear supernumerary structure that provides information on the range of the interatomic potential. Indeed, a major goal of these scattering experiments is to characterize in detail the atomic and molecular interactions, and it is the competition between repulsive and attractive portions of the potential that produces the rainbow. In addition to elastic collisions, rainbow scattering has also been observed in charge-transfer processes (e.g., Delvigne and Los (1973)).

The scattering of atoms and molecules from surfaces has also revealed rainbow effects in rotational and vibrational modes, generally for Lennard-Jones-type potentials. These observations are reviewed by Kleyn (1987), and in Nesković (1990). An excellent review of molecular scattering phenomena in general has been given by Buck (1974).

Rainbow effects have been predicted in heavy-ion collisions (Friedman *et al.* (1974)), and nuclear rainbow models developed. Clear evidence for a nuclear rainbow was found by Delbar *et al.* (1978) in the scattering of α particles by ^{40}Ca at 50 MeV, and modeled theoretically by McVoy *et al.* (1986). Similar results are obtained for scattering of α particles from ^{90}Zr (Put and Paans 1977), and Hussein and McVoy (1984) provide a general review of nuclear and rainbow scattering.

Glory scattering

As noted in Chapter 2, scattering of scalar waves from an impenetrable sphere can be interpreted as sound scattering from a perfectly soft sphere, and can also be generalized to an acoustically rigid sphere. Much of our previous analysis can be applied directly to sound waves, and we shall return to that in Chapter 8. Motivation to do that is provided by the observation of a backward glory by Marston *et al.* (1983) in the study of backscattering from an elastic sphere.

We have noted earlier the observation of both forward and backward glories in light scattering from glass spheres by Langley and Morrell (1991). Similarly, forward glories have been recorded in the same way in scattering from bubbles in liquids (Langley and Marston 1991). In this case the glory undulations provide an estimate of bubble size.

In particle–particle collisions the notion of glory scattering is somewhat other than what we have been discussing up to this point, in that the physical origins differ. For example, according to (1.13) and (1.14) the deflection function $\Theta = \pi$ only if the impact parameter is zero. The differential cross section, however, diverges as $\operatorname{cosec} \theta$ either in the forward or in the backward direction, and this can occur whenever the energy is greater than a local maximum in the potential. Interatomic potentials with both positive and negative components generate real forward glory paths that interfere with forward-diffracted trajectories arising from large impact parameters, thereby producing glory oscillations in the total cross section. Berry (1969) has developed a uniform approximation to the scattering amplitude in the glory region.

Such undulations in σ_{tot} have indeed been observed for scattering between noble gas atoms (van den Biesen, *et al.* 1982). Spacing of the oscillations is a measure of the strength-range product for the interaction potential (Bernstein 1966), and their total number can provide an estimate of the number of vibrational states for the di-atom (Kong *et al.* 1970; Mason *et al.* 1982).

Bryant and Jarmie (1968) have investigated the backscattering of α particles from spinless light nuclei at 18–50 MeV, as well as π^+ from protons at 2–8 GeV. The data are fitted quite well with a surface-wave model (see also, Brink (1985)), and exhibit a glory-like angular distribution. Further theoretical descriptions of the backward nuclear glory have been developed by Takigawa and Lee (1977).

Observation of a forward glory in heavy-ion nuclear scattering is as difficult as it is for light from water droplets. Rather than forward diffraction, the problem here is interference with Coulomb scattering, whose amplitude becomes highly singular in the forward direction. Nevertheless, the nuclear forward glory has been clearly detected in $^{12}C + ^{12}C$ scattering at a center-of-mass energy of 9.5 MeV (Ostrowski *et al.* 1989). Oscillations in the cross section with respect to angle provide the characteristic signature of the glory.

7

Scattering resonances

Given the way we seem to have continually ignored the ripple, one could be excused for thinking it nothing but a noisy nuisance. It is certainly ubiquitous, as seen in Fig. 3.4 for $Q_{\text{ext}}(\beta)$, in Fig. 6.4 for $Q_{\text{abs}}(\beta)$, in Fig. 6.6 for $Q_{\text{pr}}(\beta)$, and in Figs. 3.6 and 6.16 for backscattering. Far from being a nuisance, however, the ripple reflects a great deal of additional physics taking place within the transparent sphere, and at this point requires much closer scrutiny. Unless clearly stated otherwise we shall consider n to be real.

What is it that needs to be explained about the ripple? First and foremost, we should like to understand clearly the origin of the sharp, almost chaotic-looking peaks in these cross sections, not only mathematically, but also physically. In addition, the ripple structure appears to oscillate about the slowly varying background of Eq. (6.3), as seen in Fig. 6.1b. Why is this?

Ultimately the ripple must be linked to the behavior of the electromagnetic fields produced by the encounter of the incident plane wave with the sphere. What processes are taking place in terms of these fields that might lead to the ripple structure? Physics beyond the scattering mechanism emerges here, in the fields internal to the sphere, and a major goal in this chapter will be to explicate these phenomena.

7.1 A preliminary assessment

Initial interest in the ripple was stimulated by the 'spiky' nature of the above cross sections, which of course originates with the scattering amplitudes of Eqs. (3.95) and use of the optical theorem to yield (3.99b):

$$Q_{\text{ext}}(\beta) = \frac{2}{\beta^2} \sum_{\ell=1}^{\infty} (2\ell + 1) \, \text{Re}[a_\ell(\beta) + b_\ell(\beta)]. \tag{7.1}$$

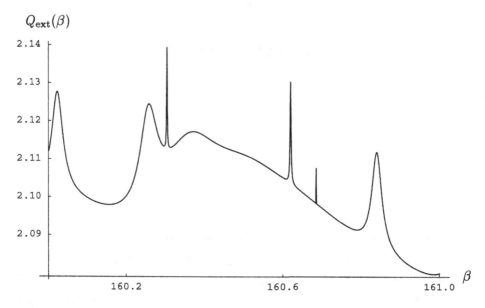

Fig. 7.1. A closeup of the extinction cross section at $n = 1.33$, illustrating very clearly the nature of the 'spikes' or ripple for β near 160.

For convenience we also rewrite here the expressions (3.88) for the partial-wave coefficients:

$$a_\ell(\beta) = \frac{\psi_\ell(\beta)\psi'_\ell(n\beta) - n\psi_\ell(n\beta)\psi'_\ell(\beta)}{\zeta^{(1)}_\ell(\beta)\psi'_\ell(n\beta) - n\psi_\ell(n\beta)\zeta^{(1)'}_\ell(\beta)}, \tag{7.2a}$$

$$b_\ell(\beta) = \frac{\psi_\ell(n\beta)\psi'_\ell(\beta) - n\psi'_\ell(n\beta)\psi_\ell(\beta)}{\zeta^{(1)'}_\ell(\beta)\psi_\ell(n\beta) - n\zeta^{(1)}_\ell(\beta)\psi'_\ell(n\beta)}. \tag{7.2b}$$

Figure 7.1 provides a closeup of $Q_{\text{ext}}(\beta)$ that nicely illustrates the structure of the ripple.

A one-to-one connection between the spikes and the coefficients a_ℓ and b_ℓ had already been suggested by Mevel (1958) and Metz and Dettmar (1963). These authors provided extensive plots to support the contention that each spike could be identified with a unique coefficient. Another 13 years passed, however, before detailed numerical confirmation of this one-to-one identification was provided by Chýlek (1976). The confirmation was extended to intervals of order $\Delta\beta \sim 10^{-7}$ by Chýlek et al. (1978a), and a detailed identification was made with the backscattering levitation experiments of Ashkin and Dziedzic (1977) at $\beta \sim 40$ (Chýlek et al. 1978b). Indeed, comparisons of measured features with Mie calculations have led to very accurate techniques for determination of size and refractive index of dielectric spheres through light scattering (Chýlek *et al.* 1983).

This early work serves to localize the origins of the spikes and indicates the need for a closer study of the structure of the partial-wave coefficients. To begin, we recall the alternative forms (3.133a) and (3.136) for a_ℓ as an example:

$$a_\ell = \frac{P_\ell^e}{P_\ell^e + iQ_\ell^e} = \frac{P_\ell^{e^2}}{P_\ell^{e^2} + Q_\ell^{e^2}} - i\frac{P_\ell^e Q_\ell^e}{P_\ell^{e^2} + Q_\ell^{e^2}}, \tag{7.3}$$

where P_ℓ^e and Q_ℓ^e are defined in (3.132). As noted in that discussion, $\mathrm{Re}\, a_\ell$ reaches its maximum value of unity when $Q_\ell^e = 0$, and its imaginary part vanishes there. According to (7.1) these must also be the local maxima in $Q_{\mathrm{ext}}(\beta)$, and thus are identifiable with the spikes. From (3.132) this criterion for a maximum provides the specific mathematical condition

$$\xi_\ell(\beta)\psi_\ell'(n\beta) - n\psi_\ell(n\beta)\xi_\ell'(\beta) = 0, \tag{7.4}$$

with a similar condition for b_ℓ obtained from the replacement $n \to n^{-1}$. We note that, for a given ℓ, this equation has infinitely many solutions at discrete values of β. While a_ℓ and b_ℓ have no poles on the real axis, the finite spikes they produce have the look of *resonances*, a characterization to be investigated further presently.

When ℓ and β are small relative to unity the quantities P_ℓ and Q_ℓ are both small in all the partial-wave coefficients, and a_ℓ and b_ℓ melt into the general background variation of $Q_{\mathrm{ext}}(\beta)$. As these variables increase a_ℓ, b_ℓ tend to sharpen up for certain values of β, and both calculation and experiment reveal that the spikes emerge in the region

$$\beta \ll \ell \ll n\beta. \tag{7.5}$$

A clear physical reason behind this condition will arise below, but its mathematical basis is readily uncovered. From the asymptotic properties of the spherical Bessel functions given in Appendix A we note that, in the region (7.5), $\psi_\ell(\beta)$ and $\psi_\ell'(\beta)$ are exponentially small, whereas $\psi_\ell(n\beta)$ and $\psi_\ell'(n\beta)$ are sinusoidal and $O(1)$; hence, P_ℓ is always small. The Neumann functions ξ_ℓ and ξ_ℓ', however, are exponentially large in the region (7.5), so that Q_ℓ is generally large. However, since $\psi_\ell(n\beta)$, $\psi_\ell'(n\beta)$ are not in phase, it is still possible for Q_ℓ to vanish at discrete values of β. Moreover, from (3.133b),

$$c_\ell = \frac{-in}{P_\ell^e + iQ_\ell^e}, \qquad d_\ell = \frac{in}{P_\ell^m + iQ_\ell^m}, \tag{7.6}$$

so that, when Q_ℓ vanishes, the smallness of P_ℓ implies that c_ℓ and d_ℓ can become *very* large, implying very large internal fields. There is clearly some physical insight to be gained here.

The foregoing analysis is only one way to examine the origin of the spikes. Although a_ℓ and b_ℓ have no poles on the *real* β-axis, they do possess singularities in the complex β-plane. From (7.2) and Appendix A the relevant conditions are

$$\frac{j'_\ell(n\beta)}{j_\ell(n\beta)} = \frac{n^2-1}{n\beta} + n\frac{h^{(1)'}_\ell(\beta)}{h^{(1)}_\ell(\beta)} \tag{7.7a}$$

for a_ℓ, and

$$n\frac{j'_\ell(n\beta)}{j_\ell(n\beta)} = \frac{h^{(1)'}_\ell(\beta)}{h^{(1)}_\ell(\beta)} \tag{7.7b}$$

for b_ℓ.

Alternatively, and equivalently, a third path is to examine the poles of the amplitudes $S^M(\lambda, \beta)$ and $S^E(\lambda, \beta)$, Eqs. (3.141), in the complex λ-plane. From (5.4), these are the roots of

$$\psi_{\lambda-1/2}(n\beta)\zeta^{(1)'}_{\lambda-1/2}(\beta) - nv_j\psi'_{\lambda-1/2}(n\beta)\zeta^{(1)}_{\lambda-1/2}(\beta) = 0, \tag{7.8}$$

with the multipole indicator v_j defined in (5.3). These roots are just the Regge poles constituting the singularities of the Debye expansion (5.27). Newton (1964) demonstrates how the poles in the two complex planes (λ and β) are related.

Equation (7.8) is actually identical to Eqs. (7.7), which in turn are equivalent to (7.4) and its companion. However, each form carries its own insights into the ripple and they complement one another. We shall examine all three views after first reviewing the connections between poles and resonances, and consider exactly what is meant by the latter.

A digression on resonances

To provide a framework and some perspective for the ripple phenomena it may be helpful to review the theory of a simple quantum-mechanical scattering problem. Although the context is quantum, the problem is still just that of scattering a scalar classical wave from an object. We follow Ohanian and Ginsburg (1974) in considering the one-dimensional potential of Fig. 7.2, which consists of an infinite hard core at the origin, an attractive rectangular well, and a repulsive rectangular barrier. This particular potential is chosen because it has most of the attributes of a central potential in three dimensions, wherein the centrifugal potential provided by angular momentum also

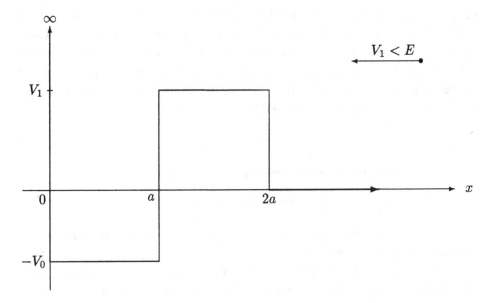

Fig. 7.2. The one-dimensional potential function of Eq. (7.9), consisting of an infinitely repulsive core at the origin, a rectangular well of depth V_0 and width a, and a rectangular barrier of height V_1 and width a.

presents a barrier. A wavefunction (or probability amplitude) describing the motion of the particle and corresponding to a stationary state of energy E must solve the stationary Schrödinger wave equation

$$\left(\frac{d^2}{dx^2} - V(x)\right)\psi(x) = E\psi(x), \tag{7.9a}$$

where

$$V(x) = \begin{cases} \infty, & x = 0, \\ -V_0, & 0 < x < a, \\ V_1, & a \leq x \leq 2a, \\ 0, & x > 2a. \end{cases} \tag{7.9b}$$

We disguise the quantum-mechanical context by employing units $\hbar^2/(2m) = 1$, $\hbar$ being Planck's constant divided by 2π, and m the mass of the particle.

Suppose for a moment that $V_0 = V_1 = 0$, so that we have an essentially free particle facing only the infinitely repulsive wall at the origin. The corresponding wavefunction satisfying (7.9a) is evidently

$$\phi(x) = \frac{1}{2i}(e^{ikx} - e^{-ikx}) = \sin(kx), \tag{7.10}$$

since we must have $\phi(0) = 0$.† The wavenumber $k = E^{1/2}$, and all time dependence is suppressed. That is, the incoming and outgoing plane waves combine to form a traveling wave with one end fixed. If $V(x) \neq 0$ for $x > 0$, but of finite range R, then the incoming wave is the same and the distortion of the outgoing wave can be described by a real phase shift δ, as discussed in Chapter 2. The wavefunction outside the potential becomes

$$\psi(x) = \frac{1}{2i}\left(e^{ikx+2i\delta} - e^{-ikx}\right), \qquad x > R. \tag{7.11}$$

One can now construct the scattered wave as the difference

$$\psi_s(x) = \psi(x) - \phi(x) = \frac{1}{2i}(e^{2i\delta} - 1)e^{ikx} = S(k)e^{ikx}, \tag{7.12}$$

defining the scattering amplitude $S(k)$, whose modulus squared measures the strength of the scattering: $\sin^2 \delta$.

An important feature of the phase shift $\delta(k)$ is the role it plays in eventually constructing a time-dependent scattered wavepacket, which we can write as

$$\psi_s(x,t) = \int_0^\infty g(k)e^{i(kx - \omega t + 2\delta)}\, dk. \tag{7.13}$$

The positive spectral function $g(k)$ is taken to have a peak at $k = k_0$, corresponding to a group velocity v_0. This peak is determined by the requirement of stationary phase at $k = k_0$, or

$$\frac{d}{dk}\left[kx + \omega t + 2\delta(k)\right]_{k=k_0} = 0, \tag{7.14}$$

which leads to the equation of motion

$$x = -v_0\left[t + 2\hbar\left(\frac{d\delta}{dE}\right)_{k=k_0}\right], \qquad v_0 \equiv \hbar k_0/m. \tag{7.15}$$

The quantity $2\hbar\, d\delta/dE$ is the *time delay* suffered by the wavepacket in crossing the region of the potential. It will be convenient here to define the *specific time delay* as

$$R^{-1}\frac{d\delta}{dk} \equiv 2\hbar\frac{d\delta}{dE}\left(\frac{\hbar k/m}{2R}\right), \tag{7.16}$$

which is just the time delay expressed in units of the transit time for the free particle to cross the distance $2R$.

Now return to the problem specified by Fig. 7.2, where $R = a$ is the range of the attractive well and $E > V_1$. The interior wavefunction is now $A\sin(k'x)$, $k' \equiv \sqrt{E + V_0}$, and the only change in the scattering solution is

---

† In this problem boundary conditions at infinity require δ-function normalization, an irrelevant detail for our purposes.

that the propagation constant for $a < x < 2a$ is $\kappa \equiv \sqrt{E - V_1}$. Constants modulating each wave in each region are determined as usual, by matching $\psi(x)$ and $\psi'(x)$ at the boundaries, conditions which determine $\delta(ka)$ and the amplitude A of the interior function. The interested reader will readily verify, after some algebra, the results

$$\delta(ka) = \frac{\pi}{2} - 2ka + \tan^{-1}\left(\frac{f_1(ka)}{f_2(ka)}\right),$$
$$A = 2ike^{-i2ka}[kf_1(ka) - i\kappa f_2(ka)], \qquad (7.17)$$

where

$$f_1(ka) = \frac{1}{2}\left(\cos(\kappa a)\sin(k'a) + \frac{k'}{\kappa}\sin(\kappa a)\cos(k'a)\right), \qquad (7.18a)$$

$$f_2(ka) = \frac{1}{2}\left(\sin(\kappa a)\sin(k'a) - \frac{k'}{\kappa}\cos(\kappa a)\cos(k'a)\right). \qquad (7.18b)$$

In Fig. 7.3 we have used these expressions to plot the relevant physical parameters for scattering when $E > V_1$, for the particular choices $V_0 = a^{-2}$ and $V_1 = 5a^{-2}$. The remarkable aspect of these plots is the event at $ka \simeq 1.8$, where: (i) the phase shift suddenly jumps by about π and passes through $-\pi/2$; (ii) the scattering strength has a sharp maximum and reaches $\sin^2\delta = 1$; (iii) the magnitude of the wave amplitude in the well possesses a sharp maximum; and, (iv) the specific time delay $a^{-1}\,d\delta/dk$ has a sharp maximum. These are the characteristics of a *resonance*, and it is necessary for *all* of them to apply to define a resonance appropriately. In particular, it is necessary that there be a sharp peak in the time delay, for it is that temporary capture of the wave or particle by the potential that is the quintessential feature of the resonance, and characterizes it also as a metastable or quasibound state. It is analogous to the classical occurrence of orbiting (Chapter 1).

A counter example, in fact, is provided by the broad peak in $\sin^2\delta$ at $ka = 1.0$. Here the specific time delay is actually negative, and δ is *not* increasing with energy. It is only a sharp increase in δ, such that it goes through $\pm\pi/2$, that constitutes a resonance at which $\sin^2\delta = 1$. A very narrow peak in $\sin^2\delta$ is also sufficient for a resonance, for that requires $|d\delta/dk|$ to be large, which can happen only if $d\delta/dk > 0$. We can see this from (7.15) which, in the units of (7.16), tells us that causality is violated unless $d\delta/dk \geq -a$; otherwise, the scattered wave left the target before the incident wave arrived.

If the barrier height is decreased, for example, the scattering will move off resonance. This entire scenario is reminiscent of the slow decay of electric current in a resonant circuit when the frequency is close to the

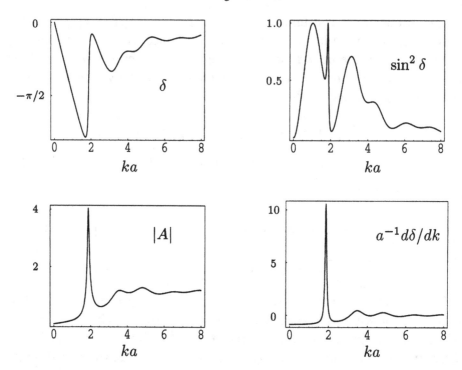

Fig. 7.3. Scattering from the one-dimensional potential of Fig. 7.2 with $V_0 = a^{-2}$ and $V_1 = 5a^{-2}$. The resonance at $ka \simeq 1.8$ is reflected in the behavior of the phase shift δ, the scattering strength $\sin^2 \delta$, the interior wave amplitude $|A|$, and the specific time delay $a^{-1} d\delta/dk$.

resonance frequency. The analogous situation in the present model arises when $0 < E < V_1$, in which case the incident wave can tunnel through the barrier, rattle around for a while inside, and tunnel back out as a decaying metastable state.

It remains to examine the scattering amplitude, or S-function, of (7.12), and hence the cross section $\sin^2 \delta$ itself. For a resonance at $k = k_0$ we know that $\delta(ka)$ increases sharply by about π near k_0, passing through $\pi/2$ (mod π) at k_0. Hence, the essential energy dependence in that neighborhood is

$$\tan \delta \simeq \frac{k_1}{k_0 - k}, \qquad 0 < k_1 \ll k_0, \qquad (7.19)$$

and one verifies that $d\delta/dk > 0$, while taking into account the multivalued nature of $\tan^{-1}$. Then the S-matrix element

$$e^{2i\delta} = \frac{1 + i \tan \delta}{1 - i \tan \delta} \simeq \frac{k - k_0 - ik_1}{k - k_0 + ik_1} \qquad (7.20)$$

possesses a pole at $k = k_0 - ik_1$, in the *lower* half-plane, and close to the real axis. The strength of the scattering is thus

$$\sin^2 \delta = \frac{1}{4}|e^{2i\delta} - 1|^2 = \frac{\frac{1}{4}\Gamma^2}{(k - k_0)^2 + \frac{1}{4}\Gamma^2}, \tag{7.21}$$

where it is customary to identify $\Gamma \equiv 2k_1$ as the full width of the resonance at half maximum when $\sin^2 \delta$ is plotted as in Fig. 7.3. Thus, the lineshape has the standard Lorentzian, or Breit–Wigner, form.

Identification of the imaginary part k_1 of the pole with the resonance width is sensible, for it is clear that the resonance becomes sharper as k_1 decreases. Because k_1 is positive the pole acts as a source of outgoing waves that diverge at infinity. Although complex energies are not physically realizable, the resonance phenomenon indicates that, at the physical energy $E_0 = \hbar^2 k_0^2/(2m)$, the wavefunction is very sensitive to small changes in the energy. If k_1 were negative the pole would appear in the upper half-plane, the wavefunction would decrease exponentially at large distances, and this would be a bound state. One readily confirms that, at resonance, $d\delta/dk = 1/k_1$, so that the narrower the resonance the longer the time delay.

7.2 Semiclassical potential analysis

With the preceding discussion as a guide we can construct an analogous picture for light scattering from a transparent sphere, much as has already been done for the impenetrable sphere in Fig. 2.11. Equation (3.80) provides a stationary 'Schrödinger' equation for each of the radial Debye potentials. In regions where n is constant we find that $\psi(r) = r\Pi$, where Π is the radial part of either Debye potential, satisfies

$$-\frac{d^2\psi(r)}{dr^2} + \left(V(r) + \frac{\ell(\ell+1)}{r^2}\right)\psi(r) = E\psi(r), \tag{7.22}$$

where $E = k^2$, $V(r) = -k^2(n^2 - 1)$, and units are again adopted such that $\hbar^2/(2m) = 1$. The parallel with the one-dimensional problem is achieved by focusing on the *effective potential*

$$U_\ell(r) = \begin{cases} k^2(1 - n^2) + \ell(\ell+1)/r^2, & r \leq a, \\ \ell(\ell+1)/r^2, & r > a, \end{cases} \tag{7.23}$$

which contains the centrifugal barrier. Although n^2 and k^2 are taken to be real and greater than unity, (7.22) is not quite the usual Schrödinger equation, for $V(r)$ depends on the energy k^2. Hence, (7.22) is strictly valid for fixed k, which must be remembered when plotting $U_\ell(r)$.

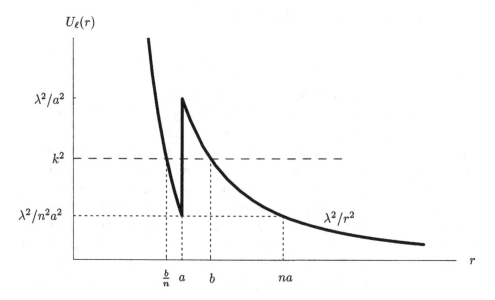

Fig. 7.4. The effective potential of Eq (7.23) for $\ell \gg 1$ and a specific value of k^2. The three radial turning points are found at $r = b/n$, a, and b.

The effective potential is depicted in Fig. 7.4 for a specific value of k^2 and $\ell \gg 1$. The local wave number

$$p_\ell \equiv \sqrt{E - U_\ell(r)} = \sqrt{k^2 n^2 - \ell(\ell+1)/r^2} \tag{7.24}$$

determines the classically forbidden regions, as well as the classical turning points,

$$r_1 = \frac{\lambda}{kn}, \qquad r_2 = \frac{\lambda}{k}, \qquad \lambda = \sqrt{\ell(\ell+1)} \simeq \ell + \tfrac{1}{2}. \tag{7.25}$$

These in turn define the top and bottom of the well:

$$\beta_T^2 = \lambda^2, \qquad \beta_B^2 = \frac{\lambda^2}{n^2}, \tag{7.26a}$$

with well depth

$$V_0 = k^2(n^2 - 1) = \frac{\lambda^2}{a^2} - \frac{\lambda^2}{n^2 a^2}. \tag{7.26b}$$

In the quantum-mechanical problem, in which $\psi(x)$ is a probability amplitude, a particle incident at energy k^2 (Fig. 7.4) can undergo *resonance penetration* (Gurney 1929) and tunnel through the classically forbidden region (barrier) into the classically allowed potential pocket. At discrete values of the energy the particle can be temporarily trapped, oscillating back and forth within the well many times before tunneling, or leaking back out. This type of time delay and sharp definite energy is just the characterization of a resonance discussed above; it is a metastable or quasibound state. More particularly, it is called a *shape resonance* (Gustafsson 1983), because its existence depends on the shape of the potential and the competition between attractive and repulsive parts (Dehmer 1984). This terminology originates in molecular scattering; in atom–atom scattering these are called *orbiting resonances*, in analogy with classical orbiting depicted in Fig. 1.5 (Toennies *et al.* 1976). In the optics literature they are often referred to as *morphology-dependent* resonances; although it is a bit extravagant sounding, the name is somewhat accurate, for they do depend on the geometry of the scatterer, and the potential pocket indeed 'morphs' with k.

Although this wave picture evolves from a quantum-mechanical analogy, the localization principle implies a direct correspondence with geometric optics and incident rays with impact parameter $b = \lambda/k$. If k^2 sits well above the barrier the ray is repelled by the inner part of the potential, the impact parameter $b < a$ is *below edge* with $\beta > \lambda = kb$, and the first refracted ray passes at a distance b/n from the center of the sphere.

Incident k^2 just at the top of the well corresponds to an *edge ray* with $b = a$ and $\lambda = \beta$; there is a single turning point at $r = a$. For incident k^2 at the bottom of the well $b = na$, $\lambda = n\beta$, and the rays are passing well clear of the sphere.

Attention is next drawn to values such that k^2 falls somewhere in the region

$$\frac{\lambda^2}{n^2 a^2} < k^2 < \frac{\lambda^2}{a^2}, \tag{7.27}$$

which is equivalent to, and therefore explains the significance of, the inequalities (7.5) for $\beta \gg 1$. There are now three turning points: $r_1 = b/n$, a, and $r_2 = b$, as indicated in Fig. 7.4 and originating from impact parameters $b = \lambda/k$. We are now in the *above-edge* region of semiclassical scattering that allows rays just outside the sphere to interact with it through diffraction – i.e., tunneling. This interpretation of diffraction as tunneling was first put forth by Nussenzveig (1989, 1992), and developed further by Johnson (1993).

As an aside, we note that the most efficient way to excite a resonance is by means of a tightly focused laser beam with the focal point positioned just above the edge (Barton *et al.* 1989). Such excitation takes place through tunneling of an incident partial wave through the centrifugal barrier, as suggested by the localization principle.

The solutions to (7.22) for the radial Debye potentials are linear combinations of the Ricatti–Bessel functions $\psi_\ell(kr)$, $\xi_\ell(kr)$ that must vanish at the origin. Matching at $r = a$, along with some algebraic reduction, yields

$$\Psi_\ell^v(r) = \begin{cases} A_\ell \left(\xi_\ell(kr) - \dfrac{Q_\ell^v}{P_\ell^v} \psi_\ell(kr) \right), & r \geq a \\ \psi_\ell(nkr), & 0 \leq r \leq a, \end{cases} \tag{7.28}$$

where $v = e, m$, and Q_ℓ^v and P_ℓ^v are defined in (3.132). Within the barrier $\Psi_\ell^v(r)$ must be exponentially decreasing, implying that $Q_\ell^e = 0$ for the TM (electric multipole) modes, and $Q_\ell^m = 0$ for TE (magnetic multipole) modes. These conditions determining the discrete 'energy levels' of a resonance are precisely those of (7.4) and its companion for b_ℓ.

A particular potential well will support J TE and J TM resonant modes, and hence J wavefunctions for each type; J is determined by β_T, Eq. (7.26a). The lowest-lying wavefunction, with k^2 near the bottom of the well, has a single peak and is the narrowest resonance since it must tunnel through the larger portion of the barrier to decay – hence, it has a longer lifetime. The next higher level has two peaks, positive and negative with one node, and is somewhat broader. The number of peaks in the classically allowed region is called the *order* number, which also corresponds to the number of peaks in $\psi_\ell(n\beta r)$ in the internal fields (see below).

As an example we consider a model with $\ell = 50$, $n = 1.5$, and units such that $a = 1.0$. The potential $U_{50}(r)$ supports four quasibound states of each type, and the TE resonances are illustrated to scale, and at their proper levels in the well, in Fig. 7.5. Note that each wavefunction, which actually defines a Debye potential, has m peaks representing its order number and $m - 1$ nodes. The width of the principal $m = 1$ resonance is about 10^{-6}, whereas that for $m = 3$ is about 0.01, as can be verified from Table 7.2 below.[†]

Figure 7.6 illustrates the behavior of the $m = 3$ wavefunction as it passes through the resonance from below. As the wave penetrates the barrier the wavefunction, and hence the field strength, increases on both sides of the spherical surface. Of particular interest are the relatively large amplitude at resonance and the phase shift of about π as ka passes through that point, much as in the scenario described in Fig. 7.3. Off resonance the amplitude is much larger outside the potential pocket. Clearly, the narrowness of these resonances implies long lifetimes and correspondingly substantial time delays.

To understand what is going on inside the sphere, consider again the scenario of Fig. 7.4, which leads to the ray description of Fig. 7.7. Upon

---

† The notation here differs from that common in the optics literature, where n is the mode or angular-momentum index and ℓ is the order number.

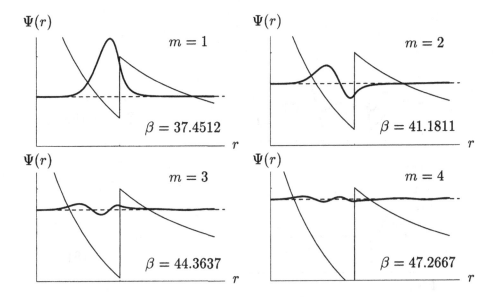

Fig. 7.5. The four TE quasibound states supported by the potential pocket of Fig. 7.4 with $\ell = 50$, $n = 1.5$, and $a = 1.0$. All are shown at their proper levels in the well.

tunneling through the barrier the ray enters the sphere and is multiply reflected in the near-surface annulus ($r = b/n$, $r = a$). These rays are almost totally internally reflected beyond the critical angle, other than small leakage at each reflection point. For the deepest, hence narrowest, resonances this leakage is small owing to the transmissivity of the barrier being small; there are thus many internal reflections with correspondingly long lifetimes. Quasibound states of light would actually be bound were there no leakage, and are the natural electromagnetic vibrational modes of the dielectric sphere found by Debye (1909b). However, they are subject to the outgoing radiation condition, so that their frequencies are always complex, yielding finite lifetimes. The analogy to semiclassical orbiting (Ford and Wheeler 1959a, b) is obvious, and presently we shall return to this interpretation in terms of the internal fields.

7.3 The complex angular-momentum plane

If we approach the resonances through Eq. (7.8) and view them as arising from the Regge poles of the Debye expansion, we are led to the notion of *Regge trajectories* discussed in connection with Fig. 2.9. In the present context we label the poles by $\lambda_{kj}(\beta)$, where j is the polarization and k is the

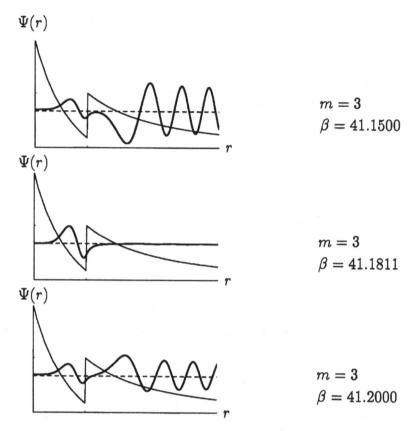

Fig. 7.6. The $\ell = 50$, $m = 3$ wavefunction as it passes through the resonance at $\beta = 41.1811$, illustrating clearly the phase shift of about π occuring on going from one side to the other.

family number. Although k plays the same role as the order m above, it here counts the number of nodes in the wavefunction, so that $k = 0, 1, 2, \ldots$.

As β varies each Regge pole $\lambda_{kj}(\beta)$ describes a trajectory in the complex λ-plane, and each trajectory is labeled by a particular family number. Thus, each trajectory passes through all resonances with the same family number, crossing a partial-wave resonance at physical values $\mathrm{Re}[\lambda_{kj}(\beta)] = \ell + \frac{1}{2}$. As usual, $\mathrm{Im}[\lambda_{kj}(\beta)]$ determines the width of the resonance, so that each trajectory provides a unified description of all resonances for a given family number.

This approach to resonances yields some additional insights. For example, their physical interpretation directly in terms of Regge poles is contained in Eq. (2.116), in which they define angular traveling waves. Here, however, the surface waves travel *inside* the spherical surface. We shall see below that this view does indeed encompass the physical character of the resonances.

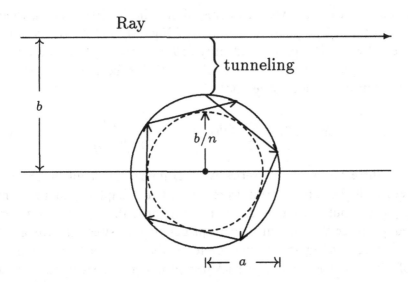

Fig. 7.7. An incident ray with impact parameter b tunnels through the centrifugal barrier to the spherical surface and is multiply reflected internally beyond the critical angle, traveling at a distance b/n from the center.

Although we shall not adopt the complex-angular-momentum scenario to locate precisely the sharp resonances of interest, there is still further information to be gleaned from this picture. Of some interest is the ability it provides us to study the contribution to the cross sections from a single pole. To pursue this, recall from Eqs. (3.99) and (3.140), and the optical theorem (3.73) that the extinction efficiency can be written

$$Q_{\text{ext}}(\beta) = \frac{1}{\beta^2} \operatorname{Re} \sum_{\ell=1}^{\infty} (2\ell + 1)[(1 - S_\ell^{\text{E}}) + (1 - S_\ell^{\text{M}})], \qquad (7.29)$$

in which we continue for the moment to consider n real. For a given multipole $j = \text{E}, \text{M}$, application of the Poisson transform provides a contribution

$$Q_{\text{ext}}^{(j)}(\beta) = \frac{2}{\beta^2} \operatorname{Re} \sum_{m=-\infty}^{\infty} e^{-im\pi} \int_0^\infty [1 - S^{(j)}(\lambda, \beta)] e^{2\pi im\lambda} \lambda \, d\lambda. \qquad (7.30)$$

Now split the m-sum and deform the contour off the real axis as in Eq. (4.74), yielding

$$Q_{\text{ext}}(\beta) = \frac{2}{\beta^2} \operatorname{Re} \left(\sum_{m=0}^{\infty} \int_0^{\infty + i\varepsilon} + \sum_{m=-1}^{-i\infty} \int_0^{\infty - i\varepsilon} \right)$$
$$\times e^{-im\pi} [1 - S^{(j)}(\lambda, \beta)] e^{2\pi im\lambda} \lambda \, d\lambda. \qquad (7.31)$$

In the present discussion we are interested only in the first integral, which is relevant to the Regge poles in region 1 of Fig. 5.1, and on the contour of this integral we can employ the identity (2.99) to perform the sum over m.

Next we introduce the Debye expansion (5.27) for the amplitude, extending it to an arbitrary number of terms P:

$$S^{(j)}(\lambda, \beta) = \sum_{p=0}^{P} S_p^{(j)}(\lambda, \beta) + \frac{S_{p+1}^{(j)}(\lambda, \beta)}{1 - R_{11}^{(j)}(\lambda, \beta)}. \qquad (7.32)$$

The first three terms of this series were examined in detail in Chapter 5, and several higher-order terms were studied in Chapter 6 in the context of the glory – but the remainder term containing the ripple was generally omitted. Displace the contour upward just enough to sweep the Regge poles between β and $n\beta$ in the remainder term, thereby acquiring $2\pi i$ times a sum of residues. The remaining background integral is essentially the same as we have always had at this stage and one proceeds as before with the leading Debye terms. For $\lambda < \beta$ outside the edge strip the contributions from $S_{p+1}^{(j)}$ are strongly damped by multiple internal reflections, so that only the residue contributions from $\lambda \geq \beta$ are significant for this term. Hence, the contribution to $Q_{\text{ext}}(\beta)$ from the Regge poles is a sum over all family numbers k, for given β, of terms

$$Q_{\text{ext}}^{(j)}(\lambda_{kj}, \beta) = -\frac{2\pi}{\beta^2} \text{Im} \left(\frac{\lambda_{kj}(\beta) e^{-i\pi \lambda_{kj}(\beta)}}{\cos[\pi \lambda_{kj}(\beta)]} \text{Res} \left[S^{(j)}(\lambda, \beta) \right]_{\lambda = \lambda_{kj}} \right), \qquad (7.33)$$

where Res[ ] denotes the residue of the entire amplitude at the pole. This form of the residue can be used because no terms in (7.32) prior to the remainder contain these poles, so that the residue formula (5.62b) implies that they make no contribution at the poles.

Since $\lambda \geq \beta \gg 1$, the contributions from (7.33) are $O(\beta^{-1})$, becoming less important relative to the geometric-optics limit of 2 as β increases. However near the physical values $\lambda = \ell + \frac{1}{2} + i\epsilon$, $\epsilon \ll 1$, the cosine term in the denominator generates a resonance peak with the well-known Lorentzian (Breit–Wigner) shape.

To examine an individual resonance (7.33) in detail it is first necessary to compute the Regge poles as functions of β for $\beta \gg 1$, and then evaluate the residues of the S-function, none of which is entirely simple. For $|\lambda| - \beta \gg \beta^{1/3}$, one employs the asymptotic expressions of Appendix C for the cylinder functions in (7.8), and writes $\lambda_{kj} \equiv \xi_{kj} + i\eta_{kj}$. Then, under the condition $|\eta_{kj}/\xi_{kj}| \ll 1$ for a well-defined resonance, considerable algebra yields the following real transcendental equation for ξ_{kj} (Nussenzveig 1969a;

Khare 1975):

$$(n^2\beta^2 - \xi_{kj})^{1/2} - \xi_{kj}\cos^{-1}[\xi_{kj}/(n\beta)] \approx k\pi + \frac{\pi}{4} + \tan^{-1}\left(\frac{1}{v_j}\frac{(\xi_{kj}^2 - \beta^2)^{1/2}}{(n^2\beta^2 - \xi_{kj}^2)^{1/2}}\right),$$

$$(7.34a)$$

and the imaginary part is

$$\eta_{kj} \approx \frac{2v_j[(n^2\beta^2 - \xi_{kj}^2)(\xi_{kj}^2 - \beta^2)]^{1/2}}{[\beta^2(n^2v_j^2 - 1) + \xi_{kj}^2(1 - v_j^2)]\cos^{-1}(\xi_{kj}/n\beta)}$$

$$\times\left(\frac{\xi_{kj}}{\beta} + \frac{(\xi_{kj}^2 - \beta^2)^{1/2}}{\beta}\right)^{-2\xi_{kj}}\exp(2\sqrt{\xi_{kj}^2 - \beta^2}). \qquad (7.34b)$$

These are now solved numerically to obtain very good approximations for $\xi_{kj}(\beta)$ and $\eta_{kj}(\beta)$.

The residues of the amplitude functions at the simple Regge poles are found by applying (5.62b) to (3.141) in the form

$$S^{(j)}(\lambda, \beta) = -\frac{D^*}{D}, \qquad (7.35)$$

where $D(\lambda, \beta)$ is given by the left-hand side of (7.8). Now express the spherical functions in terms of the cylinder functions and introduce the Schöbe expansions (C.17) to any desired order, so that $D(\lambda, \beta)$ is a series of products of Airy functions and their derivatives. The series is arranged in powers of $\gamma^2 = (2/\beta)^{2/3}$. Since $D(\lambda_{kj}, \beta) = 0$, the first term in its Taylor expansion around a pole is $\partial_\lambda D(\lambda_{kj}, \beta)(\lambda - \lambda_{kj})$, and hence the first derivative is all that contributes to the residue. Then, for real n, the residues are given by

$$\text{Res}[S^{(j)}(\lambda, \beta)]_{\lambda=\lambda_{kj}} = -\frac{D^*(\lambda_{kj})}{\gamma\dot{D}(\lambda_{kj}, \beta)}, \qquad (7.36)$$

where the superposed dot denotes differentiation with respect to λ. The pleasure of actually writing out the expressions for D and $\dot{D}$ is left to the reader, but the leading-order approximations are reasonably good.

A specific application of this approach has been provided by Guimarães and Nussenzveig (1992), who construct the Regge trajectories for $k = 0$ and 1, with $n = 1.33$ and $50 < \beta < 60$. In addition to demonstrating an excellent fit to the Mie expression for $Q_{\text{ext}}(\beta)$, they also exhibit the contribution from the $j = 1$, $k = 2$ trajectory to the TE resonance b_{61}^3 at $\ell = 61$, $\beta \simeq 58.776$. (One uses $m = 3$ rather than $k = 2$ in this notation, for it is more common). We compute this as well from (7.33) to confirm a remarkable feature of the Regge trajectories, illustrated in Fig. 7.8.

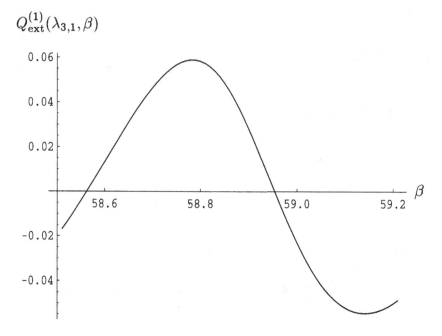

$Q_{\text{ext}}^{(1)}(\lambda_{3,1},\beta)$

Fig. 7.8. The resonant contribution to the extinction cross section from the TE resonance at $\beta \simeq 58.776$, for $\ell = 61$, $m = 3$, and $n = 1.33$. Antiresonant behavior is indicated by negative values.

Unlike the contribution to $Q_{\text{ext}}(\beta)$ from a single resonant partial wave, which is always positive, that from a Regge trajectory can turn *negative* between successive physical points, resulting in an 'antiresonant' drop of the efficiency below the background. This implies that cancelation between positive and negative contributions can arise from the phase factor in (7.29), which now explains why the background that excludes the ripple appears to provide the average about which the ripple oscillates, as in Fig. 6.1b. One might have expected this in any case from the discussion leading to Eq. (4.20). According to the associated footnote, the partial-wave coefficients tend to attain their average values of $\frac{1}{2}$ in the geometric-optics region and build a background at that level.

Guimarães and Nussenzveig have also fitted the backscattered gain in the above model while distinguishing the background from the Regge-pole contributions. The result is that, while the resonances alone make a substantial contribution to the backscattering, it is by no means a greatly dominant one. This was already evident in our study of the glory in Chapter 6.

7.4 Properties of the resonances

Despite its evident appeal, the λ-plane is not particularly well suited for a detailed study of the resonances. Rather, we return to the β-plane and first study the complex solutions to Eqs. (7.7) for fixed λ, in what amounts to a WKB approximation. These solutions are poles of the S-functions in the lower half-plane,

$$\beta_{\ell,m,j} = \beta'_{\ell,m,j} - i\beta''_{\ell,m,j}, \tag{7.37}$$

as suggested from the earlier study of the quantum-mechanical rectangular well (and as confirmed below). The real physical part β' locates the resonance on the real β-axis, while the imaginary part is $\beta'' = \Gamma/2$, one-half the full width at half maximum, or the half-width. For large β the principal resonances ($m = 1$) are close to the real axis and very sharp, leading to high Q factors:

$$Q_{\ell,m,j} \equiv \frac{\beta'_{\ell,m,j}}{2\beta''_{\ell,m,j}}, \tag{7.38}$$

which is effectively the stored energy divided by the energy lost per cycle. Locating a sharp resonance within its width can be a tricky numerical (as well as experimental) problem and will be addressed presently. For the moment it will suffice to investigate only the approximate behavior so as to appreciate the general properties. Accurate numerical computations of the complex poles in the β-plane were first carried out by Conwell *et al.* (1984).

Proceeding in much the same manner as in the preceding section, we substitute the Debye asymptotic expansions (C.2)–(C.5) into (7.7), obtaining expressions similar to those of (2.137)–(2.139). From the expansion of $J_\lambda(n\beta)$ and its derivatives we are led to define functions (Guimarães and Nussenzveig 1994)

$$\phi(\lambda, n\beta) \equiv (n^2\beta^2 - \lambda^2)^{1/2} - \lambda \cos^{-1}[\lambda/(n\beta)]$$
$$= \int_{\lambda/nk}^{a} (n^2k^2 - \lambda^2/r^2)^{1/2} \, dr, \tag{7.39a}$$

$$\psi(\lambda, \beta) \equiv -(\lambda^2 - \beta^2)^{1/2} + \lambda \cosh^{-1}(\lambda/\beta)$$
$$= -\int_{a}^{\lambda/k} (\lambda^2/r^2 - k^2)^{1/2} \, dr, \tag{7.39b}$$

which suggest the further convenient definitions

$$\delta \equiv (\lambda^2 - \beta^2)^{1/2}, \qquad \zeta \equiv (n^2\beta^2 - \lambda^2)^{1/2}. \tag{7.40}$$

Substitution of the relevant expansions for $\lambda \gg 1$ leads once more to a transcendental equation determining β', now for physical $\lambda = \ell + \frac{1}{2}$:

$$\tan\left(\phi(\lambda, n\beta') - \frac{\pi}{4}\right) \simeq \frac{\delta(\lambda, \beta')}{v_j \zeta(\lambda, \beta')} [1 - \cdots]. \qquad (7.41)$$

For our purposes it is acceptable to omit the series of higher-order terms in square brackets in (7.41). The imaginary part of the pole is then found from

$$\beta'' \simeq \frac{\beta' \delta(\lambda, \beta')}{v_j \zeta^2(\lambda, \beta')} \cos^2\left(\phi - \frac{\pi}{4}\right) e^{2\psi(\lambda, \beta')} [1 + \cdots]. \qquad (7.42)$$

In a first approximation one can set to zero the right-hand side of (7.41) and write

$$\tan\left(\phi - \frac{\pi}{4}\right) \approx 0 \longrightarrow \phi(\lambda, n\beta) = (m - \tfrac{3}{4})\pi, \qquad m = 1, 2, 3, \ldots, \qquad (7.43)$$

which is reminiscent of a Bohr–Sommerfeld quantization condition, but for the bound states of light. Expansion of ϕ in powers of $(\lambda - \beta)/\lambda$ then yields for the ℓth mode and mth order number

$$\beta'_{\ell, m, j} \approx \lambda \frac{\tan^{-1}(N)}{N} + (m - \tfrac{3}{4})\frac{\pi}{N}, \qquad N \equiv \sqrt{n^2 - 1}, \qquad (7.44)$$

where $\tan^{-1}$ is the principal value. This can now be used to iterate in (7.41) to obtain more accurate values. A polarization-independent expression of this type was first obtained by Probert-Jones (1984) and is most effectively employed near the top of the barrier for broad resonances; it turns out to be inadequate for localizing very narrow resonances. Nevertheless, it is good enough to tell us, for example, that the resonances shift to smaller β as the index of refraction increases.

The result (7.44) can now be used in (7.42) to obtain $\beta''_{\ell, m, j}$. With (7.39b), we see that this is essentially the barrier penetration factor of Eq. (2.137).

An estimate for the spacing between resonances of adjacent partial waves $(\ell, \ell + 1)$ in this approximation follows from the constancy of ϕ in (7.43). Since the total derivative vanishes we have

$$\Delta\beta = -\frac{\partial\phi/\partial\lambda}{n\, \partial\phi/\partial(n\beta)}, \qquad (7.45)$$

where we have noted that $\Delta\beta = \Delta(n\beta)/n$ and $\Delta\lambda = 1$. Then, from (7.39a),

$$\Delta_\ell\beta \equiv \beta'_{\ell+1, m, j} - \beta'_{\ell, m, j} \approx \frac{\beta}{\zeta} \tan^{-1}\left(\frac{\zeta}{\lambda}\right) \xrightarrow[\beta' \to \lambda]{} \frac{\tan^{-1}(N)}{N}. \qquad (7.46)$$

Both expressions had been reported earlier by Chýlek (1976) and Chýlek *et al.* (1990).

Similarly, the spacing between the corresponding resonances for different polarizations can be found by considering the next approximation for $\tan(\phi - \pi/4)$ in (7.41). Thus,

$$\Delta_j\beta \equiv \beta'_{\ell,m,2} - \beta'_{\ell,m,1} \approx \frac{\beta'_{\ell,m}}{\zeta(\beta'_{l,m})} \left[\tan^{-1}\left(n^2\frac{\delta}{\zeta}\right) - \tan^{-1}\left(\frac{\delta}{\zeta}\right) \right]_{\ell,m}, \qquad (7.47)$$

where $\beta'_{\ell,m} \equiv \frac{1}{2}(\beta'_{\ell,m,1} + \beta'_{\ell,m,2})$. Since the right-hand side is positive it follows that TM (a_ℓ) resonances always follow TE (b_ℓ).

In second approximation (7.41) indicates that $\phi - (m-\frac{3}{4})\pi - \tan^{-1}[\delta/(\zeta v_j)]$ is constant. A procedure similar to that yielding (7.45) then produces estimates for the spacing in order,

$$\Delta_m\beta' \equiv \beta'_{\ell,m+1,j} - \beta'_{\ell,m,j} \approx \frac{\pi\beta'_{\ell,m,j}}{\zeta(\lambda, \beta'_{\ell,m,j})}, \qquad (7.48)$$

and in refractive index,

$$\Delta_n\beta' \approx -\beta'\frac{\Delta n}{n}\left(1 - \frac{v_j\beta'^2}{\delta(\delta^2 + v_j^2\zeta^2)}\right). \qquad (7.49)$$

The effect of absorption in the form of a complex index of refraction $n = m + i\kappa$ is apparent in Fig. 6.2: as κ increases the resonances broaden and weaken until they disappear into the background.

At the top of the well $\beta' \sim \lambda$, so that (7.44) provides an approximation for the maximum order within the well:

$$m_{\max} \approx \frac{\lambda}{\pi}(N - \tan^{-1}N) + \frac{3}{4}. \qquad (7.50)$$

Therefore the number identifiable before melting into the background increases linearly with β. Hill and Benner (1986) have demonstrated numerically that the *density* of resonances (the number per unit size-parameter interval with widths less than 0.6) located below the top of the barrier increases linearly with β as well. This can be verified by integrating λ over $(\beta, n\beta)$ in (7.50) and differentiating with respect to β to obtain an approximation to the density.

For the narrower resonances we have $\beta' \simeq \lambda(1 - \epsilon)$, $\epsilon \ll 1$. From (7.39) and (7.42) we obtain a rough estimate of the Q factor,

$$Q \propto e^{\lambda\sqrt{2\epsilon}}, \qquad (7.51)$$

which increases exponentially with β. We can thus expect very high Q factors for large size parameters.

Numerical evaluations

For applications the most important property of the resonances is their precise location, without which all other parameters remain semi-quantitative; and the narrow resonances must be located with high accuracy, within their widths. In this respect we find that the most efficient procedure is to employ the real axis condition (7.4).

Recall that Eq. (7.44) followed from application of the Debye asymptotic expansions and their counterparts in the transition region $\beta \sim \lambda$, which seems like a natural procedure for the broad resonances near the top of the barrier. For the narrowest resonances, however, the Schöbe expansions of (C.17) should be employed for the Bessel functions of argument $n\beta$, and this has been done by Lam *et al.* (1992). They find that, for $\beta \gg 1$,

$$n\beta'_{\ell,m,j} \simeq \lambda + \frac{x_m^{1/3}}{2^{1/3}} - \frac{v_j}{N} + \left(\frac{3}{10}2^{-2/3}\right)\frac{x_m^2}{\lambda^{1/3}} - \frac{v_j(n^2 - 2v_j^{2/3})}{2^{1/3}N^3}\frac{x_m}{\lambda^{2/3}} + O(\lambda^{-1}), \quad (7.52)$$

where again x_m is defined by $\mathrm{Ai}(-x_m) = 0$. A similar expression had also been found by Schiller and Byer (1991), which was subsequently developed to higher order by Schiller (1993). It is the latter result that we use as a seed to a numerical root finder, resulting in very accurate resonance locations. Again, for the broader resonances (7.44) appears to be more accurate.

From (7.48) we immediately obtain an expression for the spacing in size parameter:

$$n\Delta_\ell\beta \simeq 1 + \frac{2^{-1/3}}{3}x_m\lambda^{-2/3} - \frac{2^{-2/3}}{10}x_m^2\lambda^{-4/3}$$
$$\times \left(\frac{2^{2/3}}{3}\frac{v_j(n^2 - 2v_j^2/3)}{N^{4/3}} - \frac{2^{-1/3}}{9}\right)\frac{x_m}{\lambda^{5/3}} + O(\lambda^{-2}), \quad (7.53)$$

which is somewhat more accurate than (7.46). If we employ just the first two terms on the right-hand side of (7.49) and take $\ell = 100$, $m = 4$, and $n = 1.33$, we find that

$$\Delta\beta_{100,1,j} \simeq 0.814. \quad (7.54)$$

This corresponds to the main quasiperiod (6.33) in the backscattered intensity. In Fig. 6.18, say, the sharpest resonances apparently are not resolved. Indeed, in this case we should be using (7.42), which yields $\Delta_\ell\beta \simeq 0.821$. More accurate spacing can be obtained, of course, by employing the roots determined numerically.

To find the widths of the resonances it is instructive to return to Eqs. (7.7) and examine the complex roots in the β-plane. Using the TE mode as an example, we rewrite (7.7b) as

$$j_\ell(nz)h_\ell(z)M_\ell(z) = 0, \tag{7.55}$$

with $z = \beta' - i\beta''$, and

$$M_\ell(z) = \frac{h_\ell^{(1)'}(z)}{h_\ell^{(1)}(z)} - n\frac{j_\ell'(z)}{j_\ell(nz)}. \tag{7.56}$$

The resonance therefore occurs at $M_\ell(z_0) = 0$, $z_0 = \beta_0' - i\beta_0''$.

For any real size parameter β on the real axis but near the pole z_0 a Taylor expansion yields

$$M_\ell(\beta) \simeq (\beta - z_0)M_\ell'(z_0). \tag{7.57a}$$

However, use of the differential equation (A.1) for the spherical Bessel functions shows that, at the pole, $M_\ell'(z_0) = n^2 - 1$ (modulo some algebra), and hence

$$M_\ell(\beta) \simeq (\beta - z_0)(n^2 - 1). \tag{7.57b}$$

For $\ell > \beta \gg 1$ Appendix A confirms that $|j_\ell(\beta)| \ll |y_\ell(\beta)|$, so that in (7.56)

$$\frac{h_\ell^{(1)'}(z)}{h_\ell^{(1)}(z)} \simeq \frac{y_\ell'(z)}{y_\ell(z)}\left(1 + i\frac{j_\ell(z)y_\ell'(z) - j_\ell'(z)y_\ell(z)}{y_\ell(z)y_\ell'(z)}\right) = \frac{y_\ell'(z)}{y_\ell(z)}\left(1 + \frac{i}{z^2 y_\ell(z)y_\ell'(z)}\right), \tag{7.58}$$

where we have employed the Wronskian of Eq. (A.5a).

Now, nothing in the preceding argument requires the refractive index to be real, so we can just as easily write $n = m + i\kappa$ everywhere. Thus, from (7.56)–(7.58)

$$M_\ell(\beta') \simeq -\left[\frac{y_{\ell+1}(\beta')}{y_\ell(\beta')} - \mathrm{Re}\left(\frac{nj_{\ell+1}(n\beta')}{j_\ell(n\beta')}\right)\right] + i\left[\frac{1}{\beta'^2 y_\ell^2(\beta')} + \mathrm{Im}\left(\frac{nj_{\ell+1}(n\beta')}{j_\ell(n\beta')}\right)\right]$$

$$= i\beta'(m^2 - \kappa^2 - 1). \tag{7.59}$$

For n real $\mathrm{Re}[M_\ell(\beta')]$ satisfies the condition (7.7b), and the imaginary part gives the full width at half maximum:

$$\Gamma_{\ell,m,1} = \frac{2}{m^2 - \kappa^2 - 1}\left[\frac{1}{\beta_{\ell,m,1}'^2 y_\ell^2(\beta_{\ell,m,1}')} + \mathrm{Im}\left(\frac{nj_{\ell+1}(n\beta_{\ell,m,1}')}{j_\ell(n\beta_{\ell,m,1}')}\right)\right]$$

$$\xrightarrow[\kappa\to 0]{} \frac{2}{(n^2 - 1)\beta_{\ell,m,1}'^2 y_\ell^2(\beta_{\ell,m,1}')}, \tag{7.60}$$

whose accuracy is essentially that of $\beta_{\ell,m,1}'$. The reader should verify that the corresponding result for TM modes is obtained through the following replacements:

$$\beta'_{\ell,m,1} \to \beta'_{\ell,m,2}, \qquad n^2 - 1 \to (n^2 - 1)\left[\frac{\lambda^2}{\beta'^2_{\ell,m,2}}\left(1 + \frac{1}{n^2}\right) - 1\right]. \qquad (7.61)$$

Note that the associated internal TE partial-wave coefficients in (7.6) can be written as

$$d_\ell(\beta) = \frac{i/\beta^2}{h_\ell^{(1)}(\beta)j_\ell(n\beta)M_\ell(\beta)}, \qquad (7.62)$$

and near a resonance

$$d_\ell(\beta) \simeq \frac{y_\ell(\beta')}{j_\ell(n\beta')}\frac{\Gamma_\ell/2}{(\beta - \beta') + i\Gamma_\ell/2}, \qquad (7.63)$$

where near the resonance we have replaced β by β' in the spherical Bessel functions. This expression provides a measure of the strength of the resonance in the internal fields, as well as exhibiting the characteristic Lorentzian line shape. A similar form can be found for the TM coefficient c_ℓ.

Because the internal coefficients at resonance are very much larger than the scattering coefficients, the resonance strengths of the latter are correspondingly different. A similar lineshape for the latter can be developed by referring to the form (7.3), for example, and rewriting it as

$$ia_\ell = \frac{P_\ell^e}{Q_\ell^e - iP_\ell^e} = \frac{|P_\ell^e|}{G_\ell^e - i|P_\ell^e|}, \qquad (7.64)$$

where $G_\ell^e \equiv Q_\ell^e P_\ell^{e*}/|P_\ell^e|$. This has the merit of enfolding all of the complex behavior into the function $G_\ell^e(\beta)$ if we allow n to be complex. The real part of G_ℓ^e vanishes at a resonance, so we replace it near there with the first nonzero term in its Taylor expansion. Some algebra then reduces (7.64) to

$$a_\ell \simeq i\frac{C\Gamma_\ell/2}{(\beta - \beta'_\ell) + i\Gamma/2}, \qquad (7.65)$$

where

$$\frac{\Gamma_\ell}{2} \equiv \frac{|P_\ell^e| - \mathrm{Im}\,G_\ell^e}{[-d\mathrm{Re}G_\ell^e/d\beta]_{\beta=\beta'}}, \qquad (7.66a)$$

and

$$C \equiv \frac{|P_\ell^e|}{|P_\ell^e| - \mathrm{Im}\,G_\ell^e} \xrightarrow[n\text{ real}]{} 1 \qquad (7.66b)$$

is the strength, or *coupling*. The presence of absorption leads to a considerable reduction in C.

Table 7.1. *Resonance locations, widths, and strengths for selected TE modes,*
n = 1.45

ℓ	$\beta'_{\ell,1,1}$	$\Gamma(\kappa = 0)/\beta'$	$\Gamma(\kappa = 10^{-8})/\beta'$	C
50	38.706	2.133×10^{-7}	2.147×10^{-7}	9.938×10^{-1}
100	74.406	7.881×10^{-15}	1.360×10^{-9}	5.794×10^{-6}
200	144.902	2.421×10^{-30}	1.370×10^{-9}	1.767×10^{-21}
300	214.942	3.455×10^{-46}	1.372×10^{-9}	2.516×10^{-37}
400	284.764	3.391×10^{-62}	1.375×10^{-9}	2.467×10^{-53}
500	354.453	2.641×10^{-78}	1.376×10^{-9}	1.920×10^{-69}

At this point it is useful to examine the perspective provided by some typical numbers generated from numerical computation. In Table 7.1 we exhibit a computational sampling of principal resonance locations and widths over a broad range of TE mode numbers ℓ, obtained using the above procedures. Although the locations have been determined to many more significant figures, they have been rounded off here for convenience of tabulation. The third column lists the resonance widths for real index of refraction $n = 1.45$, indicating the extreme narrowness attained for large size parameters. Clearly, it is out of the question to locate sharp principal resonances within their widths for size parameters exceeding a few hundred, either theoretically or experimentally. There is, however, always some absorption present, and the fourth column gives the widths for $n = 1.45 + i10^{-8}$, while the fifth exhibits the reduced strengths, which are no longer unity. The locations are essentially unaffected by this tiny amount of absorption, but we see that the widths have broadened considerably. Thus, the experimental possibilities for observing high-Q fields become much more promising, as will be discussed further below.

Table 7.2 tabulates representative results for the principal resonance corresponding to $\ell = 50$, $n = 1.5 + i10^{-8}$ studied earlier, along with the first four secondary resonances and their widths. The rapid broadening with increasing order is evident.

Finally, we note that Guimarães and Nussenzveig (1994) have developed a uniform approximation scheme for locating the resonances in the complex β-plane. For β not too large, these algorithms allow for determination of β' within the width β'' with very small error. Ultimately, however, one must resort to numerical techniques for extracting complex roots of transcendental equations. There are excellent routines available for this, along with increas-

Table 7.2. *Locations and widths for low-order TE resonances, l = 50 and*
n = 1.45 + *i*10^{-8}

m	1	2	3	4
$\beta'_{50,m,1}$	37.4512	41.1811	44.3637	47.2667
$\Gamma_{50,m,1}$	8.8675 × 10^{-7}	2.8013 × 10^{-4}	1.1048 × 10^{-2}	1.0164 × 10^{-1}

ingly fast processors, so that one may as well attack the exact equations directly. Some of the computational issues are addressed in Appendix F.

7.5 Internal and near fields

It should be abundantly clear at this point that the quintessential physical attribute of the resonances, shared by their varied manifestations, is some kind of capture and time delay. For electromagnetic scattering from a spherically symmetric, nonabsorbing target, the time delay has been studied by Nussenzveig (1997) as an average over all scattering angles and the incident energy spectrum. This average time delay is most effectively expressed as a spectral average over the incident wavepacket of the *spectral time delay* $\Delta\tau(\omega)$, $\omega \equiv kc$. In particular, for a dielectric sphere of radius *a* Nussenzveig shows that this quantity has the partial-wave expansion

$$\Delta\tau(\omega) = \frac{a}{\beta c}\sum_{\ell=1}^{\infty}(2\ell+1)\left(\frac{d\delta_\ell^m}{d\beta} + \frac{d\delta_\ell^e}{d\beta}\right), \qquad (7.67)$$

where the phase shifts δ_ℓ^m and δ_ℓ^e are defined in (3.134). With this result one can in principle compute the time delays for individual partial-wave resonances.

These remarks suggest that in the case of electromagnetic resonances of a dielectric sphere the principal physical features to be understood are the behaviors of the near and internal fields. The former are given explicitly by Eqs. (3.90), whereas the latter are obtained from these by making the replacements noted following (3.91):

$$E_r = iE_0\cos\phi\sum_\ell i^\ell(2\ell+1)c_\ell(\beta)\frac{\psi_\ell(Kr)}{(Kr)^2}P_\ell^1(\cos\theta),$$

$$E_\theta = -iE_0\cos\phi\sum_\ell \varepsilon_\ell\left(c_\ell(\beta)\frac{\psi_\ell'(Kr)}{Kr}\tau_\ell(\cos\theta) + d_\ell(\beta)\frac{i\psi_\ell(Kr)}{Kr}\pi_\ell(\cos\theta)\right),$$

$$E_\phi = iE_0\sin\phi\sum_\ell \varepsilon_\ell\left(c_\ell(\beta)\frac{\psi_\ell'(Kr)}{Kr}\pi_\ell(\cos\theta) + d_\ell(\beta)\frac{i\psi_\ell(Kr)}{Kr}\tau_\ell(\cos\theta)\right); \quad (7.68a)$$

$$H_r = iE_0 \sin\phi \sum_\ell i^\ell (2\ell + 1) d_\ell(\beta) \frac{\psi_\ell(Kr)}{(Kr)^2} P_\ell^1(\cos\theta),$$

$$H_\theta = -iE_0 n \sin\phi \sum_\ell \varepsilon_\ell \left(c_\ell(\beta) \frac{i\psi_\ell(Kr)}{Kr} \pi_\ell(\cos\theta) + d_\ell(\beta) \frac{\psi_\ell'(Kr)}{Kr} \tau_\ell(\cos\theta) \right),$$

$$H_\phi = -iE_0 n \cos\phi \sum_\ell \varepsilon_\ell \left(c_\ell(\beta) \frac{i\psi_\ell(Kr)}{Kr} \tau_\ell(\cos\theta) + d_\ell(\beta) \frac{\psi_\ell'(Kr)}{Kr} \pi_\ell(\cos\theta) \right),$$

$$(7.68b)$$

where $K \equiv nk$ and c_ℓ and d_ℓ are given in (7.6). Inside the sphere the refractive index n is taken to be that of the dielectric alone.

We see that the internal fields are governed, aside from the angular dependence, by c_ℓ or d_ℓ, and by a Ricatti–Bessel function or its derivative,

$$\psi_\ell(nkr) = \psi_\ell(n\beta\rho), \qquad \rho \equiv r/a. \qquad (7.69)$$

Thus, when $Q_\ell^{e,m} = 0$, as at a resonance, these fields can be very large; this will be made explicit presently. The internal fields are still reasonably small until ρ approaches $\ell/(n\beta')$, at which point one of them increases rapidly, the other remaining small. This oscillating field propagates outward while inducing the other, and as the spherical surface is approached the two fields combine to form a spherical electromagnetic standing wave. At the point $\rho = \ell/n\beta'$ we have $2\pi a = \ell(\lambda/n)$, where here λ is the wavelength, so that precisely a resonant-mode number of wavelengths fits around the surface of the sphere.

Quantitatively, consider as an example a TM resonance in a_ℓ, so that $|d_\ell| \ll |c_\ell|$. As $n\beta\rho$ exceeds ℓ the Ricatti–Bessel functions become oscillatory, as in (A.16a):

$$\psi_\ell(n\beta\rho) \approx \cos[n\beta\rho - (\ell+1)\pi/2],$$
$$\psi_\ell'(n\beta\rho) \approx -\sin[n\beta\rho - (\ell+1)\pi/2],$$
$$\psi_\ell''(n\beta\rho) \approx -\psi_\ell(n\beta\rho). \qquad (7.70)$$

Combination of these approximations with the differential equation (A.18) yields

$$\frac{\psi_\ell(n\beta\rho)}{(n\beta\rho)^2} = \frac{1}{\ell(\ell+1)}[\psi_\ell''(n\beta\rho) + \psi_\ell(n\beta\rho)] \approx 0, \qquad (7.71)$$

asymptotically. In addition, for $\ell \sim n\beta\rho$ we find that ψ_ℓ dominates ψ_ℓ'. On applying these results to (7.64), we see that the radial components of the internal fields vanish in this approximation, and, for the dominant resonance mode, the remaining components are

$$E_\theta \approx \frac{iE_0}{2} \cos\phi \, \frac{2\ell+1}{\ell(\ell+1)} \frac{c_\ell \tau_\ell}{n\beta\rho} \left[e^{in\beta\rho} + (-1)^\ell e^{-in\beta\rho} \right],$$

$$E_\phi \approx \frac{-iE_0}{2} \sin\phi \, \frac{2\ell+1}{\ell(\ell+1)} \frac{c_\ell \pi_\ell}{n\beta\rho} \left[e^{in\beta\rho} + (-1)^\ell e^{-in\beta\rho} \right], \qquad (7.72a)$$

$$H_\theta \approx \frac{-iE_0}{2} n \sin\phi \, \frac{2\ell+1}{\ell(\ell+1)} \frac{c_\ell \pi_\ell}{n\beta\rho} \left[e^{in\beta\rho} - (-1)^\ell e^{-in\beta\rho} \right],$$

$$H_\phi \approx \frac{-iE_0}{2} n \cos\phi \, \frac{2\ell+1}{\ell(\ell+1)} \frac{c_\ell \tau_\ell}{n\beta\rho} \left[e^{in\beta\rho} - (-1)^\ell e^{-in\beta\rho} \right]. \qquad (7.72b)$$

Hence, for the outgoing wave

$$e^{in\beta\rho}: \qquad \frac{E_\theta}{H_\phi} = -\frac{E_\phi}{H_\theta} = \frac{1}{n}, \qquad (7.73a)$$

and for the incoming wave

$$e^{-in\beta\rho}: \qquad -\frac{E_\theta}{H_\phi} = \frac{E_\phi}{H_\theta} = \frac{1}{n}, \qquad (7.73b)$$

so that we have simple electromagnetic standing waves with ℓ nodes around the surface.

For the very narrow low-order resonances these high-intensity waves have often been called 'whispering gallery modes', after the analogous acoustic effect in curved rooms or galleries. The most famous of these is the walkway below the dome of St Paul's cathedral in London: if a person speaks softly close to and along the wall, a listener anywhere else along the wall hears the speech almost perfectly. A proper explanation takes much the form suggested in Fig. 7.7 and was put forth by Rayleigh in 1878 (e.g., Walker (1978)).

To describe the behavior of the fields at resonance more precisely it is necessary to return to the source function $S(r)$, defined in Eq. (3.108) as the angle-averaged relative electric intensity. Consider a TE mode near resonance, for example, so that (3.109b) reduces to

$$S(r) \equiv \int \frac{E \cdot E^*}{|E_0|^2} \, d\Omega \simeq \frac{2\ell+1}{2} |d_\ell|^2 |j_\ell(n\beta'\rho)|^2. \qquad (7.74)$$

However, at the resonance d_ℓ has the form (7.63), with the half-width $\Gamma/2$ given by (7.60). Substitution into (7.74) provides an alternative form of $S(r)$ at a TE resonance:

$$S_{res}(r) \approx \left(\frac{2\ell+1}{2} \frac{1}{\beta'^2(n^2-1)} \frac{1}{j_\ell^2(n\beta')} \right) \frac{|j_\ell(n\beta'\rho)|^2}{\Gamma_\ell/2}. \qquad (7.75)$$

Thus, for very narrow resonances the internal field intensity is enormous and located very near the surface. This assertion is verified by noting that the location of the maximum in $S_{res}(r)$ is determined solely by the spherical Bessel function. One finds that this is the solution of

$$\frac{j_{\ell+1}(n\beta'\rho)}{j_\ell(n\beta'\rho)} = \frac{\ell+1}{n\beta'\rho}. \tag{7.76}$$

Although this is readily solved numerically for ρ, it is instructive to find an analytic approximation to the location of the maximum, which begins with the asymptotic form (C.6a):

$$j_\ell(n\beta'\rho) = \left(\frac{2}{\pi n\beta'\rho}\right)^{1/2} J_\lambda(n\beta'\rho)$$

$$\simeq \left(\frac{2}{\pi n\beta'\rho}\right)^{1/2} \left(\frac{2}{a}\right)^{1/2} \text{Ai}\left(2^{1/3}\frac{n\beta'-\lambda}{\lambda^{1/3}}\right), \tag{7.77}$$

and again $\lambda = \ell + \frac{1}{2}$. When $n\beta'\rho \geq \lambda$ the Airy function vanishes just inside the surface, at a zero $-x_m$ of $\text{Ai}(x)$. This occurs at

$$\rho_{max} \simeq 1 - \frac{x_m}{n\beta'_{\ell,m,j}}\left(\frac{\lambda}{2}\right)^{1/3}, \tag{7.78}$$

where we have employed the first approximation in (7.52). For the principal TE resonance at $\ell = 564$, $n = 1.45$, say, Eq. (7.77) yields $\rho_{max} \simeq 0.9735$, compared with the solution $\rho \simeq 0.9873$ of (7.75). The two increase and come closer together as β increases, while exhibiting only a mild sensitivity to changes in refractive index.

Figure 7.9 presents the exact Mie results for the source function at a TE resonance for $\ell = 50$, $n = 1.45$. The intensity distribution of the principal resonance located at $\rho = 0.9549$ is depicted in Fig. 7.9a, whereas Fig. 7.9b illustrates the distribution associated with the broader resonance of order $m = 3$. In the latter case we note the presence of exactly $m = 3$ maxima, corresponding to the radial local maxima of the spherical Bessel function in (7.71).

It was noted earlier that, as β increases, the low-order resonances become so sharp that it eventually becomes impossible to locate them within their widths, either theoretically or experimentally. While this is certainly true for the external elastic scattering resonances in a_ℓ and b_ℓ, the situation is somewhat better internally. The elastic scattering resonances produce integrated intensities from $S(r)$ that vanish rapidly as Q increases, because $|a_\ell|^2$ and $|b_\ell|^2$ are bounded above by unity. In contrast, c_ℓ and d_ℓ can be as large as the input parameters dictate, so that high Q can yield large observable intensities in which broadening does *not* reduce the sharpness.

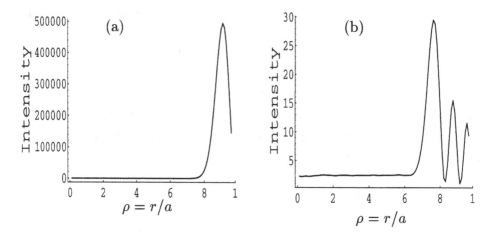

Fig. 7.9. The source function $S(r)$ for a TE resonance at $\ell = 50$, $n = 1.45$. (a) The principal resonance, $m = 1$. (b) The resonance of order $m = 3$.

For example, consider a TM resonance for $\ell = 100$, $n = 1.45$, which occurs at $\beta' \simeq 74.89$. We compute the line width to be about 1.38×10^{-12} and $|c_{100}|^2 \sim 4.89 \times 10^{12}$. Clearly, the integrated intensity remains significant. Hill and Benner (1986) have pointed out that it is just this effect that provides the optical feedback required to produce stimulated Raman scattering and lasing in microspheres (see below).

Figures 7.10 and 7.11 provide a broader view of the field intensity within the sphere, as well as just outside the surface. Here the intensity is computed along the optical (z) axis and in the plane $\phi = \pi/2$ directly from the expressions (7.67) and (3.90). In Fig. 7.10, where β is near but not on a resonance, our attention is drawn to the fields just outside the illuminated side of the surface and just outside the shadow side. These are simply the expected intensity enhancements due to axial focusing. At a resonance, however, as illustrated in Fig. 7.11, the internal fields just under the surface increase enormously, dwarfing the near external fields, and the total volume-integrated intensity of Eq. (3.110) increases by orders of magnitude. Figure 7.10 exhibits a very rapid oscillation of the internal intensity, a result of interference between refracted and internally reflected rays.

Corresponding contour plots of the field intensities shown in Figs. 7.10 and 7.11 are presented in Fig. 7.12. The standing–wave pattern around the inner surface is particularly clear in Fig. 7.12b and leaves little doubt as to its 'whispering-gallery' character.

A general feature revealed by the preceding figures is that the strong fields, both on and off resonance, are confined to the regions very near to the inner

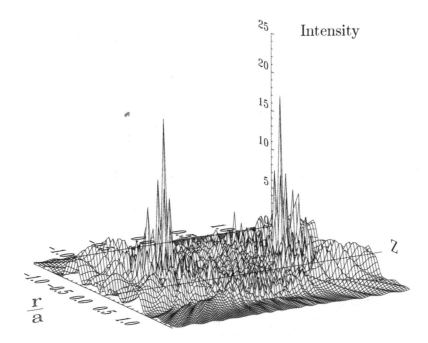

Fig. 7.10. The field intensity along the optical axis and in the plane $\phi = \pi/2$ near the principal TE resonance at $l = 50$, $n = 1.5$, just below resonance at $\beta' = 37.45$. Negative values of z and r/a serve only as directional references with respect to the spherical center.

and outer sides of the surface. As suggested in Figs. 7.4 and 7.7, the near-edge rays producing the whispering-gallery modes have impact parameters in the range $a < b < na$, corresponding to $\beta < \mathrm{Re}\,\lambda < n\beta$. In turn, the associated domain of standing waves within the sphere is $a > r > a/n$. The two points $r = a/n$ and $r = na$ are known as the *aplanatic points* of the sphere $r = a$ (e.g., Born and Wolf (1975, p. 149)); the spheres they define are perfect stigmatic images of one another, in the sense that spherical aberration is absent.

Calculations and experimental observations of both external and near fields, both in cylinders and in spheres, have been carried out by Benincasa *et al.* (1987). Photographs of these intensity distributions are included, and are recommended viewing for the reader.

Because of the relatively large size parameters being considered, one would expect geometric optics to provide a good approximation to the distribution of electromagnetic energy within the sphere. This expectation has been verified by a number of authors (Hovenac and Lock 1992; Chowdhury *et al.* 1992; Velesco *et al.* 1997) for size parameters $\beta \sim 500$, which demonstrate

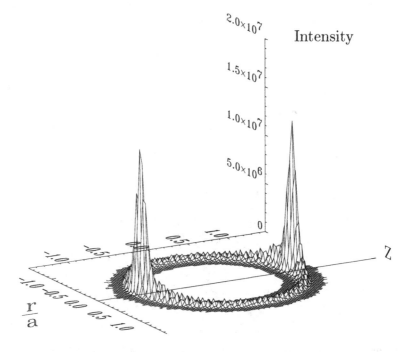

Fig. 7.11. The field intensity along the optical axis and in the plane $\phi = \pi/2$ near the principal TE resonance at $\ell = 50$, $n = 1.5$, on resonance at $\beta' = 37.451\,166\,659\,082\,52$. Negative values of z and r/a serve only as directional references with respect to the spherical center.

(a)   (b)

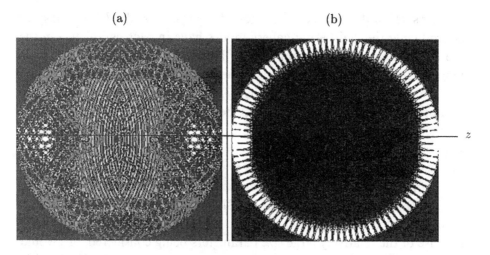

Fig. 7.12. Contour plots of the field intensities shown in Figs. 7.10 and 7.11, in which lighter shades correspond to higher intensity. (a) Figure 7.10, just below resonance. (b) Figure 7.11, on resonance and exhibiting the whispering-gallery standing waves.

that the ray description provides a reasonable approximation to the energy-density distribution within a lossless dielectric sphere. When phase effects of different rays are included there is also good agreement with the Mie calculations.

While providing an attractive physical interpretation of field effects within the sphere, however, these ray-tracing techniques are not capable of describing resonance effects in any detail. A more productive way to approach this viewpoint is through the Debye expansion of Chapter 5. Equations (5.23) can be substituted directly into (7.68) to yield the Debye expansion of the fields, and the contributions near a resonance can be isolated in the same way as in (7.33). In a now-familiar way one again introduces the modified Watson transformation, followed by continuation into the complex λ-plane, and evaluates the resulting contour integrals asymptotically.

Although we shall not pursue this development here, preferring instead to utilize the direct Mie computations on the real β-line, such calculations have been carried out by Guimarães (1991, 1993). For $\beta \leq 100$ excellent fits to the exact Mie resonances are obtained by including about the first 18 terms in the Debye expansion (contrasted with only a few for off-resonance fields). In this sense it is the semiclassical theory, not geometric optics alone, that matches the Mie results almost exactly, and provides an intuitively pleasing physical interpretation.

This brings us back to the motivation expressed at the beginning of this chapter. The ripple observed in the cross sections, which are far-field quantities, is the signature of resonant fields within the sphere as β is varied. The observed spikes are not *caused* by small denominators, for that is simply a mathematical manifestation. Rather, they are a result of electromagnetic fields in resonance. In common with resonant electric circuits, dielectric spheres in resonance provide new tools for probing other physical phenomena.

7.6 Observations and applications

We have already noted an early application of the resonance structure of Mie scattering to optical levitation through radiation pressure by Ashkin and colleagues (e.g., Ashkin (1980)). These authors used the sharp resonances to detect fractional changes in droplet radius of 1 part in 10^6, thereby demonstrating how smooth the surface of a droplet is maintained through surface tension. Their techniques led to an accurate method for determining the absolute size and refractive index of spheres by comparing computed shapes and locations of the resonances with measurements (Chýlek *et al.*

1983). Subsequently, a sizing algorithm was developed based solely on the alignment of resonances (Hill *et al.* 1985). More recently direct methods based on the angular distribution of scattering have been developed and these laser diagnostics for characterizing droplets have been reviewed by Chen *et al.* (1996).

Owen *et al.* (1982) first observed resonances in Raman scattering from dielectric microspheres, and they were also observed in fluorescence emission from small spheres on substrates (Hill *et al.* 1984). They have also been observed in the Raman spectra of optically levitated liquid droplets of water and glycerol (Thurn and Kiefer 1985). Stimulated Raman scattering (SRS) has been observed by Snow *et al.* (1985) in mixtures of water and ethanol, in which the resonant field intensities from first-order Stokes emission can act as a pump for multiorder Stokes emission. Up to 14 orders of Stokes peaks in CCl_4 have been reported (Qian and Chang 1986). In addition, Q factors on the order of 10^7 have been seen in SRS (Zhang *et al.* 1988; Hsieh *et al.* 1988; Pinnick *et al* 1988).

Although frequency splitting of degenerate azimuthal modes is expected when droplets depart from spherical, and has been observed (Chen *et al.* 1991), such splitting of high-Q modes has also been seen in undeformed silica microspheres. With observed Q values $\gtrsim 10^8$, each resonance is split into a doublet and it is shown that the splitting arises from *internal* backscattering that couples the clockwise and counterclockwise modes and lifts the degeneracy (Weiss, *et al.* 1995). The requisite backscattering apparently arises from the presence of a number of independent microscopic dipoles much smaller than a wavelength, and the consequent Rayleigh scattering couples the two modes.

Lasing in microdroplets, in which the usual role of mirrors is played by near-total internal reflection, as in Fig. 7.7, has been reported by Tzeng *et al.* (1984). Optical third-harmonic generation enhanced by Mie resonances has been observed by Acker *et al.* (1989) and studied further by Hill *et al.* (1993). One gains the impression that the potential for further applications of high-Q resonances in microspheres is enormous.

8

Extensions and further applications

The preceding chapters have been concerned primarily with explicating the physical features of scattering electromagnetic plane waves from dielectric spheres, particularly for large size parameters. Much of the theoretical development is strongly dependent upon the maximal symmetry of this scenario, as well as upon the idealization of an incident plane wave. Although the spherical target is a good approximation to many of those encountered in important physical problems, there exist many other situations in which departures from sphericity cannot be ignored. Moreover, an infinite plane wave is clearly a fiction, albeit a very useful one, and in reality we have only the approximation of a *locally* plane wave. In many applications the incident radiation is provided by a tightly focused laser beam that may, but need not, satisfy this criterion. Thus, while the bulk of the work presented here provides a sound basis for understanding the basic scattering problem, there is a large body of physical applications in which one or more of the idealizations inherent in our model may fail to be realized. In this final chapter we shall attempt a brief and necessarily incomplete survey of some of the ways in which the fundamental model and its analysis must be altered in these situations. For the most part derivations and extensive mathematical expressions are omitted.

To this point our study can be summarized as pertaining to the

> scattering of an **electromagnetic plane wave** from **a single homogeneous dielectric** ($n > 1$) **sphere**.

Each of the terms emphasized in boldface type is capable of modification in one way or another. We shall consider each possibility in turn, as well as

267

various other modifications of this basic scenario, and in doing so we hasten to point out that each constitutes a very active area of research at present. Almost all of these extensions provide fertile ground for application of the asymptotic methods espoused in earlier chapters to models with large size parameters.

8.1 Inhomogeneous dielectric spheres

Although we have taken our spherical targets to be homogeneous and isotropic, it is clear that this is a valid presumption only on a particular scale. The scale chosen, and hence the actual size of the spheres under consideration, is such that the material properties of the sphere can be described by a macroscopic relative index of refraction n. On this same scale, however, it is certainly possible for n to vary spatially in an arbitrary manner. For example, the dielectric sphere could contain a number of inclusions with different refractive indices and nonspherical geometries. While various numerical techniques have been developed to attack such problems (e.g., Michel (1997)), an analytic treatment is virtually out of the question. As with the study of nonequilibrium fluid flows, significant progress toward understanding the underlying physical mechanisms is achieved by restricting the type of inhomogeneity to simpler and more systematic variations.

We review here, in a general way, electromagnetic scattering from spherical models in which the refractive index is allowed to vary radially. That is, a spherical dielectric that is optically isotropic with $n = n(r)$.

The coated sphere

This model of two concentric spheres, a homogeneous core of radius b and refractive index n_1 enclosed by a homogeneous shell of radius a and refractive index n_2, is perhaps the simplest of inhomogeneous models, primarily because spherical symmetry is maintained. Throughout this chapter we take the refractive index of the embedding medium to be unity, and all three media are presumed nonmagnetic. The coated sphere so described is depicted in Fig. 8.1.

Plane-wave scattering from this target is described in exactly the same way as in Chapter 3, Eqs. (3.74)–(3.90), except that now there are three spatial regions to consider and matching is to be done at two separate boundaries. Moreover, in the shell $b < r < a$ the Ricatti–Bessel functions ψ_ℓ and ξ_ℓ must be included independently as solutions of the radial equation. There

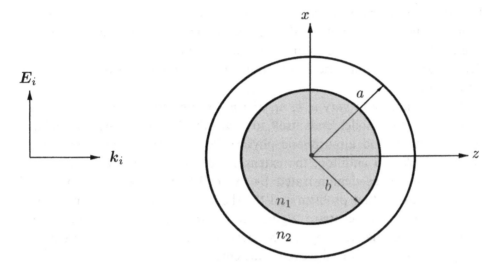

Fig. 8.1. A coated sphere consisting of two concentric spheres of different refractive indices. The shell has radius a and index n_2, while the core radius is b with index n_1. In this scenario the target is illuminated by a plane wave linearly polarized along the x-axis.

are then eight independent partial-wave coefficients determined by a set of eight linear equations of the form (3.87).

We adopt the notation $\alpha \equiv kb$, $\beta \equiv ka$ for the size parameters, and let a_ℓ and b_ℓ be the external scattering coefficients; c_ℓ and d_ℓ refer to the core; and e_ℓ, f_ℓ, u_ℓ, and v_ℓ refer to the shell. We are primarily interested in the scattering coefficients alone, which were first obtained by Aden and Kerker (1951). As expected, they are a straightforward generalization of those for the homogeneous sphere, Eqs. (3.88), and are functions of n_1, n_2, α and β. In a slightly different notation we find that†

$$a_\ell = \frac{\psi_\ell'(\beta)[\psi_\ell'(n_2\beta) - A_\ell\xi_\ell'(n_2\beta)] - n_2\psi_\ell'(\beta)[\psi_\ell(n_2\beta) - A_\ell\xi_\ell(n_2\beta)]}{\zeta^{(1)}(\beta)[\psi_\ell'(n_2\beta) - A_\ell\xi_\ell'(n_2\beta)] - n_2\zeta_\ell^{(1)'}(\beta)[\psi_\ell(n_2\beta) - A_\ell\xi_\ell(n_2\beta)]},$$

$$b_\ell = \frac{\psi_\ell'(\beta)[\psi_\ell(n_2\beta) - B_\ell\xi_\ell(n_2\beta)] - n_2\psi_\ell(\beta)[\psi_\ell'(n_2\beta) - B_\ell\xi_\ell'(n_2\beta)]}{\zeta^{(1)'}(\beta)[\psi_\ell(n_2\beta) - B_\ell\xi_\ell(n_2\beta)] - n_2\zeta_\ell^{(1)}(\beta)[\psi_\ell'(n_2\beta) - B_\ell\xi_\ell'(n_2\beta)]}; \quad (8.1)$$

$$A_\ell \equiv \frac{n_2\psi_\ell(n_2\alpha)\psi_\ell'(n_1\alpha) - n_1\psi_\ell'(n_2\alpha)\psi_\ell(n_1\alpha)}{n_2\xi_\ell(n_2\alpha)\psi_\ell'(n_1\alpha) - n_1\xi_\ell'(n_2\alpha)\psi_\ell(n_1\alpha)},$$

$$B_\ell \equiv \frac{n_2\psi_\ell(n_1\alpha)\psi_\ell'(n_2\alpha) - n_1\psi_\ell'(n_1\alpha)\psi_\ell(n_2\alpha)}{n_2\xi_\ell'(n_2\alpha)\psi_\ell'(n_1\alpha) - n_1\xi_\ell(n_2\alpha)\psi_\ell'(n_1\alpha)}. \quad (8.2)$$

† Note that, although Aden and Kerker, and others, define the Ricatti–Neumann functions with an additional minus sign, the signs in (8.1) are unaffected.

Equations (8.1) obviously reduce to (3.88) for $n_1 = n_2$, and Kerker (1969) demonstrates that a similar reduction occurs when the two radii approach one another. Similar expressions are readily obtained for the internal coefficients, and stable numerical algorithms have been developed by Toon and Ackerman (1981).

With the ability to vary a, b, n_1, and n_2, where the refractive indices can be complex, this model lends itself to a wealth of applications, particularly in astrophysics and atmospheric physics. Some of these are described by Kerker (1969). In addition, the extension to multiple layers is completely straightforward, as demonstrated by Wait (1963). He also exploited the analogy between that problem and the theory of non-uniform transmission lines to develop a recursive algorithm for solving it. In the same vein, Bhandari (1985) has developed a recursive numerical algorithm for efficient computation of the partial-wave coefficients.

Continuous nonuniformity

The mathematical problem in the event that $n(r)$ varies continuously has been analyzed in detail by Wyatt (1962). Although Wait (1963) suggested that one might approach this problem by taking appropriate limits in the above multilayered model, the proper approach is to make the replacement $k \to k(r)$ in the wave equation (3.76). The immediate complication here is that the two Debye potentials (3.75) now satisfy different wave equations within the sphere:

$$\nabla^2 \Pi_1 - \frac{2}{kr} \frac{\partial k}{\partial r} \frac{\partial(r\Pi_1)}{\partial r} + k^2 \Pi_1 = 0,$$
$$\nabla^2 \Pi_2 + k^2 \Pi_2 = 0. \tag{8.3}$$

(Note that $k(r)$ is simply the constant wavenumber of the incident wave times $n(r)$.) These separate into two identical angular equations but different radial equations. We write $W(r) \equiv rR_1(r)$ for the electric potential, and $G(r) \equiv rR_2(r)$ for the magnetic, which then satisfy

$$\frac{d^2 W_\ell}{dr^2} - \frac{2}{k} \frac{dk}{dr} \frac{dW_\ell}{dr} + \left(k^2 - \frac{\ell(\ell+1)}{r^2} \right) W_\ell = 0, \tag{8.4a}$$

$$\frac{d^2 G_\ell}{dr^2} + \left(k^2 - \frac{\ell(\ell+1)}{r^2} \right) G_\ell = 0. \tag{8.4b}$$

As usual, these interior solutions must be chosen to be regular at the origin.

Because k depends on the independent variable these equations for W_ℓ and G_ℓ are generally a good deal more complicated than those considered earlier, and possess closed-form solutions only for some special functional forms of $n(r)$. The scattering coefficients, however, have simple forms similar to those of (3.88):

$$a_\ell = \frac{\psi_\ell(\beta)W_\ell'(\beta) - n^2 W_\ell(\beta)\psi_\ell'(\beta)}{\zeta_\ell^{(1)}(\beta)W_\ell'(\beta) - n^2 W_\ell(\beta)\zeta_\ell^{(1)'}(\beta)}, \tag{8.5a}$$

$$b_\ell = \frac{\psi_\ell(\beta)G_\ell'(\beta) - G_\ell(\beta)\psi_\ell'(\beta)}{\zeta_\ell^{(1)}(\beta)G_\ell'(\beta) - G_\ell(\beta)\zeta_\ell^{(1)'}(\beta)}, \tag{8.5b}$$

where $n = n(a)$, and normalization is provided by the limiting case of constant n:

$$G_\ell(\beta) = W_\ell(\beta) = \psi_\ell(n\beta),$$
$$G_\ell'(\beta) = W_\ell'(\beta) = n\psi_\ell'(n\beta). \tag{8.6}$$

One example of an exact solution arises when n has a power-law dependence on r (Kerker 1969). Let

$$n = A\rho^p, \qquad \rho \equiv kr, \tag{8.7}$$

where A and p are (possibly complex) constants. Solutions of (8.4) regular at the origin are verified to be

$$W_\ell(\rho) = \rho^{p+1}\left(\frac{\pi}{2\rho}\right)^{1/2} J_{\pm\nu}(X), \tag{8.8a}$$

$$G_\ell(\rho) = \rho\left(\frac{\pi}{2\rho}\right)^{1/2} J_{\pm\mu}(X), \tag{8.8b}$$

where $X = A\rho^{p+1}/(p+1)$, and

$$\nu = \frac{[\ell(\ell+1) + (p+\frac{1}{2})^2]^{1/2}}{p+1}, \qquad \mu = \frac{2\ell+1}{2(p+1)}. \tag{8.9}$$

These reduce to the Ricatti–Bessel functions for $p = 0$ (Eqs. (A.2) and (A.11a)).

Anisotropy

We have agreed to restrict our discussion to isotropic dielectric spheres, primarily because the completely arbitrary problem of scattering from an optically anisotropic sphere is virtually intractable. In the guise of inhomogeneity, however, some progress has been made.

One such scenario is that in which the spheres of the coated-sphere problem are not concentric, so that the sphere is eccentrically stratified, as

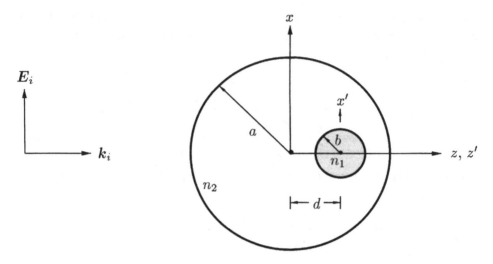

Fig. 8.2. An eccentrically stratified sphere in which the core of Fig. 8.1 is offset a distance d along the z-axis.

in Fig. 8.2. Essentially, the core of Fig. 8.1 has been reduced and translated radially along the z-axis. There is actually an exact solution to this problem that can be constructed as follows (Fikioris and Uzunoglu 1979).

Return to a description of the spherically symmetric scattering problem in terms of vector spherical harmonics $X_{\ell m}$, in which the fields are expanded as in Eq. (3.12). Although complete spherical symmetry is absent here, one can still employ such expansions fruitfully. To apply the interior boundary conditions properly, however, it is necessary to relate the two spherical coordinate systems by a translation transformation, and this is most efficiently carried out in terms of an alternative representation for the vector spherical harmonics. We shall describe such a formulation briefly below, in Eqs. (8.40), and simply note here that this description readily leads to expansions of the scattered fields in the coordinate system of the large sphere. The expansion coefficients are determined by an infinite set of simultaneous equations that, to say the least, presents a challenging computational problem, though it is amenable to approximation techniques. Fikioris and Uzunoglu have studied the scenario in which $(n_2 - n_1)kb \ll 1$ – that is, a small tenuous inclusion. The resulting perturbation expressions can provide some insight into the general properties of the system, and a related model will be studied in Section C below.

Roth and Dignam (1973) have investigated a scenario in which the outer shell in Fig. 8.1 is optically anisotropic. The physical model envisions the coating as an oriented molecular layer in which the local optical axis is coincident with the local normal to the particle surface. This is realized by specifying unequal tangential (n_t) and radial (n_r) components of the relative refractive index in the shell $b < r < a$, or region 2.

Remarkably, the overall spherical symmetry leads once again to the Mie solution, but with rather complicated partial-wave coefficients. The complication arises because, while the magnetic (TE) Debye potential satisfies the same equation as in the isotropic case, with n_2 replaced by n_t, the electric (TM) equation contains the ratio (n_r/n_t). Thus, the radial solutions for the two multipoles differ and are found to be

$$\psi_\ell^v(z) = \left(\frac{\pi z}{2}\right)^{1/2} J_w(z),$$

$$\chi_\ell^v(z) = -\left(\frac{\pi z}{2}\right)^{1/2} J_{-w}(z), \tag{8.10}$$

where $z \equiv k n_t r$, $w \equiv S_\ell^v + \frac{1}{2}$, $v = e, m$, and

$$S_\ell^e \equiv \left[\ell(\ell+1)\left(\frac{n_t}{n_r}\right)^2 + \frac{1}{4}\right]^{1/2} - \frac{1}{2},$$

$$S_\ell^m \equiv \ell. \tag{8.11}$$

In an isotropic medium these solutions have the same form for both multipoles and reduce to Ricatti–Bessel functions.

The scattering coefficients are again ratios of sums of products of the radial solutions evaluated at $r = a$ and b, and we shall refer to Roth and Dignam (1973) for their explicit forms. Applications have been made in the thin-film limit, $(a - b)/a \ll 1$, by expanding in powers of this parameter. Calculations of this kind have been made through second order and have been found useful in biochemical applications (Hahn and Aragón 1994).

Finally, although it is isotropic, the spherical medium may be able to support longitudinally polarized waves and, for example, allow for resonant excitation of bulk plasmons in metal spheres. We suppose the dielectric constant to have a longitudinal component $\epsilon_L(k_L, \omega)$, and a transverse component $\epsilon_T(k_T, \omega)$, with $n = 1$ in the external medium. A longitudinal plasma wave will have the dispersion relation $\epsilon_L(k_L, \omega) = 0$, from which k_L is determined, whereas for transverse propagation

$$k_T^2 = \epsilon_T(k_T, \omega)\omega^2/c^2. \tag{8.12}$$

One now proceeds just as in the original Mie problem, noting that there is now an additional electric field associated with longitudinal plasma modes, though there is no magnetic field (e.g., Jackson (1975, p. 492)). Hence, there is an additional internal partial-wave coefficient e_ℓ, along with an additional boundary condition. This condition can be taken as a requirement that the radial components of the total external electric field be equal at $r = a$. One now finds the scattering coefficients given by

$$a_\ell = \frac{g_\ell \psi_\ell(\beta) + \beta j'_\ell(\beta_L)[\psi_\ell(\beta)\psi'_\ell(\beta_T) - n_T\psi_\ell(\beta_T)\psi'_\ell(\beta)]}{g_\ell \zeta_\ell^{(1)}(\beta) + \beta j'_\ell(\beta_L)[\zeta_\ell^{(1)}(\beta)\psi'_\ell(\beta_T) - n_T\psi_\ell(\beta_T)\zeta_\ell^{(1)'}(\beta)]}, \quad (8.13a)$$

$$b_\ell = \frac{\psi_\ell(\beta_T)\psi'_\ell(\beta) - n_T\psi'_\ell(\beta_T)\psi_\ell(\beta)}{\zeta_\ell^{(1)'}\psi_\ell(\beta_T) - n_T\zeta_\ell(\beta)\psi'_\ell(\beta_T)}, \quad (8.13b)$$

where $\beta = ka$, $\beta_L = k_L a$, $\beta_T = k_T a$, and

$$g_\ell \equiv \ell(\ell + 1)\frac{j_\ell(\beta_L)j_\ell(\beta_T)}{\beta_L}(n_T^2 - 1). \quad (8.14)$$

The magnetic coefficient b_ℓ is exactly the same as the original, (3.88b), with n_T identified as the usual index of refraction. Clearly, the excitation of magnetic modes is not affected by the presence of longitudinal fields. However, the electric coefficient a_ℓ is quite different, which indicates that a simultaneous excitation of transverse and longitudinal electric modes occurs. When $g_\ell = 0$ the coefficients a_ℓ reduce to those of (3.88a).

Ruppin (1975) has employed this model in a study of the optical properties of small metal spheres. One interesting new result for sodium emerges in the extinction cross section as a function of frequency, wherein a series of absorption peaks appears just above the plasma frequency.

Resonances and perturbation theory

We have already noted that small departures from sphericity lead to observable line splittings by lifting the azimuthal degeneracy. One might, therefore, expect similar changes in the resonance structure to arise from inhomogeneity, and this is indeed the case.

A Rayleigh–Schrödinger type of perturbation method has been constructed to assess the effects of a small inhomogeneity in the form $n + \delta n(r)$ (Lai et al. 1991; Leung and Pang 1996; Lee, et al. 1998). The effects on resonance widths and Q factors were addressed in particular and it was concluded that, except in some cases of spherically symmetric perturbation, the widths always increase.

The exceptional cases have been studied nonperturbatively, but numerically, by Chowdhury, et al. (1991) by employing the results of Eqs. (8.5). These authors study two forms of radial inhomogeneity – one in which $n(r)$

rolls off smoothly to a fixed value near the surface, and another in which it increases smoothly near the surface. The numerical results show that, as the refractive index decreases smoothly near the surface, resonant size parameters increase and the Q factors decrease. A spherically symmetric increase in refractive index, on the other hand, results in a decrease in the resonance location, while Q tends to increase. The shift in resonance location in both cases is exactly what is predicted by Eq. (7.49).

Mazumder *et al.* (1992) have also examined the effects on resonances of eccentric spherical inclusions of the type considered above in Fig. 8.2, in which the refractive index differs little from that within the host sphere. They find that the resonant size parameter always decreases from the homogeneous value, but that the Q factors can either decrease or increase depending on the original Q value and the location and shape of the perturbation.

The coated sphere of Fig. 8.1 possesses a particularly interesting resonance structure, in that it is possible for both concentric surfaces to support whispering gallery types of standing waves. Criteria for determining the outer resonances follow, as in Chapter 3, from setting $\mathrm{Re}(a_\ell)$ and $\mathrm{Re}(b_\ell)$ to unity in Eqs. (8.1) and (8.2):

$$A_\ell(n_1, n_2, \alpha) = \frac{\xi_\ell(\beta)\psi'_\ell(n_2\beta) - n_2\xi'_\ell(\beta)\psi_\ell(n_2\beta)}{\xi_\ell(\beta)\xi'_\ell(n_2\beta) - n_2\xi'_\ell(\beta)\xi_\ell(n_2\beta)},$$

$$B_\ell(n_1, n_2, \alpha) = \frac{\xi'_\ell(\beta)\psi_\ell(n_2\beta) - n_2\xi_\ell(\beta)\psi'_\ell(n_2\beta)}{\xi'_\ell(\beta)\xi_\ell(n_2\beta) - n_2\xi_\ell(\beta)\xi'_\ell(n_2\beta)}. \tag{8.15}$$

To find the core resonances one must first find the core coefficients c_ℓ and d_ℓ, which we shall not do here, and then examine the denominators to obtain the appropriate criteria.

Hightower and Richardson (1988) have carried out a numerical study of these resonances for the coated sphere, including the case where the core is a cavity, and they illustrate how the resonance locations and widths change as the two relative indices and the core radius are varied. In addition, Hightower *et al.* (1988) have performed measurements on glass spheres ($n = 1.51$) coated with glycerine ($n = 1.473$) for which $b = 4.0$ μm and shell thickness varied between 0 and 2.5 μm. They find excellent agreement with the numerical results.

8.2 Incident beams and wavepackets

Although many, if not most, modern light-scattering experiments utilize a laser beam as source, quite often the beam width is very much larger than the particle diameter so that the incident illumination is effectively a plane wave. However, this is not always the case, for the incident beam can sometimes be focused down to a transverse focal waist on the order of a wavelength. Consequently, these and other deviations from a locally plane wave as incident radiation have come to be of some importance in light scattering.

Evanescent waves

When an electromagnetic wave is totally reflected at the interface between two media with different refractive indices there will be, in contrast to geometric optics, penetration into the optically thinner medium, and a wave will propagate along the interior surface with energy that decreases exponentially with increasing distance from the interface. This *evanescent wave* can be observed to scatter from objects near the interface and thus be used as a probe with which to explore various phenomena. Such a technique has found application in near-field optics (e.g., Pohl and Courjon (1993)), as well as in instruments for studying viruses, where the evanescent wave actually enhances the illumination intensity over that of the incident beam (Hirschfeld, *et al.* 1977).

An extension of the Mie theory to include illumination by an incident evanescent wave was first developed by Chew, *et al.* (1979), and corrected and further extended subsequently (Liu, *et al.* 1995; Zvyagin and Goto 1998). The model is depicted in Fig. 8.3, where a plane wave is incident from below in medium n with wavevector k in the xz-plane of incidence, and the angle of incidence with the x-axis is θ_i. If $\theta \to \theta_c = \sin^{-1}(n'/n)$, only an evanescent wave will be present in medium $n' < n$ with wavevector $k' = n'k$. This wave will be incident along the z-axis upon a dielectric sphere of radius a and refractive index n_{sp}, situated a distance d from the interface, and all media are considered nonmagnetic. We choose d to be a number of wavelengths, such that boundary conditions at the interface and on the spherical surface are independent, yet not so large that the incident wave has decayed too much to be effective. The y- and y'-axes are perpendicular to and positive out of the page, and, when θ_i exceeds the critical angle, $\theta_{k'}$ will be imaginary. The evanescent wave is then incident as indicated. Once again it is most effective to formulate the scattering problem in terms of vector spherical harmonics and to expand the fields as in Eq. (3.12). Moreover, polarization of the incident wave plays a more sensitive role than usual in determining that of the scattered wave, for the exponential damping destroys the symmetry responsible for preventing cross polarization in the ordinary plane-wave case. It is customary, then, to restrict the incident plane-wave polarization to s-type (E_{inc} along the y-axis), and to p-type (E_{inc} in the plane

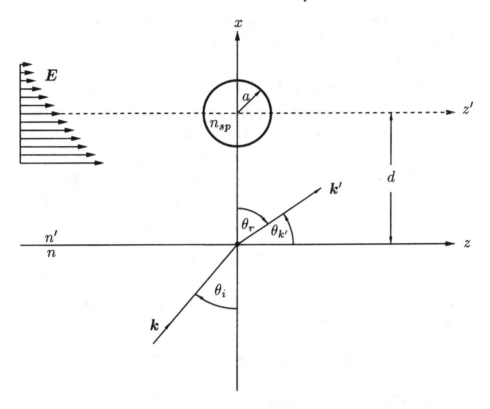

Fig. 8.3. A plane wave incident at an interface lying in the yz-plane produces an evanescent wave that illuminates a dielectric sphere lying beyond but near the interface.

of incidence). For s polarization the incident and evanescent waves are

$$E_{\text{inc}} = \hat{y} E_0 e^{i\boldsymbol{k} \cdot \boldsymbol{r}},$$

$$E_{\text{ev}} = \hat{y} E_0' e^{i\boldsymbol{k}' \cdot \boldsymbol{r}}, \qquad E_0' \equiv \frac{2n \cos \theta_i}{n \cos \theta_i + n' \cos \theta_r} E_0, \qquad (8.16)$$

and for p polarization

$$E_{\text{inc}} = (\sin \theta_i, 0, -\cos \theta_i) E_0 e^{i\boldsymbol{k} \cdot \boldsymbol{r}},$$

$$E_{\text{ev}} = (\sin \theta_r, 0, -\cos \theta_r) E_0' e^{i\boldsymbol{k}' \cdot \boldsymbol{r}}, \qquad E_0' \equiv \frac{2n \cos \theta_i}{n \cos \theta_i + n' \cos \theta_r} E_0, \quad (8.17)$$

where we have introduced the Fresnel transmission coefficient. In both cases note that

$$e^{i\boldsymbol{k}' \cdot \boldsymbol{r}} = e^{-\gamma x} e^{ik'(n/n') \sin \theta_i}, \qquad (8.18)$$

where $\gamma \equiv k'(n^2 \sin^2 \theta_i / n'^2 - 1)^{1/2}$ is the damping constant.

Employing the orthogonality properties of the vector spherical harmonics (Appendix D), one finds the following expressions for the multipole coefficients (Liu *et al.* 1995):

$$\alpha_M^s(\ell,m) = i^{\ell+1}\eta_{\ell,m}[P_\ell^{m+1} - (\ell+m)(\ell-m+1)P_\ell^{m-1}]E_0',$$

$$\alpha_E^s(\ell,m) = i^{\ell-2}n'\eta_{\ell,m}[P_{\ell-1}^{m+1} + (\ell+m)(\ell+m-1)P_{\ell-1}^{m-1}]E_0'; \quad (8.19)$$

$$\alpha_M^p(\ell,m) = i^\ell\eta_{\ell,m}\{[P_\ell^{m+1} + (\ell+m)(\ell-m+1)P_\ell^{m-1}]\cos\theta_{k'}$$
$$-2m\sin\theta_{k'}\,P_\ell^m\}E_0',$$

$$\alpha_E^p(\ell,m) = i^{\ell-1}\eta_{\ell,m}\{[(\ell+m)(\ell+m-1)P_{\ell-1}^{m-1} - P_{\ell-1}^{m+1}]\cos\theta_{k'}$$
$$+2(\ell+m)\sin\theta_{k'}\,P_{\ell-1}^m\}E_0', \quad (8.20)$$

where $P_\ell^m(\cos\theta_{k'})$ is the associated Legendre polynomial, and

$$\eta_{\ell,m} \equiv \left(\frac{\pi(2\ell+1)(\ell-m)!}{\ell(\ell+1)(\ell+m)!}\right)^{1/2}. \quad (8.21)$$

Note carefully that $\theta_{k'}$ here is not restricted to $[0,\pi]$; it is imaginary for the evanescent wave. The scattered field in the far zone is then

$$E_{sca}^v(r) \simeq \frac{e^{ik'r}}{k'r}\sum_{\ell,m}(-1)^{\ell-1}\left(\frac{1}{n'}B_E^v(\ell,m)\hat{r}\times X_{\ell m} - B_M^v(\ell,m)X_{\ell m}\right), \quad (8.22)$$

where the vector spherical harmonics $X_{\ell m}$ are defined in Appendix D, $v = s, p$, and the partial-wave coefficients are

$$B_E^v(\ell,m) = \frac{n'^2\psi_\ell(\beta')\psi_\ell'(n_{sp}\beta') - n_{sp}\psi_\ell(n_{sp}\beta')\psi_\ell'(\beta')}{n_{sp}\psi_\ell(n_{sp}\beta')\zeta_\ell^{(1)'}(\beta') - n'^2\zeta_\ell^{(1)}(\beta')\psi_\ell'(n_{sp}\beta')}e^{-\gamma d}\alpha_E^v(\ell,m),$$

$$B_M^v(\ell,m) = \frac{\psi_\ell(n_{sp}\beta')\psi_\ell'(\beta') - n_{sp}\psi_\ell(\beta')\psi_\ell'(n_{sp}\beta')}{n_{sp}\psi_\ell'(n_{sp}\beta')\zeta_\ell^{(1)}(\beta') - \zeta_\ell^{(1)'}(\beta')\psi_\ell(n_{sp}\beta')}e^{-\gamma d}\alpha_M^v(\ell,m), \quad (8.23)$$

for size parameter $\beta' = k'a$.

With these results one readily computes the various cross sections. In particular, the differential scattering cross section has been computed by Liu, *et al.* (1995) for $n = 1.5$ and various angles of incidence at and above the critical angle, and over a range of size parameters β'. A first observation is that one sees a higher density of resonances than for plane-wave incidence, and the principle resonances have both decreasing width and decreasing strength as the angle of incidence increases. In addition, the axial symmetry of the internal energy distribution is lost as θ_i increases, and the bulk of the energy is coupled into the sphere at the bottom, which is to be expected from Fig. 8.3. The standing wave pattern around the inside surface at resonance (e.g., Fig. 7.12b) also washes out with increasing angle of incidence. In both cases one understands these results by noting that the two counter-propagating surface waves of Eqs. (2.116), and mentioned again in Chapter 7, have increasingly different amplitudes as θ_i increases – the wave induced at the top of the sphere is smaller than that at the bottom, and so the proper interference pattern cannot be sustained.

Gaussian beams

When the radiation incident upon a dielectric sphere is a narrow beam whose half-width is within a few sphere radii the assumption of an incident plane wave breaks down. Indeed, if the beam axis does not coincide with an axis through the center of the sphere, then a great deal of symmetry is removed from the problem. An immediate consequence is that Eqs. (3.12) must be employed in their full form, for their reduction to (3.22) is no longer valid and all azimuthal eigenvalues m must be included.

Efforts to generalize the Mie theory to include irradiation by focused laser beams have been under way for over 30 years, and the early work includes that of Morita et $al.$ (1968), Tsai and Pogorzelski (1975), Tam and Corriveau (1978), and Kim and Lee (1983). Subsequently, a full extension of the ordinary Mie theory was developed by Gouesbet et $al.$ (1985, 1988), and by Barton et $al.$ (1988); the latter is a straightforward extension of the work in Chapter 3.

We begin again with the exposition in terms of the Debye potentials in Eqs. (3.74)–(3.84), noting that now the partial-wave expansions are to be generalized. As an example, the potentials (3.81) for the incident wave become

$$\Pi_1^{inc} = \sum_{\ell=0}^{\infty} \sum_{m=-\ell}^{\ell} i^\ell \frac{2\ell+1}{2\ell(\ell+1)} A_{\ell m}\, j_\ell(kr) P_\ell^{|m|}(\cos\theta)e^{im\phi},$$

$$\Pi_2^{inc} = \sum_{\ell=0}^{\infty} \sum_{m=-\ell}^{\ell} i^\ell \frac{2\ell+1}{2\ell(\ell+1)} B_{\ell m}\, j_\ell(kr) P_\ell^{|m|}(\cos\theta)e^{im\phi}. \qquad (8.24)$$

For an arbitrary incident electromagnetic wave the expansion coefficients follow from orthogonality, as in (3.13), and from the expressions (3.78):

$$\begin{Bmatrix} A_{\ell m} \\ B_{\ell m} \end{Bmatrix} = \frac{(-i)^{\ell-1}}{2\pi} \frac{kr}{j_\ell(kr)} \frac{(\ell-|m|)!}{(\ell+|m|)!}$$

$$\times \int_0^\pi d\theta \int_0^{2\pi} d\phi \sin\theta\, P_\ell^{|m|}(\cos\theta)e^{-im\phi} r \cdot \begin{Bmatrix} E_{inc}(r) \\ H_{inc}(r) \end{Bmatrix}. \qquad (8.25)$$

These are usually evaluated at $r = a$, since in any event they must be independent of r. Utilization of $|m|$ in the associated Legendre polynomials possesses some computational advantages (Lock 1993).

Proceeding exactly as in Chapter 3, we find that the plane-wave expansions (3.95) for the scattering amplitudes are generalized to

$$S_1(\beta, \theta, \phi) = \frac{1}{E_0} \sum_{\ell m} \frac{2\ell + 1}{2\ell(\ell + 1)} [-ima_\ell(\beta) A_{\ell m} \pi_\ell^{|m|} + b_\ell(\beta) B_{\ell m} \tau_\ell^{|m|}], \quad (8.26a)$$

$$S_2(\beta, \theta, \phi) = \frac{1}{E_0} \sum_{\ell m} \frac{2\ell + 1}{2\ell(\ell + 1)} [imb_\ell(\beta) A_{\ell m} \pi_\ell^{|m|} + a_\ell(\beta) B_{\ell m} \tau_\ell^{|m|}], \quad (8.26b)$$

where a_ℓ and b_ℓ are just the partial-wave coefficients of (3.88), and

$$\pi_\ell^{|m|}(\theta, \phi) \equiv \frac{P_\ell^{|m|}(\cos \theta)}{\sin \theta} e^{im\phi}, \quad (8.27a)$$

$$\tau_\ell^{|m|}(\theta, \phi) \equiv \frac{dP_\ell^{|m|}(\cos \theta)}{d\theta} e^{im\phi}. \quad (8.27b)$$

Expressions containing the internal fields, such as that for the source function of (3.109), will likewise contain products $c_\ell A_{\ell m}$ and $d_\ell B_{\ell m}$. The various cross sections now follow in the usual way, though one is advised to proceed via the definitions (3.26) in this scenario.

As a check on our generalization, recall from (3.15) that the radial components of a plane electromagnetic wave incident along the z-axis with linear polarization of the electric field along the x-axis are

$$\hat{r} \cdot E_{\text{inc}} = E_0 e^{ikr \cos \theta} \sin \theta \cos \phi,$$

$$\hat{r} \cdot H_{\text{inc}} = E_0 e^{ikr \cos \theta} \sin \theta \sin \phi. \quad (8.28)$$

Substitution into (8.25) and reference to Eqs. (A.15a) and (D.12a), along with an integration by parts, leads to the reduction

$$A_{\ell m} = \begin{cases} E_0, & m = \pm 1 \\ 0, & \text{otherwise,} \end{cases}$$

$$B_{\ell m} = \begin{cases} \mp i E_0, & m = \pm 1 \\ 0, & \text{otherwise.} \end{cases} \quad (8.29)$$

Thus, for incident plane waves the scattering amplitudes (8.26) reduce to the set (3.95).

At this point we must be more specific, and it is most useful to envision the incident radiation as a focused Gaussian beam. To do that we need an explicit description of such beams, and it is common to adopt that of Davis (1979). Presume the Gaussian beam to be traveling in the z direction with the focal waist of half-width w_0 at the origin. The Davis formalism then allows one to express the incident fields as a series expansion in powers of the transverse beam-confinement parameter $s \equiv 1/(kw_0)$. For short wavelengths, or $s \ll 1$, the first-order result (TEM$_{00}$ mode) is adequate and

$$E_{\mathrm{inc}}^{(1)} = \frac{E_0}{D} e^{i(kz-\omega t)} e^{-R^2/(w_0^2 D)} \left(\hat{x} - \frac{2isx}{w_0 D} \hat{z} \right),$$

$$H_{\mathrm{inc}}^{(1)} = \frac{E_0}{D} e^{i(kz-\omega t)} e^{-R^2/(w_0^2 D)} \left(\hat{y} - \frac{2isy}{w_0 D} \hat{z} \right), \qquad (8.30)$$

where $R^2 \equiv x^2 + y^2$, $D \equiv 1 + 2isz/w_0$, and the harmonic time dependence will be omitted subsequently. Retention of only the first (or a few) terms in the expansion means that the fields (8.30) satisfy Maxwell's equations only approximately and, although the approximation appears to be rather good for w_0 much greater than a wavelength, this should be kept clearly in mind when considering further developments of the incident-beam scenario.

The beam described by (8.30) is quite general in this approximation and, although the beam axis is parallel to the z-axis through the center of the target sphere, it can be offset arbitrarily from that center in the xy-plane. For illustrative purposes here we shall consider the beam to be on-axis; that is, the symmetry axis of the beam is taken to coincide with the z-axis containing the center of the particle. In this case the formalism simplifies considerably and the radial components of the incident fields can be written as

$$\hat{r} \cdot E_{\mathrm{inc}}^{(1)} = E_0 e^{ikr \cos \theta} f(kr, \theta) \sin \theta \cos \phi,$$

$$\hat{r} \cdot H_{\mathrm{inc}}^{(1)} = E_0 e^{ikr \cos \theta} f(kr, \theta) \sin \theta \sin \phi, \qquad (8.31)$$

where $f(kr, \theta)$ is a smoothly varying function that can be identified from (8.30). For $f = 1$ these fields reduce to the plane-wave fields (8.28).

When the incident Gaussian beam is on-axis azimuthal symmetry is restored and the scattering amplitudes take the simple forms

$$S_1(\beta, \theta) = \sum_{\ell=1}^{\infty} \frac{2\ell + 1}{\ell(\ell + 1)} g_\ell [a_\ell(\beta) \pi_\ell(\cos \theta) + b_\ell(\beta) \tau_\ell(\cos \theta)], \quad (8.32\mathrm{a})$$

$$S_2(\beta, \theta) = \sum_{\ell=1}^{\infty} \frac{2\ell + 1}{\ell(\ell + 1)} g_\ell [a_\ell(\beta) \tau_\ell(\cos \theta) + b_\ell(\beta) \pi_\ell(\cos \theta)], \quad (8.32\mathrm{b})$$

where the *beam-shape coefficients* are

$$g_\ell \equiv \frac{(-i)^{\ell-1}}{2\ell(\ell + 1)} \frac{kr}{j_\ell(kr)} \int_0^\pi \sin^2 \theta \, f(kr, \theta) e^{ikr \cos \theta} P_\ell^1(\cos \theta) \, d\theta. \qquad (8.33)$$

Thus, the cross sections of the plane-wave theory are modified by simply inserting a factor $|g_\ell|^2$ in the partial-wave sums (or $g_\ell g_{\ell+1}^*$ for the radiation pressure). Although the integrals (8.33) are still formidable, Grehan *et al.* (1986) have employed the localization principle to obtain a rather good leading-order approximation for small s:

$$g_\ell \simeq e^{-s^2(\ell+1/2)^2}. \tag{8.34}$$

Detailed justification for this approximation is discussed by Lock and Gouesbet (1994), and Gouesbet and Lock (1994).

We shall restrict further comments on this extension of the Mie theory to two particular results. The first concerns the issue of what we mean by extinction and its relation to the optical theorem when a dielectric sphere is illuminated by a Gaussian beam; and for this discussion we presume that we have an on-axis incident beam. When the beam half-width $w_0 \gg a$ we expect the beam to closely resemble a locally plane wave and extinction to occur just as we have considered it heretofore. In the geometric-optics limit of large size parameters the efficiency obtained from the optical theorem is $Q_{ext}(\beta) \simeq 2$ and, as we have discussed in Chapter 4 and subsequently, roughly half of this is due to reflection and half to diffraction.

If the beam is now narrowed, so that $w_0 < a$, we should expect diffraction to make a much smaller contribution to $Q_{ext}(\beta)$, and for $w_0 \ll a$ we would not be surprised to see the large-β limit approach unity. Lock (1995) has examined this question in some detail and demonstrated that the extinction efficiency as usually defined does *not* behave in this way. Rather, as w_0/a decreases $Q_{ext}(\beta)$ starts oscillating about the value 2 (rather than above it, as is the case for an incident plane wave), and the amplitude of the oscillations actually *increases* with β! It is clear that what is going on in the shadow of the sphere has been altered significantly and that this has a strong effect on the scattering in the far field. Equally clear is the need for new observables to describe how the energy is distributed in the case of narrow-beam illumination, as well as some experimental guidance, and Lock has made a start on the former.

Closely related to this question is that of the relation of σ_{ext} to the optical theorem, Eq. (3.73). This has been addressed by Lock *et al.* (1995), who begin by expanding various quantities in powers of the parameter s to obtain a .wide-beam approximation:

$$\sigma_{ext} \simeq \frac{4\pi}{k^2} \operatorname{Re} S(0) \left(1 - \frac{s^2 |S(0)|^2}{\operatorname{Re} S(0)} \right). \tag{8.35}$$

That is, the optical theorem itself is only an approximation that becomes exact in the limit of plane-wave scattering. We note, however, that (8.35) is *not* the beginning of a series expansion of $\mathrm{Re}\,S(0)$, but only the stated approximation.

Unless the frequency is very high the expression (8.35) is not useful when $w_0 < a$, and from the above discussion it is not at all clear that even the scattering cross section for a nonabsorbing particle can be related to $S(0)$. One can still argue that, for $\beta \gg 1$, most of the scattered intensity is in the near-forward direction, and Lock *et al.* (1995) find an approximation to Q_{ext} in that way that has roughly the same *form* as the plane-wave result (6.3), though with somewhat different behavior for large β. Thus, the leading factor of 2 is still present and the difficulties of interpretation raised above remain. We conclude that the field of narrow-Gaussian-beam scattering is still developing, with important insights yet to be gained.

Despite these problems, the present theory has also yielded some very positive results in terms of well-understood observables. Barton, *et al.* (1988) have carried out a numerical study of the source function both for on-axis and for off-axis incidence of a narrow ($w_0/a = 0.8$) Gaussian beam (TEM$_{00}$). The requisite scattered and internal fields are readily obtained from the Debye potentials as indicated in Eq. (3.78). As expected, the beam provides a much stronger channeling of the incident energy through the center of the particle. In addition, as the position of the focal point is moved off center in the xy-plane, the internal energy captured is reduced and it is concentrated to the side where the focal point is positioned. Thus, the distribution of energy within the sphere seems well understood.

8.3 Proximate spheres

Throughout the preceding chapters we have been concerned only with scattering from a single macroscopic particle. There is nothing to prevent us from immediately applying these results to a collection of particles, however, as long as multiple-scattering effects are negligible; that is, as long as the density of the medium is low enough to ensure that there is a sufficient average distance between particles that they can be considered independent, meaning that they do not interact appreciably with one another via their radiation fields when they are illuminated by an incident electromagnetic field. In this event both the forward scattering amplitude and the extinction efficiency of the medium are simple sums of those for each particle.

By considering a rectangular slab of N particles constituting such a material, say, one concludes that the overall effect is to provide the medium with a complex index of refraction (e.g., van de Hulst (1957)). The real part

$$n = 1 + 2\pi N k^{-3} \operatorname{Im}[S(0)] \tag{8.36}$$

affects the phase of a wave traveling through the medium, while the imaginary part provides the total extinction efficiency:

$$\Sigma_{\text{ext}} = 4\pi N k^{-2} \operatorname{Re}[S(0)] = N\sigma_{\text{ext}}. \tag{8.37}$$

Identification of this refractive index depends on it not differing too much from unity. Rayleigh had arrived at (8.36) already in 1871 (Strutt 1899) and, although it was not written explicitly, deduced (8.37). Further advances in studying this system are discussed in Meeten (1997).

What happens in the above system as the interparticle distances are decreased? For the moment we shall focus on a pair of possibly dissimilar dielectric spheres, a problem first studied quantitatively by Trinks (1935), and extended by Levine and Olaofe (1968). These authors employed the Debye potentials in much the same way as we have done in Chapter 3, but it happens that a description in terms of vector spherical harmonics, as also discussed there, is much more efficient and more amenable to generalization to systems of many particles.

In Chapter 3 we chose to employ the vector spherical harmonics $X_{\ell m}$, whose properties are discussed in Appendix D, because these have broader utility in some other areas of physics. Owing to the availability of important transformation theorems, it is more useful here to introduce an alternative representation of the divergenceless solutions to the vector wave equation. This representation is characterized by the linearly independent vector functions

$$M_{\ell m}^{(j)}(r) \equiv \nabla \times r z_{\ell m}^{(j)}, \qquad N_{\ell m}^{(j)}(r) \equiv \frac{1}{k} \nabla \times M_{\ell m}^{(j)}(r), \tag{8.38}$$

where $\{z_{\ell m}^{(j)}\}$ form a complete set of solutions to the scalar Helmholtz equation (3.7):

$$z_{\ell m}^{(1)} = j_\ell(kr) \left(\frac{(2\ell + 1)(\ell - m)!}{4\pi(\ell + m)!} \right)^{1/2} P_\ell^m(\cos\theta) e^{im\phi},$$

$$z_{\ell m}^{(3)} = h_\ell^{(1)}(kr) \left(\frac{(2\ell + 1)(\ell - m)!}{4\pi(\ell + m)!} \right)^{1/2} P_\ell^m(\cos\theta) e^{im\phi}, \tag{8.39}$$

and $\{z_{\ell m}^{(2)}, z_{\ell m}^{(4)}\} \propto \{y_\ell, h_\ell^{(2)}\}$ will not be needed here. A description in terms of $M_{\ell m}$ and $N_{\ell m}$ is completely equivalent to one in terms of $X_{\ell m}$. As usual, $0 \le \ell \le \infty$, $-\ell \le m \le \ell$.

The general scattering problem, independent of any symmetries possessed by the target, is described in terms of the implicitly time-harmonic incident, scattered, and internal electric fields,

$$E_{\text{inc}} = \sum_{\ell,m} \left[p_{\ell m} N^{(1)}_{\ell m}(\mathbf{r}) + q_{\ell m} M^{(1)}_{\ell m}(\mathbf{r}) \right], \tag{8.40a}$$

$$E_{\text{sca}} = \sum_{\ell,m} \left[a_{\ell m} N^{(3)}_{\ell m}(\mathbf{r}) + b_{\ell m} M^{(3)}_{\ell m}(\mathbf{r}) \right], \tag{8.40b}$$

$$E_{\text{int}} = \sum_{\ell,m} \left[c_{\ell m} N^{(1)}_{\ell m}(\mathbf{r}) + d_{\ell m} M^{(1)}_{\ell m}(\mathbf{r}) \right], \tag{8.40c}$$

and the magnetic fields are proportional to $\nabla \times E$. Application of the boundary conditions at the spherical surface allows one to determine the coefficients $a_{\ell m}$ and $b_{\ell m}$ in terms of those of the incident field, $p_{\ell m}$ and $q_{\ell m}$. Scattering amplitudes are identified, and cross sections are calculated in the usual way, though now one must deal with angular functions

$$\pi_{\ell m}(\theta, \phi) \equiv \frac{m}{\sin\theta} \left(\frac{(2\ell+1)(\ell-m)!}{4\pi(\ell+m)!} \right)^{1/2} P^m_\ell(\cos\theta) e^{im\phi}, \tag{8.41a}$$

$$\tau_{\ell m}(\theta, \phi) \equiv \left(\frac{(2\ell+1)(\ell-m)!}{4\pi(\ell+m)!} \right)^{1/2} \frac{d}{d\theta} P^m_\ell(\cos\theta) e^{im\phi}. \tag{8.41b}$$

Envision two spheres close enough together with a distance d between their centers so that they can no longer be considered independent, and let a linearly polarized plane wave be incident along the z-axis, as in Fig. 8.4. Because the targets possess spherical symmetry in their own coordinate systems the problem simplifies somewhat, but not entirely. Although the total scattered field E_{sca} is just a sum of the fields scattered from each sphere, this is a bit deceiving, for each of those fields is a functional of the other. The problem is highly nonlinear owing to the multiple scattering between spheres.

This system was first solved formally by Bruning and Lo (1971), by utilizing the translation and rotation addition theorems for vector spherical harmonics (Stein 1961; Cruzan 1962). In this way the expansions (8.40) for the individual particles can be combined into total fields by well-defined transformations. This results in an effective linearization of the problem in the form of an infinite set of simultaneous linear equations determining the expansion coefficients, and is complicated by the additional presence of transformation coefficients characterizing the translation. A first approach is to truncate these in an appropriate way and employ matrix-inversion techniques, yielding a manageable computational task, at least for size parameters of $O(1)$.

Kattawar and Dean (1983) carried out such a program for two identical spheres and obtained excellent agreement with earlier experimental results

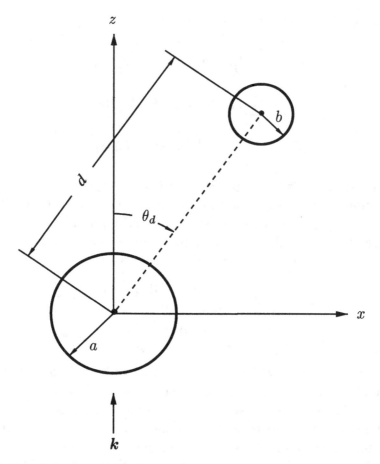

Fig. 8.4. Two dielectric spheres close enough to interact via multiple scattering upon illumination by a plane wave incident along the z-axis.

of Wang *et al.* (1981). Of some interest is the observation of an apparent resonance in the scattered intensity when the axis connecting the two spheres is orthogonal to the incident wavevector. In addition, the data seem to indicate that this type of multiple scattering becomes important when the spheres are within about ten sphere diameters of one another. Both the numerical and the experimental work have been confirmed independently by means of an integral-equation approach (Hall and Mao 1995), which also suggests that, for other than close proximity ($kd \lesssim 4$), feedback from multiple scattering has little effect on the extinction efficiency. That is, the observed oscillations in Q_{ext} as a function of the separation are due primarily to interference between the scattered fields from the two spheres, which approaches the constant sum of the extinction efficiencies for two single spheres at $kd \simeq 20$ and $\beta = 0.9283$.

Scattering from two coated spheres of the type shown in Fig. 8.1 has been investigated by Hamid, *et al.* (1992), who provide representative numerical results for various cross sections. The behavior of resonances in the forward scattered intensity for two identical spheres has been investigated numerically by Fuller (1989). When the spheres are very close together and $\beta \simeq 30$ the single-sphere resonances are shown to split and broaden, and it is clear that the multiple scattering affects the resonance spectrum significantly. Finally, a system consisting of two conducting osculating, or overlapping spheres has been studied by Videen *et al.* (1996), a geometry intermediate between those of Figs. 8.2 and 8.4. The backscattered intensity as a function of the orientation angle has been computed and compared with experiment for $\beta \simeq 1.6$ and $kd \simeq 2.68$, with good general agreement. These various scenarios and their good agreement with experiment lend strong support to the theoretical framework for multiple scattering developed by Bruning and Lo (1971).

Many spheres

Extension of the preceding formalism to a system of N spheres is quite straightforward, the feasibility having been demonstrated long ago in a rather formal way by Twersky (1967). A first extension was carried out by Borghese, *et al.* (1979), who actually employed the vector spherical harmonic representation $X_{\ell m}$.

Subsequently the multipole expansions were worked out and applied in a different way by Fuller and Kattawar (1988a, b) and used to study clusters of three and four identical spherical particles. For $kd = 2ka = 6.228$ and $n = 1.366 + 0.005i$ very good agreement with known data is achieved. In this work the scattering coefficients were derived using a method alternative to matrix inversion, and which nicely parallels the physical process. This is called an *order-of-scattering* (OS) technique, since it is organized according to the successive reflections from one sphere to another that eventually lead to the total scattered field. It is a very efficient method that can reduce the computation time by an order of magnitude in some cases.

The matrix-inversion approach has been further refined by Hamid *et al.* (1990), and an effective group-theoretical method is presented by Borghese *et al.* (1984). More recently, the entire multiple-sphere formalism has been carefully re-analyzed by Mackowski (1991, 1994), from which a number of fresh analytic techniques and efficient computational algorithms have emerged. Despite this progress, however, the general problem for many spheres and large size parameters remains computationally difficult, and provides fertile ground for asymptotic analyses.

8.4 Bubbles and $n < 1$

Although most of our experience with light scattering is from targets that are optically more dense than the surrounding medium, there is certainly no fundamental difficulty in turning the situation around. Indeed, this is a scenario that plays right into our hands, for the major application one envisions for relative refractive indices less than unity is to gas bubbles in a liquid, and bubbles are almost always much larger than the wavelength of light incident upon them. Thus, we note immediately two features of scattering scenarios with $n < 1$ that should dominate the physics: total reflection at the spherical surface can play a significant role, and the effects of semiclassical scattering should be very important.

In the ray description of geometric optics, discussed in Chapter 4, a few new wrinkles arise when the relative refractive index is less than unity; it will be convenient to take $n = 0.75$ for most of this section. Figure 8.5a illustrates the existence of a minimum scattering angle θ_m, below which all rays are totally reflected and none is transmitted into the sphere. The corresponding angle of incidence is the critical angle θ_c, so that

$$\theta_m = \pi - 2\theta_c = 2\cos^{-1} n,$$
$$\theta_c = \sin^{-1} n, \qquad n < 1. \tag{8.42}$$

With the above value of n we have $\theta_c \simeq 48.59°$ and $\theta_m \simeq 82.82°$ for air bubbles in water. Total reflection occurs for all angles of incidence $\theta_i > \theta_c$, thereby defining an additional shadow region along with the geometric shadow.

Figure 8.5b shows how several rays can contribute to the scattering in a specific direction ($\theta = 60°$ here), and these are chosen to illustrate a particular point below. For the moment we note only that they correspond to different terms in the Debye expansion, which formally proceeds as earlier in the case $n > 1$. Here, however, all terms in the expansion have the same shadow boundary defined by $\theta = \theta_m$, while the Regge poles corresponding to narrow resonances are to be found in the region $\text{Re}\,\lambda < n\beta$, and now $\beta \gg n\beta \gg 1$. The effective potential for $n < 1$ corresponds to a *barrier* (e.g., Eq. (7.23)), so that neither resonances nor ripple are to be expected in the cross sections.

Davis (1955) appears to be the first to have studied this system to any extent, carrying out a detailed geometric optics analysis. Among other things, he uncovered a singularity in the derivative of the scattered intensity with respect to the scattering angle $\theta = \theta_m$, indicating that there is a sharp decrease in scattered intensity at larger angles. Such a discontinuity originates with

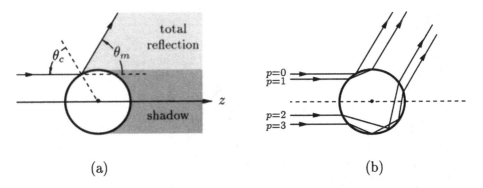

Fig. 8.5. Scattering from a dielectric sphere whose refractive index is *less* than that of the surrounding medium. (a) An illustration of the critical angle of incidence θ_c and the minimum scattering angle θ_m below which there is only total reflection. (b) Scattering of rays corresponding to various terms in the Debye expansion.

the Fresnel reflectivities (1.37) and (1.38), which increase very rapidly to unity as the angle of incidence approaches the critical angle θ_c. The derivatives of $R_\parallel$ and $R_\perp$ contain a factor $(1 - n^{-1} \sin^2 \theta_i)^{-1/2}$, providing an infinite slope at $\theta_i = \theta_c$, and total reflection beyond. The corresponding critical scattering angle θ_m represents the minimum deviation of partially reflected rays.

This enhancement of the scattered intensity for $\theta < \theta_m$ is referred to as *near-critical scattering* and was actually reported long ago by Pulfrich (1888), who observed it in the scattering of natural light from a cloud of air bubbles in water. It is also observed clearly in bubble-chamber measurements (Glaser 1958), and seen commonly by scuba divers.

Near-critical scattering is contained in the Mie partial-wave series, of course, and Fig. 8.6 exhibits the behavior of the gain factor G_2 of Eq. (3.125) near the critical scattering angle. Several features of the phenomenon are to be noted here, the first of which is the major background oscillation of the intensity in the region of total reflection that decays rapidly in the vicinity of $\theta = \theta_m$. This oscillation is a result of diffraction, which also smooths the discontinuity of geometric optics in the near-critical region; it is quite reminiscent of the Airy rainbow curve of Fig. 6.9 with 'supernumerary' oscillations and a rapid decay to the 'shadow'. A physical-optics model similar in spirit to Airy's theory and that includes these diffraction effects has been constructed by Marston (1979), and is extended in Marston and Kingsbury (1984). From this model we learn that the background oscillation arises from interference between the direct-reflection ($p = 0$) and direct-transmission ($p = 1$) rays shown at the top of the sphere in Fig. 8.5b – once again we encounter a near-edge effect. Note the growth in amplitude

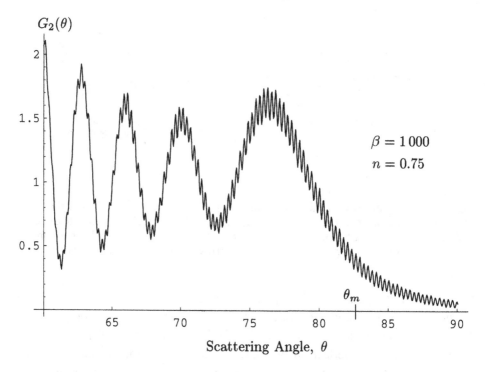

Fig. 8.6. The gain factor G_2 for $n = 0.75$, Eq. (3.125), as a function of scattering angle over a range including the critical scattering angle θ_m. Evident are the broad background oscillations increasing in amplitude into the total reflection region, the rapid dropping off into the region of partial reflection, and the high-frequency fine structure.

as one moves below the critical scattering angle and further into the region of total reflection. This increase reflects the increasing amplitudes of directly transmitted rays toward the forward direction.

A second feature of Fig. 8.6 is the fine structure in the form of very rapid oscillations superimposed on the background. The above model shows that this fine structure arises from further interference between the $p = 0$ ray and rays such as those shown at the bottom of the sphere. (Of course, all this is symmetric and both sets of rays can also be found in the exchanged positions and contributing to the scattering below the optical axis.) Further experimental work (Langley and Marston 1984) illustrates how very well the Mie curve fits the data, and that the physical-optics model does a reasonable job of explicating the general physical situation.

While this model provides a good overall qualitative picture of near-critical-angle scattering, and even quantitative agreement with the Mie curves for $\theta_i \lesssim \theta_c$, it does not give a very accurate account of the intensities for θ

close to θ_m. As earlier, one needs to introduce a detailed asymptotic analysis of the scattering amplitudes in the region of an angle of minimum deviation, so that the effects of diffraction at a curved surface are made manifest. Such a study has been carried out by Fiedler-Ferrari *et al.* (1991), who first remove the fine structure by subtracting out the rays indicated at the bottom of the sphere in Fig. 8.5b, which are the source of these rapid oscillations. Clearly, these rays have little to do with the near-critical-angle effect, though they do contribute to the overall intensity. This operation, which requires removal of some 100 Debye terms, then reveals accurately the main background curve.

Physically, direct reflection dominates close to the critical angle, modified by the effects of surface curvature, though the contributions to partial and total reflection are treated separately. In the latter an interesting polarization-dependent shift of the scattering angle is uncovered in the planar (i.e., $\beta \gg 1$) limit. This is a leading-order effect of surface curvature, rather than a dynamical effect, arising from a spread in the angles of incidence that is absent for a plane interface. One finds that (8.42) for the scattering angle is modified to

$$\theta = \pi - 2\theta_c - \delta\theta_j, \qquad \delta\theta_j \simeq -\frac{v_j}{\beta}\sqrt{\frac{2n}{\epsilon}}\frac{1}{(1-n^2)^{3/4}}, \qquad (8.43)$$

where ϵ is defined by $\theta_i = \theta_c + \epsilon$, and v_j is the multipole indicator. That is, an incident ray just above the critical ray tunnels into the sphere and, rather than being reflected immediately, travels an additional arc length before re-emerging at the same reflection angle. The effect is illustrated in Fig. 8.7, in which we presume that $|\delta\theta_j| \ll \epsilon$. A shift of this type in total reflection from a plane interface was first noticed by Goos and Hänchen (1947).

The result of the asymptotic analysis in the complex angular-momentum plane is in excellent agreement with that of the Mie series throughout the near-critical region, and extending well beyond it on either side. Numerous examples of comparison curves, for both polarizations and their phase difference, are provided by Fiedler-Ferrari, *et al.* (1991) for $n = 0.75$ and $\beta = 5000$.

Earlier we noted that the forward glory has been observed in scattering from bubbles (Langley and Marston 1991) and is described by the physical-optics model. The spacing of the minima in the ring pattern is proportional to the inverse of the bubble radius, and thus provides a means for sizing the bubbles.

More recently, scattering of light from bubbles has proved important in studying the fascinating and still incompletely understood phenomenon

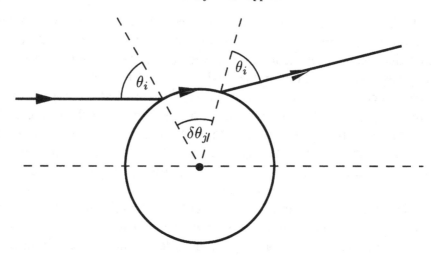

Fig. 8.7. An incident ray just above the critical ray tunnels into the sphere and, rather than being reflected immediately, travels an additional arc length determined by the Goos–Hänchen angular displacement $\delta\theta_j$ before re-emerging at the same reflection angle.

of sonoluminescence (e.g., Crum (1994)). It is basically a cavitation phenomenon, which was first observed by Frenzel and Schultes (1934). Under certain circumstances sound waves can cause bubbles in a water tank to collapse extremely rapidly and generate ultrashort (50 ps) light flashes. The bubbles act as transducers of sound energy to light energy. By noting from the geometric theory that the scattered intensity is proportional to the square of the bubble radius (e.g., Davis (1955)), one can illuminate the bubble with ultrashort laser pulses at different times during its implosion and determine the speed of its collapse. Such an experiment has found this speed to be about $\sim 1500\,\text{m s}$ (Weninger, *et al.* 1997), and provides good evidence for a shock-wave model of the phenomenon. Lentz *et al.* (1995) have used a similar technique to measure the size of sonoluminescing bubbles.

Thin films

By now the reader may already have remarked that the preceding discussion does not appear to encompass those bubbles with which we are perhaps most familiar, such as the soap bubbles in air that delight all children at one time or another. Indeed, such bubbles do not immediately fit into the above discussion, but they can be studied by returning to the coated-sphere model and adapting it to this notion. Moreover, it is worth doing so for several reasons, not the least of which is that the potential applications go well beyond soap bubbles to various objects of biochemical interest.

Let the outer shell in Fig. 8.1 be reduced to a small thickness δ and presume the core to be of the same medium as that outside the concentric spheres. The refractive index n of the shell is then relative both to the core and to the external medium, and will be taken to be *greater* than unity. In (8.1) and (8.2) we then make the replacements $n_1 \rightarrow 1$ and $n_2 \rightarrow n$ and, for $\delta = \beta - \alpha \ll 1$, we can expand these partial-wave coefficients in powers of δ. This is accomplished by expanding all the Riccati–Bessel functions of argument $\beta = \alpha + \delta$ and retaining only the leading-order terms in δ. Some algebraic reduction yields the approximations

$$a_\ell \simeq -i(n^2 - 1)\left([\psi_\ell'(\alpha)]^2 + \frac{\ell(\ell+1)}{n^2\alpha^2}\psi_\ell^2(\alpha) \right)\delta + O(\delta^2),$$
$$b_\ell \simeq -i(n^2 - 1)\psi_\ell^2(\alpha)\delta + O(\delta^2), \tag{8.44}$$

where $\alpha = kb$ is the size parameter of the cavity.

These are the thin-film partial-wave coefficients, first worked out by Aragón and Elwenspoek (1982). It is somewhat enjoyable to pursue this development a bit further here, because this is one of those very rare problems in scattering theory that can actually be solved exactly in closed form! The authors accomplish this with a clever algebraic device that we now outline.

Prior to substituting (8.44) into (3.95) for the scattering amplitudes, let us agree to work with normalized quantities $\bar{a}_\ell \equiv ia_\ell/(n^2 - 1)\delta$, etc., and to further consider temporarily only the difference quantities $\Delta\bar{a}_\ell \equiv \bar{a}_\ell(n) - \bar{a}_\ell(n = 1)$, etc. (According to Eq. (4.8) this is tantamount to subtracting out the Rayleigh–Gans limit). With this in mind, and reference to (D.43b) for $\tau_\ell(x)$, $x = \cos\theta$, we have for polarization 1

$$\Delta\bar{S}_1 = -\frac{n^2 - 1}{n^2}\frac{d}{dx}\sum_{\ell=0}^{\infty}(2\ell + 1)\frac{\psi_\ell^2(\alpha)}{\alpha^2}P_\ell(x). \tag{8.45}$$

Now, this sum is exactly in the form of a well-known addition theorem for spherical Bessel functions (Abramowitz and Stegun 1964, 10.1.45) and has the value $j_0(qa)$, where $q = 2k\sin(\theta/2)$. That is, q is the magnitude of the elastic momentum transferred from the incident to the scattered wave. From the recurrence formulas the derivative is then $[\alpha^2/(qa)]j_1(qa)$. To obtain $\bar{S}_1$ itself we now need $\bar{S}_1(n = 1)$, and this requires substantial manipulation. However, the sum can again be written in closed form and is found by the above authors to be

$$\bar{S}_1(n=1) = \sum_{\ell=0}^{\infty} (2\ell+1)\psi_\ell^2(\alpha)P_\ell(\cos\theta) = \alpha^2 j_0(qa). \tag{8.46}$$

Thus, finally,

$$S_1(\beta,\theta) = -i\delta\alpha^2(n^2-1)\left(j_0(qa) - \frac{n^2-1}{n^2}\frac{j_1(qa)}{qa}\right), \tag{8.47}$$

and similarly

$$S_2(\beta,\theta) = -i\delta\alpha^2(n^2-1)\left[j_0(qa)\cos\theta + \frac{(n^2-1)}{n^2}\frac{j_1(qa)}{qa}\right.$$
$$\left. + \frac{(n^2-1)}{n^2}\frac{1+\cos\theta}{2}\left(\frac{j_1(qa)}{qa} - j_0(qa)\right)\right]. \tag{8.48}$$

Note that these amplitudes suffer the same shortcoming as those for Rayleigh scattering, Eqs. (4.1), and for the Rayleigh–Gans theory, Eqs. (4.9), in that they cannot be used in the optical theorem to find the extinction efficiency. To do so one would have to continue the calculation to at least $O(\delta^2)$. One can integrate their squares, of course, to obtain the scattering cross section, but that appears to be analytically intractable. The intensities themselves can be studied readily, however, as can the polarization. Indeed, Aragón and Elwenspoek (1982) show that the depolarization ratio $|S_2/S_1|^2$ for the bubble is almost a factor of three greater than that for the corresponding solid sphere; the scattered intensity in polarization 2 is much greater for the bubble.

A similar closed-form analysis has been applied to the anisotropic model of Eqs. (8.10) and (8.11) by Lange and Aragón (1990). It may also be of interest to note that, although it does not result in a completely closed form, the theory has been extended to second order in δ by Hahn and Aragón (1994).

8.5 Nonspherical particles

The final departure from our standard model that we shall address here is perhaps also the most important to many areas of research involving light scattering from particles. An extension of the model to some geometrically regular, but noncircular particle cross sections was discussed briefly in Section 4.2 in connection with the anomalous-diffraction approximation. However, a complete theoretical development for systems lacking spherical symmetry has proved to be very difficult to uncover, though in some cases computational models have provided some insights.

In principle it is not at all difficult to formulate the scattering problem for particles of arbitrary shape and finite size – this was already done in Chapter 3. The incident, scattered, and internal fields are given by Eqs. (8.40), with the help of (8.38) and (8.39). There are now generally four scattering amplitudes to be considered, as in (3.31), and (3.95) is replaced by a set of four equations involving the angular functions (8.41) and the partial-wave scattering coefficients $a_{\ell m}$ and $b_{\ell m}$ that must be determined from boundary conditions. This stipulation presents an intractable problem for completely arbitrary particles (e.g., ones with various facets and edges). There are actually four possible coefficients $a_{\ell m}^{(1,2)}$ and $b_{\ell m}^{(1,2)}$ necessary in order to account for cross polarization.

It is well known that the 3-dimensional wave equation is separable in only 11 curvilinear coordinate systems. These are illustrated and discussed in detail by Morse and Feshbach (1953, p. 655). Boundary matching is thus severely limited to only a few geometries other than spherical for which we can obtain exact solutions in the form of partial-wave sums.

Solvable geometries

Four of the 11 coordinate systems mentioned above are cylindrical, and special interest has long been shown in the circular cylinder of infinite length; (infinite so as to avoid messy end effects). These can serve as models of various fibers, as well as for biological objects such as viruses, and in many cases the particle is long enough to appear effectively infinite. Over a century ago Rayleigh solved the problem for a plane wave incident perpendicular to the cylinder axis (Strutt 1881, 1918), well before the sphere had been treated properly. Only many years later was a complete and correct solution for oblique incidence obtained (Wait 1955, 1965).

The procedure for constructing the general solution for the circular cylinder, analogous to that in Chapter 3 for the sphere, is outlined both by van de Hulst (1957) and Kerker (1969). As is expected, the scattering coefficients are composed of cylinder functions, or ordinary Bessel functions of integer order. For example, in a cylindrical geometry with z-axis along the symmetry axis of the cylinder and transverse coordinates (r, φ), a plane wave incident along the positive x-axis has the expansion (e.g., Morse and Feshbach (1953, p. 828))

$$e^{ikx} = \sum_{m=0}^{\infty} \epsilon_m i^m \cos(m\varphi) \, J_m(kr), \qquad \epsilon_m \equiv \begin{cases} 1, & m = 0 \\ 2, & m \neq 0, \end{cases} \qquad (8.49)$$

a cylindrical analog of the spherical expansion (2.16). In the case of oblique incidence the expressions for the partial-wave coefficients are somewhat cumbersome, but can be found explicitly in Bohren and Huffman (1983). When the incident wavevector is perpendicular to the cylinder axis they reduce to forms similar to those for the sphere:

$$a_\ell = \frac{J_\ell(\beta)J'_\ell(n\beta) - nJ'_\ell(\beta)J_\ell(n\beta)}{J'_\ell(n\beta)H^{(1)}_\ell(\beta) - nJ_\ell(n\beta)H^{(1)'}_\ell(\beta)},$$

$$b_\ell = \frac{J'_\ell(\beta)J_\ell(n\beta) - nJ_\ell(\beta)J'_\ell(n\beta)}{J_\ell(n\beta)H^{(1)'}_\ell(\beta) - nJ'_\ell(n\beta)H^{(1)}_\ell(\beta)}. \tag{8.50}$$

Oblique incidence results in a significant cross-polarized component in the scattered field, which is absent at normal incidence. Cross-polarization requires four partial-wave coefficients, but the two that vanish at perpendicular incidence differ only in sign (Kerker *et al.* 1966), a consequence of time-reversal invariance (Cohen and Acquista 1982). Cross-polarized components also vanish if the cylinder is perfectly conducting. When incidence is perpendicular the scattered radiation propagates as a cylindrical wave, but at oblique incidence it is propagated along the surface of a cone centered on the cylindrical axis, as illustrated in Fig. 8.8. As the angle of incidence ϕ decreases to zero the cone opens up into a plane perpendicular to the cylinder.

Scattering from circular cylinders is also amenable to application of some of our earlier analytic techniques for large size parameters. In this vein, Guimarães and Mendonça (1997) have performed a semiclassical analysis of resonant modes and extended it to coated cylinders as well. The Debye expansion has been employed by Lock and Adler (1997) to analyze the primary rainbow produced by oblique incidence, and they have observed it experimentally.

Aside from the approximation of infinite length, practical applications of scattering from cylinders also face an ambiguity as to orientation. Collections of fibers, say, will rarely have all their components uniformly aligned, so that one must inevitably average over all orientations. Although it is finite in extent, this problem persists also for oblate and prolate spheroids, which have discrete symmetry axes. Nevertheless, this scenario of scattering from spheroids has been solved exactly (Asano and Yamamoto 1975; Asano and Sato 1980). Because it provides a model for nonspherical particles on the level of the Mie solution, it is one badly in need of asymptotic analysis on the level of that developed for the sphere. This has yet to be done.

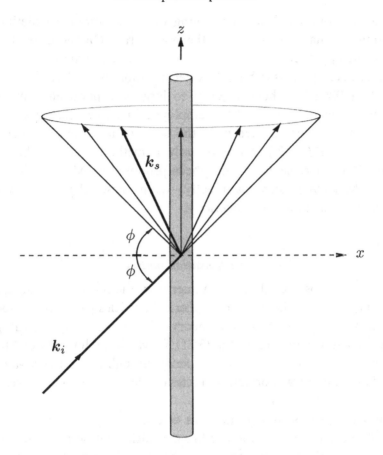

Fig. 8.8. The conical pattern for scattering of a plane wave from an infinite circular cylinder. The angle of oblique incidence ϕ with respect to the xy-plane is the same as that of the exterior apex angle of the cone.

Perturbation theory

An obvious approach to the study of light scattering from particles that depart only slightly from spherical is through perturbation methods. Raindrops distorted by gravity, for example, provide a possible application. The main developments of this sort have focused on boundary perturbations, of both the boundary conditions and of the physical boundary itself. The earliest work in this area seems to be that of Yeh (1964, 1965), who employed Taylor expansions in the perturbation parameter characterizing the departure from sphericity, and then retained only the first terms. These ideas were later extended to all orders in a series of articles by Erma (1968a, b, 1969).

Although asymptotic series need not converge in the ordinary sense, to be useful their successive terms should decrease rapidly up to a certain point.

Thus, on the one hand, a shortcoming of the perturbation method is its restriction to small distortions at the boundaries. On the other hand, the merit of the approach is its applicability to *arbitrary* shapes.

Further developments in boundary perturbation theory have been reported by Krebs (1982a, b), who extends it to large size parameters by utilizing some of the techniques we have employed in earlier chapters. The type of distortion considered here is in the form of a variable particle radius $r(\theta) = a[1 + \epsilon P_2(\cos \theta)]$. More recently a method valid in the opposite domain of size parameters $\lesssim 1$ has been proposed by Martin (1993), which perturbs the Debye potentials directly. Again, these schemes apply only to quite small departures from sphericity.

The transition matrix

Currently, the most popular, and numerically the most powerful, approach to the study of scattering from nonspherical particles parallels methods long used extensively in the quantum theory of scattering. Indeed, the general notion is already expressed in Eq. (3.31), in which the linearity of Maxwell's equations relates scattered to incident fields through the scattering-amplitude matrix S. A similar transformation relates incident to scattered Stokes vectors via the transfer matrix in (3.36).

In his exhaustive symmetry analysis of electromagnetic scattering Waterman (1971) extended the notion of linear transformation to a direct relation between the partial-wave expansion coefficients of the incident and scattered waves. Irrespective of particle shape and composition the various fields can always be expanded as in (8.40), and the T-matrix is introduced by asserting that

$$\begin{pmatrix} a \\ b \end{pmatrix} = \begin{pmatrix} T^{11} & T^{12} \\ T^{21} & T^{22} \end{pmatrix} \begin{pmatrix} p \\ q \end{pmatrix}. \tag{8.51}$$

For spherical particles T is diagonal.

Aside from its applicability to arbitrary shapes, a major attribute of the T-matrix is that it is independent of the incident and scattered fields, depending only upon the size parameter, refractive index, and particle orientation. In addition, its components are analytically well behaved and it is readily averaged over all orientations. Once T has been determined one can easily compute the elements of the scattering-amplitude matrix, which completely determines the scattering.

The next step is to compute the T-matrix elements for nonspherical particles, and this remains a formidable problem. For single particles the standard approach is the so-called *extended boundary condition method* (EBCM) introduced by Waterman. This technique first relates the coefficients of the incident fields to those of the internal fields, and thence the scattered fields to the internal fields. The elements of the matrices involved here are two-dimensional integrals of products of the vector spherical harmonics (8.38) taken over the surface of the particle, and some matrix algebra allows one to identify the elements of the *T*-matrix. This procedure is in general rather complicated, but is simplified considerably if the particles possess an axis of symmetry. Indeed, almost all applications of the method have been to objects of revolution, or axisymmetric particles, and in these cases it provides a very efficient computational tool.

Ström (1975) has provided a nice explication of the *T*-matrix method, and a recent review (Mishchenko, *et al.* 1996) provides a broad overview of its current status. An early, yet representative, application to small homogeneous spheroids and finite cylinders can be found in Barber and Yeh (1975), and smooth deformations in terms of Chebyshev polynomials have been treated by .Wiscombe and Mugnai (1980).

There are numerous other extensions of the basic model that could be discussed profitably – optically active particles (Bohren and Huffman 1983), acoustic scattering (Williams and Marston 1985), rotating spheres (de Zutter 1980), and glare spots (Lock 1987), to name just a few – but those above will have to suffice. In most of the extensions discussed here we see a common thread; namely, a necessary reliance on numerical analysis, even for relatively small size parameters. The physical picture of scattering from homogeneous spheres that emerged in earlier chapters is not yet as well developed in these other scenarios. To make it so, particularly for large size parameters, will require adaptation of asymptotic techniques for summing the various partial-wave series, along with other analytic approaches. Rather than consider this a negative outlook, we take it to herald an exciting future for research in scattering theory, and some efforts along these lines have already begun. For example, these techniques have been extended to resonant light scattering from a dielectric cylinder with an eccentric inclusion (Simão *et al.* 1999). Much more is to be expected.

Mathematical appendices

Appendices A–D define and describe properties of the various special functions employed in the main text, although we have included for the most part only those properties directly relevant to present needs. Authoritative general references to the behavior of these functions include the *Handbook of Mathematical Functions* edited by Abramowitz and Stegun (1964), which we usually abbreviate as HMF; Gradshteyn and Ryzhik (1980), *Table of Integrals, Series, and Products*; Watson's *Theory of Bessel Functions* (1995); Whittaker and Watson, *A Course of Modern Analysis* (1963); the Bateman-manuscript project's *Higher Transcendental Functions* (Erdélyi *et al.* 1953); and Szegö's *Orthogonal Polynomials* (1959). The reader is referred to these sources for all proofs – none is given here, though occasionally some are suggested. Additional appendices provide reference to mathematical and numerical techniques of value in studying scattering processes.

Appendix A
Spherical Bessel functions

The second-order differential equation

$$\left(\frac{d^2}{dz^2} + \frac{2}{z}\frac{d}{dz} + 1 - \frac{\ell(\ell+1)}{z^2} \right) u_\ell(z) = 0, \quad \ell = 0, \pm 1, \pm 2, \dots, \qquad \text{(A.1)}$$

possesses four separate solutions in terms of Bessel functions of half-integer index. The spherical Bessel functions of the first and second kinds, respectively, are special cases of the corresponding cylindrical functions:

$$j_\ell(z) \equiv \sqrt{\pi/(2z)}\, J_{\ell+1/2}(z), \qquad \text{(A.2)}$$

$$y_\ell(z) \equiv \sqrt{\pi/(2z)}\, Y_{\ell+1/2}(z), \qquad \text{(A.3)}$$

and the second is often called a Neumann function. These are linearly independent solutions for every ℓ. The spherical Hankel functions of the first and second kinds,

$$h_\ell^{(1)}(z) \equiv j_\ell(z) + i y_\ell(z) = \sqrt{\pi/(2z)}\, H_{\ell+1/2}^{(1)}(z), \qquad \text{(A.4a)}$$

$$h_\ell^{(2)}(z) \equiv j_\ell(z) - i y_\ell(z) = \sqrt{\pi/(2z)}\, H_{\ell+1/2}^{(2)}(z), \qquad \text{(A.4b)}$$

also form a pair of complex conjugate linearly independent solutions. The Wronskian of two such solutions is defined as $W\{f(x), g(x)\} \equiv f(x)g'(x) - f'(x)g(x)$.

Wronskians

$$W\{j_\ell(z), y_\ell(z)\} = z^{-2}, \qquad \text{(A.5a)}$$

$$W\{h_\ell^{(1)}(z), h_\ell^{(2)}(z)\} = -2i z^{-2}. \qquad \text{(A.5b)}$$

Cross products

$$j_\ell(z)y_{\ell-1}(z) - j_{\ell-1}(z)y_\ell(z) = z^{-2}, \qquad\qquad \text{(A.6a)}$$

$$j_{\ell+1}(z)y_{\ell-1}(z) - j_{\ell-1}(z)y_{\ell+1}(z) = (2\ell+1)z^{-3}. \qquad \text{(A.6b)}$$

Ascending series

$$j_\ell(z) = \frac{z^\ell}{(2\ell+1)!!}\left(1 - \frac{z^2/2}{1!(2\ell+3)} + \frac{(z^2/2)^2}{2!(2\ell+3)(2\ell+5)} - \cdots\right), \qquad \text{(A.7)}$$

$$y_\ell(z) = -\frac{(2\ell-1)!!}{z^{\ell+1}}\left(1 - \frac{z^2/2}{1!(1-2\ell)} + \frac{(z^2/2)^2}{2!(1-2\ell)(3-2\ell)} - \cdots\right), \qquad \text{(A.8)}$$

where $(2\ell+1)!! \equiv 1\cdot 3\cdot 5\cdots(2\ell+1)$, for example.

Explicit forms

$$j_0(z) = \frac{\sin z}{z} \qquad\qquad\qquad\qquad\qquad\qquad \text{(A.9a)}$$

$$j_1(z) = \frac{\sin z}{z^2} - \frac{\cos z}{z}, \qquad\qquad\qquad\qquad \text{(A.9b)}$$

$$j_2(z) = \left(\frac{3}{z^3} - \frac{1}{z}\right)\sin z - \frac{3}{z^2}\cos z. \qquad\qquad \text{(A.9c)}$$

$$y_0(z) = -j_{-1}(z) = -\frac{\cos z}{z}, \qquad\qquad\qquad \text{(A.10a)}$$

$$y_1(z) = j_{-2}(z) = -\frac{\cos z}{z^2} - \frac{\sin z}{z}, \qquad\qquad \text{(A.10b)}$$

$$y_2(z) = -j_{-3}(z) = \left(-\frac{3}{z^3} + \frac{1}{z}\right)\cos z - \frac{3}{z^2}\sin z. \qquad \text{(A.10c)}$$

Some low-order functions are shown explicitly in Fig. A.1.

Interrelations

$$y_\ell(z) = (-1)^{\ell+1}j_{-\ell-1}(z), \qquad \ell = 0, \pm1, \ldots, \qquad \text{(A.11a)}$$

$$h^{(1)}_{-\ell-1}(z) = i(-1)^\ell h^{(1)}_\ell(z), \qquad \ell = 0, 1, \ldots, \qquad \text{(A.11b)}$$

$$h^{(2)}_{-\ell-1}(z) = -i(-1)^\ell h^{(2)}_\ell(z), \qquad \ell = 0, 1, \ldots. \qquad \text{(A.11c)}$$

Spherical Bessel Functions

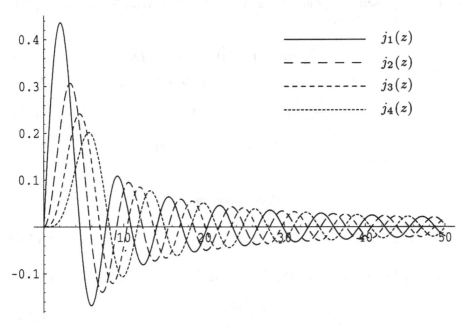

Spherical Neumann Functions

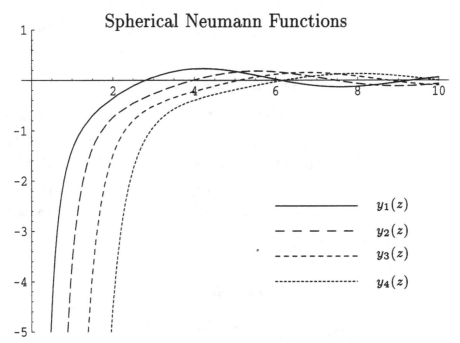

Fig. A.1. Representative behavior of the spherical Bessel and Neumann functions for $\ell = 1$–4.

Recurrence relations

Let $f_\ell(z)$ represent any one of the functions $j_\ell(z)$, $y_\ell(z)$, $h_\ell^{(1)}(z)$ and $h_\ell^{(2)}(z)$, for $\ell = 0, \pm 1, \ldots$ Then

$$f_{\ell-1}(z) + f_{\ell+1}(z) = \frac{(2\ell + 1)}{z} f_\ell(z), \tag{A.12a}$$

$$\ell f_{\ell-1}(z) - (\ell + 1) f_{\ell+1}(z) = (2\ell + 1) f_\ell'(z), \tag{A.12b}$$

$$\frac{\ell + 1}{z} f_\ell(z) + f_\ell'(z) = f_{\ell-1}(z), \tag{A.12c}$$

$$\frac{\ell}{z} f_\ell(z) - f_\ell'(z) = f_{\ell+1}(z). \tag{A.12d}$$

Differentiation

$$\left(\frac{1}{z}\frac{d}{dz}\right)^m \left[z^{\ell+1} f_\ell(z)\right] = z^{\ell-m+1} f_{\ell-m}(z), \tag{A.13a}$$

$$\left(\frac{1}{z}\frac{d}{dz}\right)^m \left[z^{-\ell} f_\ell(z)\right] = (-1)^m z^{-\ell-m} f_{\ell+m}(z), \qquad m = 1, 2, 3, \ldots; \tag{A.13b}$$

$$j_\ell(z) = z^\ell \left(-\frac{1}{z}\frac{d}{dz}\right)^\ell \frac{\sin z}{z}, \tag{A.13c}$$

$$y_\ell(z) = -z^\ell \left(-\frac{1}{z}\frac{d}{dz}\right)^\ell \frac{\cos z}{z}, \qquad \ell = 0, 1, 2, \ldots. \tag{A.13d}$$

Analytic continuation

$$j_\ell(ze^{im\pi}) = e^{im\ell\pi} j_\ell(z), \tag{A.14a}$$

$$y_\ell(ze^{im\pi}) = (-1)^m e^{im\ell\pi} y_\ell(z), \tag{A.14b}$$

$$h_\ell^{(1)}(ze^{i(2m+1)\pi}) = (-1)^\ell h_\ell^{(2)}(z), \tag{A.14c}$$

$$h_\ell^{(2)}(ze^{i(2m+1)\pi}) = (-1)^\ell h_\ell^{(1)}(z), \tag{A.14d}$$

$$h_\ell^{(j)}(ze^{i2m\pi}) = h_\ell^{(j)}(z), \tag{A.14e}$$

for $j = 1, 2$ and $m, \ell = 0, 1, 2, \ldots.$

Integrals

$$j_\ell(z) = \frac{1}{2i^\ell} \int_{-1}^{+1} e^{izt} P_\ell(t) \, dt. \tag{A.15a}$$

$$\frac{2k^2}{\pi} \int_0^\infty r^2 j_\ell(kr) j_\ell(k'r) \, dr = \delta(k - k'). \tag{A.15b}$$

The first is Poisson's integral representation in terms of Legendre polynomials, and the second is the orthogonality relation. Integrals of the latter type with unequal indices have been studied by Maximon (1991).

Asymptotic values

These follow directly from the expressions for the corresponding cylindrical functions in Appendix C, or HMF, but we provide the leading terms here for $x \gg \ell$;

$$j_\ell(x) \approx \frac{1}{x} \sin(x - \ell\pi/2), \tag{A.16a}$$

$$y_\ell(x) \approx -\frac{1}{x} \cos(x - \ell\pi/2). \tag{A.16b}$$

The asymptotic behavior of the spherical Hankel functions is contained explicitly in the exact expressions

$$h_\ell^{(1,2)}(x) = \frac{e^{\pm ix}}{x} i^{\mp(\ell+1)} \sum_{k=0}^{\ell} \frac{(\ell+k)!}{k!(\ell-k)!}(\mp 2ix)^{-k}$$

$$\xrightarrow[x \gg 1]{} (\mp i)^{\ell+1} \frac{e^{\pm ix}}{x}. \tag{A.17}$$

The asymptotic behavior as $\ell \to \infty$ for fixed x is best obtained from that of the cylindrical functions in Appendix C, but a qualitative feel for this can be gleaned from Fig. A.2.

Riccati–Bessel functions

The second-order equation

$$\left(\frac{d^2}{dz^2} + 1 - \frac{\ell(\ell+1)}{z^2}\right) w_\ell(z) = 0, \qquad \ell = 0, \pm 1, \pm 2, \ldots, \tag{A.18}$$

has the following pairs of linearly independent solutions:

$$\begin{aligned} \psi_\ell(z) &\equiv z j_\ell(z), & \xi_\ell(z) &\equiv z y_\ell(z); \\ \zeta_\ell^{(1)}(z) &\equiv z h_\ell^{(1)}(z), & \zeta_\ell^{(2)}(z) &\equiv z h_\ell^{(2)}(z), \end{aligned} \tag{A.19}$$

known as Riccati–Bessel functions. Substitution into (A.18) leads back to (A.1).

All properties of these functions follow directly from those of the spherical Bessel functions, although it may be useful to record the Wronskians

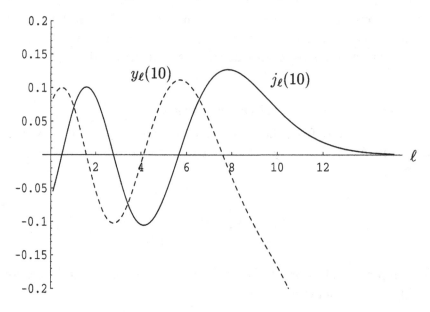

Fig. A.2. The spherical Bessel functions $j_\ell(x)$ and $y_\ell(x)$ for $x = 10$, illustrating the asymptotic behavior when $\ell > x$.

explicitly:

$$W\{\psi_\ell(z), \xi_\ell(z)\} = 1, \qquad W\left\{\zeta_\ell^{(1)}(z), \zeta_\ell^{(2)}(z)\right\} = -2i, \qquad \ell = 0, 1, \dots. \quad \text{(A.20)}$$

If $g_\ell(z) = z f_\ell(z)$ is any Riccati–Bessel function, it follows from the recurrence relations above that the logarithmic derivative is

$$D_\ell(z) \equiv \frac{g_\ell'(z)}{g_\ell(z)} = \frac{\ell+1}{z} - \frac{f_{\ell+1}(z)}{f_\ell(z)} = \frac{\ell+1}{z} - \frac{g_{\ell+1}(z)}{g_\ell(z)}$$

$$= \frac{\ell+1}{z} - \frac{1}{D_{\ell+1}(z) + (\ell+1)/z}. \quad \text{(A.21)}$$

Appendix B
Airy functions

The differential equation

$$\left(\frac{d^2}{dz^2} - z\right)w(z) = 0 \qquad (B.1)$$

has three pairs of linearly independent solutions:

$$\text{Ai}(z), \ \text{Bi}(z); \qquad (B.2a)$$

$$\text{Ai}(z), \ \text{Ai}(ze^{2\pi i/3}); \qquad (B.2b)$$

$$\text{Ai}(z), \ \text{Ai}(ze^{-2\pi i/3}), \qquad (B.2c)$$

called Airy functions. Their explicit definitions are either by integral representation or in terms of cylindrical functions:

$$\text{Ai}(z) = \frac{3^{1/3}}{\pi} \int_0^\infty \cos(t^3 + 3^{1/3}zt)\, dt$$

$$= \frac{\sqrt{z}}{3}[I_{-1/3}(\zeta) - I_{1/3}(\zeta)], \qquad (B.3a)$$

$$\text{Bi}(z) = \frac{3^{1/3}}{\pi} \int_0^\infty \left[e^{-t^3 + 3^{1/3}zt} + \sin(t^3 + 3^{1/3}zt)\right] dt$$

$$= \sqrt{\frac{z}{3}}[I_{-1/3}(\zeta) + I_{1/3}(\zeta)], \qquad (B.3b)$$

where $I_\nu(z)$ is the modified Bessel function of the first kind (HMF), and we have introduced what turns out to be an ubiquitous variable in the study of Bessel and related functions,

$$\zeta \equiv \tfrac{2}{3}z^{3/2}. \qquad (B.4)$$

Alternatively,

$$\text{Ai}(-z) = \frac{\sqrt{z}}{3}[J_{1/3}(\zeta) + J_{-1/3}(\zeta)], \qquad (B.5a)$$

$$Bi(-z) = \sqrt{\frac{z}{3}}\left[J_{-1/3}(\zeta) - J_{1/3}(\zeta)\right]. \tag{B.5b}$$

Wronskians

$$W\{Ai(z), Bi(z)\} = \pi^{-1}, \tag{B.6a}$$

$$W\{Ai(z), Ai(ze^{\pm 2\pi i/3})\} = \frac{1}{2\pi}e^{\mp i\pi/6}, \tag{B.6b}$$

$$W\{Ai(ze^{2\pi i/3}), Ai(ze^{-2\pi i/3})\} = \frac{i}{2\pi}. \tag{B.6c}$$

Note that the derivatives are with respect to z; for example,

$$\frac{d}{dz}Ai(ze^{2\pi i/3}) = e^{2\pi i/3}\,Ai'(ze^{2\pi i/3}).$$

Ascending series

Let

$$f(z) = \sum_{n=0}^{\infty} 3^n \left(\tfrac{1}{3}\right)_n \frac{z^{3n}}{(3n)!},$$

$$g(z) = \sum_{n=0}^{\infty} 3^n \left(\tfrac{2}{3}\right)_n \frac{z^{3n+1}}{(3n+1)!}, \tag{B.7}$$

where $(a)_n \equiv a(a+1)\cdots(a+n-1) = \Gamma(a+n)/\Gamma(a)$ is Pockhammer's symbol. Then,

$$Ai(z) = c_1 f(z) - c_2 g(z), \tag{B.8a}$$

$$Bi(z) = \sqrt{3}[c_1 f(z) + c_2 g(z)], \tag{B.8b}$$

where

$$c_1 \equiv Ai(0) = 3^{-1/2}\,Bi(0) = 3^{-2/3}/\Gamma(2/3), \tag{B.8c}$$

$$c_2 \equiv -Ai'(0) = 3^{-1/2}\,Bi'(0) = 3^{-1/3}/\Gamma(1/3). \tag{B.8d}$$

Interrelations

$$Bi(z) = e^{i\pi/6}\,Ai(ze^{2\pi i/3}) + e^{-i\pi/6}\,Ai(ze^{-2\pi i/3}), \tag{B.9a}$$

$$Ai(ze^{\pm 2\pi i/3}) = \tfrac{1}{2}e^{\pm i\pi/3}[Ai(z) \mp i\,Bi(z)], \tag{B.9b}$$

$$Ai(z) + e^{-2\pi i/3}\,Ai(ze^{-2\pi i/3}) + e^{2\pi i/3}\,Ai(ze^{2\pi i/3}) = 0. \tag{B.9c}$$

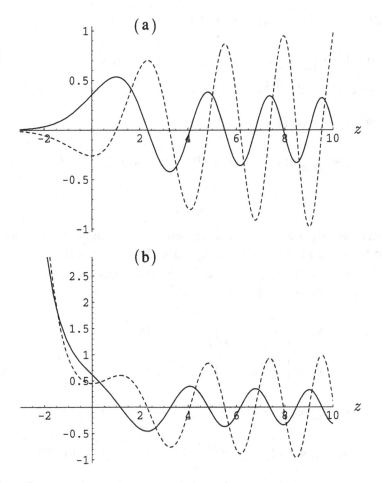

Fig. B.1. (a) The Airy function Ai($-x$) (solid line) and its derivative Ai$'$($-x$) (dashed line). (b) the Airy function Bi($-x$) (solid line) and its derivative Bi$'$($-x$) (dashed line).

Figure B.1 illustrates the behavior of the Airy functions and their derivatives both for positive and for negative values of their arguments.

Asymptotic behavior ($|z| \gg 1$)

$$\text{Ai}(z) \approx \frac{e^{-\zeta}}{2\pi^{1/2}z^{1/4}}[1 + O(\zeta^{-1})], \quad |\arg z| < \pi, \tag{B.10a}$$

$$\text{Ai}(-z) \approx \pi^{-1/2}z^{-1/4}\sin(\zeta + \pi/4) + O(\zeta^{-1}), \quad |\arg z| < 2\pi/3, \tag{B.10b}$$

$$\text{Ai}'(z) \approx -\frac{z^{1/4}e^{-\zeta}}{2\pi^{1/2}}[1 + O(\zeta^{-1})], \quad |\arg z| < \pi, \tag{B.10c}$$

$$\mathrm{Ai}'(-z) \approx -\frac{z^{1/4}}{\pi^{1/2}}\cos(\zeta + \pi/4) + O(\zeta^{-1}), \quad |\arg z| < 2\pi/3, \qquad \text{(B.10d)}$$

$$\mathrm{Bi}(z) \approx \frac{e^{\zeta}}{\pi^{1/2}z^{1/4}}[1 + O(\zeta^{-1})], \quad |\arg z| < \pi/3, \qquad \text{(B.10e)}$$

$$\mathrm{Bi}(-z) \approx \pi^{-1/2}z^{-1/4}\cos(\zeta + \pi/4) + O(\zeta^{-1}), \quad |\arg z| < 2\pi/3, \qquad \text{(B.10f)}$$

$$\mathrm{Bi}'(z) \approx \frac{z^{1/4}}{\pi^{1/2}}e^{\zeta}[1 + O(\zeta^{-1})], \quad |\arg z| < \pi/3, \qquad \text{(B.10g)}$$

$$\mathrm{Bi}'(-z) \approx \frac{z^{1/4}}{\pi^{1/2}}\sin(\zeta + \pi/4) + O(\zeta^{-1}), \quad |\arg z| < 2\pi/3. \qquad \text{(B.10h)}$$

Zeros

The functions $\mathrm{Ai}(z)$ and $\mathrm{Ai}'(z)$ have zeros on the negative real axis only, denoted by $-x_n$ and $-x_n'$, respectively. Table B.1 provides the first 25 positive zeros of $\mathrm{Ai}(-x)$ and $\mathrm{Ai}'(-x)$, and Table B.2 gives the evaluations $a_n \equiv \mathrm{Ai}(-x_n')$ and $a_n' \equiv \mathrm{Ai}'(-x_n)$. Asymptotically, for $n \gg 1$,

$$-x_n \approx \left[\frac{3\pi}{2}\left(n - \frac{1}{4}\right)\right]^{2/3}[1 + O(n^{-2})], \qquad \text{(B.11a)}$$

$$-x_n' \approx \left[\frac{3\pi}{2}\left(n - \frac{3}{4}\right)\right]^{2/3}[1 + O(n^{-2})], \qquad \text{(B.11b)}$$

$$\mathrm{Ai}'(x_n) \approx (-1)^{n-1}\pi^{-1/2}\left[\frac{3\pi}{2}\left(n - \frac{1}{4}\right)\right]^{1/6}[1 + O(n^{-2})], \qquad \text{(B.11c)}$$

$$\mathrm{Ai}(x_n') \approx (-1)^{n-1}\pi^{-1/2}\left[\frac{3\pi}{2}\left(n - \frac{3}{4}\right)\right]^{1/6}[1 + O(n^{-2})]. \qquad \text{(B.11d)}$$

Further properties of the Airy functions are given in HMF and in Watson (1995).

Table B.1. *Zeros of* Ai$(-x)$ *and* Ai$'(-x)$

n	x_n	x'_n
1	2.33810741046	1.01879297165
2	4.08794944413	3.24819758218
3	5.52055982810	4.82009921118
4	6.78670808996	6.16330735564
5	7.94413358713	7.37217725505
6	9.02265085334	8.48848673403
7	10.0401743416	9.53544905244
8	11.0085243037	10.5276603970
9	11.9360155632	11.4750566335
10	12.8287767529	12.3847883718
11	13.6914890352	13.2622189617
12	14.5278299518	14.1115019705
13	15.3407551360	14.9359371967
14	16.1326851569	15.7382013737
15	16.9056339974	16.5205038254
16	17.6613001057	17.2846950502
17	18.4011325992	18.0323446225
18	19.1263804742	18.7647984377
19	19.8381298917	19.4832216566
20	20.5373329077	20.1886315095
21	21.2248299436	20.8819227555
22	21.9013675956	21.5638877232
23	22.5676129175	22.2352322853
24	23.2241650011	22.8965887389
25	23.8715644555	23.5485262959

Table B.2. $a_n = \text{Ai}(-x'_n)$ and $a'_n = \text{Ai}'(-x_n)$

n	a_n	a'_n
1	0.535656656016	0.701210822721
2	−0.419015478033	−0.803111369655
3	0.380406468628	0.865204025894
4	−0.357907943712	−0.91085073705
5	0.342301244412	0.947335709442
6	−0.330476229148	−0.977922808569
7	0.321022288195	1.00437012266
8	−0.313185390979	−1.02773868882
9	0.306517293883	1.04872064859
10	−0.300730829323	−1.06779385916
11	0.295631481002	1.08530283135
12	−0.291081677204	−1.10150457028
13	0.286980706999	1.11659617793
14	−0.283252736125	−1.13073231049
15	0.27983930536	1.14403667327
16	−0.276694445069	−1.15660984912
17	0.273781385647	1.16853478449
18	−0.271070278577	−1.17988072987
19	0.268536578282	1.19070613116
20	−0.266159868216	−1.20106079152
21	0.263922992961	1.21098751487
22	−0.261811405695	−1.2205233739
23	0.259812670151	1.22970070151
24	−0.257916075333	−1.23854787533
25	0.25611233378	1.24708994526

Appendix C
Asymptotic properties of cylinder functions

Almost all properties of the Bessel functions $J_\lambda(z)$, $Y_\lambda(z)$, and $H_\lambda^{(1,2)}(z)$ can be found in HMF or Watson (1995). Specifically, if λ is fixed and $|z| \to \infty$, we have the well-known oscillatory behavior

$$J_\lambda(z) \simeq \sqrt{\frac{2}{\pi z}} \, \cos\left(z - \frac{\lambda\pi}{2} - \frac{\pi}{4}\right) + O(z), \qquad \text{(C.1a)}$$

$$Y_\lambda(z) \simeq \sqrt{\frac{2}{\pi z}} \, \sin\left(z - \frac{\lambda\pi}{2} - \frac{\pi}{4}\right) + O(z). \qquad \text{(C.1b)}$$

When z is fixed and λ becomes large, however, the asymptotic forms can be more complicated. We begin with the simplest forms and work toward the most elaborate, culminating with uniform representations.

Large real order

The leading behaviors of the Debye expansions (Debye, 1909a) are given by HMF with the definition

$$\tanh\alpha \equiv \sqrt{1 - (z/\lambda)^2}, \qquad \text{(C.2)}$$

in which $\alpha > 0$. This assertion is verified by noting that

$$\alpha \xrightarrow[z \ll \lambda]{} \ln(\lambda/z) \gg 1.$$

Then, for $z < \lambda$,

$$J_\lambda(z) \simeq \frac{e^{\lambda(\tanh\alpha - \alpha)}}{\sqrt{2\pi\lambda \tanh\alpha}} \xrightarrow[z \ll \lambda]{} \frac{(ez/\lambda)^\lambda}{2^\lambda \sqrt{2\pi\lambda}}, \qquad \text{(C.3a)}$$

$$Y_\lambda(z) \simeq -\frac{2e^{\lambda(\alpha - \tanh\alpha)}}{\sqrt{2\pi\lambda \tanh\alpha}} \xrightarrow[z \ll \lambda]{} \frac{2^{\lambda+1}}{\sqrt{2\pi\lambda}}(ez/\lambda)^{-\lambda}. \qquad \text{(C.3b)}$$

313

In the opposite situation, $\lambda < z$, the implication is that both z and λ are large. Here we employ the definition

$$\tan \eta \equiv \sqrt{(z/\lambda)^2 - 1}, \qquad 0 < \eta < \pi/2, \tag{C.4}$$

so that

$$J_\lambda(z) \simeq \sqrt{\frac{2}{\pi \lambda \tan \eta}} \, \cos\left(\lambda \tan \eta - \lambda \eta - \frac{\pi}{4} \right), \tag{C.5a}$$

$$Y_\lambda(z) \simeq \sqrt{\frac{2}{\pi \lambda \tan \eta}} \, \sin\left(\lambda \tan \eta - \lambda \eta - \frac{\pi}{4} \right), \qquad \lambda < z. \tag{C.5b}$$

Of considerable importance is the transition region in which z and λ are both large and $z \simeq \lambda$. In this event the appropriate expansions are given in terms of Airy functions and we find that (HMF)

$$J_\lambda(z) \simeq (2/\lambda)^{1/3} \, \text{Ai}(2^{1/3} c)[1 + O(\lambda^{-2/3})], \tag{C.6a}$$

$$Y_\lambda(z) \simeq -(2/\lambda)^{1/3} \, \text{Bi}(2^{1/3} c)[1 + O(\lambda^{-2/3})], \tag{C.6b}$$

where

$$c \equiv \frac{\lambda - z}{\lambda^{1/3}}, \qquad z \simeq \lambda. \tag{C.7}$$

If $c \ll 1$ we can employ the series expansions of the Airy functions (Appendix B), but then one must account for terms already omitted from Eqs. (C.6). These series have been developed extensively by Schöbe (1954), and, to leading order,

$$J_\lambda(z) \simeq \left(\frac{2}{z}\right)^{1/3} \frac{3^{-2/3}}{\Gamma(2/3)} - \frac{3^{-1/3}}{\Gamma(1/3)} \left(\frac{2}{z}\right)^{1/3} \frac{\lambda - z}{(\lambda/2)^{1/3}}, \tag{C.8a}$$

$$Y_\lambda(z) \simeq -\left(\frac{2}{z}\right)^{1/3} \frac{3^{-1/6}}{\Gamma(2/3)} - \frac{3^{1/6}}{\Gamma(1/3)} \left(\frac{2}{z}\right)^{1/3} \frac{\lambda - z}{(\lambda/2)^{1/3}}, \qquad z \simeq \lambda \gg 1. \tag{C.8b}$$

Of course, because λ and z are both very large in the transition region they do not have to differ greatly before c becomes large. In this case we employ the asymptotic expansions for the Airy functions in Appendix B, and for c positive we find

$$J_\lambda(z) \simeq \frac{(2^{1/3} c)^{-1/4}}{2^{2/3} \pi^{1/2} \lambda^{1/3}} \exp\left(-\frac{2}{3}(2^{1/3} c)^{3/2}\right) [1 - O(z^{-3/2})], \tag{C.9a}$$

$$Y_\lambda(z) \simeq -\frac{2^{1/3} (2^{1/3} c)^{-1/4}}{\pi^{1/2} \lambda^{1/3}} \exp\left(\frac{2}{3}(2^{1/3} c)^{3/2}\right) [1 - O(z^{-3/2})], \qquad \lambda > z. \tag{C.9b}$$

On the other side, $c < 0$, we have

$$J_\lambda(z) \simeq \frac{2^{1/3}(2^{1/3}c)^{-1/4}}{\pi^{1/2}\lambda^{1/3}} \sin\left[\tfrac{2}{3}(2^{1/3}c)^{3/2} + \pi/4\right] [1 + O(z^{-3/2})], \tag{C.10a}$$

$$Y_\lambda(z) \simeq -\frac{2^{1/3}(2^{1/3}c)^{-1/4}}{\pi^{1/2}\lambda^{1/3}} \cos\left[\tfrac{2}{3}(2^{1/3}c)^{3/2} + \pi/4\right] [1 + O(z^{-3/2})], \quad \lambda < z. \tag{C.10b}$$

Complex order, $|\lambda| \gg 1$

Primary interest throughout the text lies with the asymptotic behavior of the cylinder functions $Z_\lambda(z)$ in the complex λ-plane, for $|\lambda| \gg 1$ and $z > 0$. As suggested in the previous section, this behavior is generally described in terms of auxiliary functions of the form

$$A(\lambda, \beta) \equiv (2/\pi)^{1/2}(\lambda^2 - \beta^2)^{-1/4}, \tag{C.11a}$$

$$v(\lambda, \beta) \equiv (\lambda^2 - \beta^2)^{1/2} - \lambda \ln\left(\frac{\lambda}{\beta} + \frac{(\lambda^2 - \beta^2)^{1/2}}{\beta}\right), \tag{C.11b}$$

where now we adopt the notation for the size parameter that is used throughout the text. The branch of the square root in (C.11) is specified by the condition

$$(\lambda^2 - \beta^2)^{1/2} \xrightarrow[|\lambda| \to \infty]{} \lambda = |\lambda|e^{i\varphi}, \qquad -\pi < \varphi < \pi. \tag{C.12}$$

Then,

$$A \xrightarrow[|\lambda| \to \infty]{} (2/\pi\lambda)^{1/2}, \tag{C.13a}$$

$$e^v \xrightarrow[|\lambda| \to \infty]{} (e\beta/2\lambda)^\lambda, \tag{C.13b}$$

reminiscent of Eqs. (C.3); this is a common feature of the cylinder functions in the complex plane.

The situation for the Hankel functions is complicated by the appearance of the Stokes phenomenon in their behavior along the curves p_1 and p_{-1} of Fig. 2.10. Nussenzveig (1965) has carried out a careful study of the Hankel functions as $|\lambda| \to \infty$, the results of which are indicated schematically in Fig. C.1. We recall that the λ-zeros of $H_\lambda^{(1)}(\beta)$ are asymptotically located on the curves $p_{\pm 1}$; similarly, the zeros of $H_\lambda^{(2)}(\beta$ are found asymptotically along $p_{\pm 2}$. The portions of the real axis labeled j and j' contain the zeros of $J_\lambda(\beta)$. These six curves divide the λ-plane into five regions, A–E, that summarize the asymptotic behavior of $H_\lambda^{(1,2)}(\beta)$ and $J_{\pm\lambda}(\beta)$, where we note that, when λ is not an integer, J_λ and $J_{-\lambda}$ are linearly independent. The leading behavior

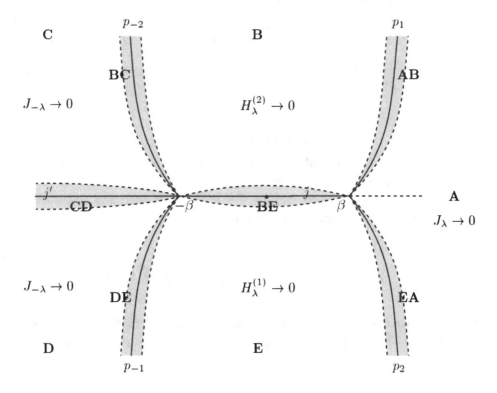

Fig. C.1. Regions of various asymptotic behaviors of the cylinder functions as $|\lambda| \to \infty$ in the complex λ-plane. Those functions that vanish in this limit are indicated in each region.

of each function in each region is given in Table C.1, and, in employing these results, it is useful to note the reflection properties

$$H^{(1,2)}_{-\lambda}(\beta) = e^{\pm i\pi\lambda} H^{(1,2)}_{\lambda}(\beta). \tag{C.14}$$

The four shaded regions which contain the zeros of the Hankel functions, and which leave the real axis at an angle of $\pi/3$, have angular width $(|\lambda|\ln|\lambda|)^{-1}$ with arc length tending to zero as $(\ln|\lambda|)^{-1}$. Within these regions we have the representations for $|\lambda| \to \infty$

$$H^{(1)}_{\lambda}(\beta) \simeq 2Ae^{i\pi/4}\sinh(v - i\pi/4), \qquad \text{in AB;}$$
$$\simeq -2Ae^{-i\pi\lambda - i\pi/4}\sinh(v - i\pi\lambda + i\pi/4), \qquad \text{in DE;} \tag{C.15a}$$
$$H^{(2)}_{\lambda}(\beta) \simeq -2Ae^{+i\pi\lambda + i\pi/4}\sinh(v - i\pi\lambda - i\pi/4), \qquad \text{in BC,}$$
$$\simeq 2Ae^{-i\pi/4}\sinh(v + i\pi/4), \qquad \text{in EA.} \tag{C.15b}$$

Table C.1. *The cylinder functions as* $|\lambda| \to \infty$

Region	A	B	C	D	E
	$\mathrm{Re}\,\nu < 0$	$\mathrm{Re}\,\nu > 0$	$\mathrm{Re}\,\nu > 0$	$\mathrm{Re}\,\nu > 0$	$\mathrm{Re}\,\nu > 0$
$H_\lambda^{(1)}$	$-iAe^{-\nu}$	A^ν	Ae^ν	$-Ae^{\nu-2i\pi\lambda}$	$-iA^{-\nu}$
$H_\lambda^{(2)}$	$iAe^{-\nu}$	$iAe^{-\nu}$	$-Ae^{\nu+2i\pi\lambda}$	Ae^ν	Ae^ν
J_λ	$\frac{1}{2}Ae^\nu$	$\frac{1}{2}Ae^\nu$	$\frac{1}{2}Ae^\nu$	$\frac{1}{2}Ae^\nu$	$\frac{1}{2}Ae^\nu$
$J_{-\lambda}$	—	$\frac{i}{2}Ae^{-\nu-i\pi\lambda}$	$\frac{i}{2}Ae^{-\nu-i\pi\lambda}$	$-\frac{i}{2}Ae^{-\nu+i\pi\lambda}$	$-\frac{i}{2}Ae^{-\nu+i\pi\lambda}$

The asymptotic forms provided in Fig. C.1 and Table C.1 are valid for $|\lambda| \gg \beta$, but for smaller values of $|\lambda|$ one needs finer results. In particular, for the region BE in a neighborhood of the real axis, we can employ the Debye asymptotic expansions (HMF) in the forms

$$H_\lambda^{(1,2)}(\beta) \simeq (2/\pi)^{1/2}(\beta^2 - \lambda^2)^{-1/4} \exp\left\{\pm i\left[(\beta^2 - \lambda^2)^{1/2} - \lambda\cos^{-1}(\lambda/\beta) - \pi/4\right]\right\}$$

$$\times \left[1 \mp \frac{i}{8(\beta^2 - \lambda^2)^{1/2}}\left(1 + \frac{5}{3}\frac{\lambda^2}{(\beta^2 - \lambda^2)} + \cdots\right) + \cdots\right], \qquad \text{(C.16)}$$

for $-\beta < \lambda < \beta$, with $(\beta^2 - \lambda^2)^{-1/4} > 0$ and $0 < \cos^{-1}(\lambda/\beta) < \pi/2$.

These expansions fail in the neighborhood of $\lambda = \pm\beta$, so that, if $|\lambda - \beta| = O(\beta^{1/3})$, we turn to the complete expansions of Schöbe (1954). For $|\lambda| \gg 1$,

$$J_\lambda(\beta) \simeq \left(\frac{2}{\beta}\right)^{1/3} \sum_{n=0}^\infty (-1)^n \left(\frac{2}{\beta}\right)^{2n/3} [A_n(z)\,\mathrm{Ai}(z) + B_n(z)\,\mathrm{Ai}'(z)], \qquad \text{(C.17a)}$$

$$J_\lambda'(\beta) \simeq -\left(\frac{2}{\beta}\right)^{2/3} \sum_{n=0}^\infty (-1)^n \left(\frac{2}{\beta}\right)^{2n/3} [\bar{A}_n(z)\,\mathrm{Ai}(z) + \bar{B}_n(z)\,\mathrm{Ai}'(z)], \qquad \text{(C.17b)}$$

$$H_\lambda^{(1)}(\beta) \simeq 2e^{-i\pi/3}\left(\frac{2}{\beta}\right)^{1/3}$$

$$\times \sum_{n=0}^\infty (-1)^n \left(\frac{2}{\beta}\right)^{2n/3} [A_n(\xi)\,\mathrm{Ai}(-\xi) - e^{-i\pi/3}B_n(\xi)\,\mathrm{Ai}'(-\xi)], \qquad \text{(C.17c)}$$

$$H_\lambda^{(1)'}(\beta) \simeq -2e^{-i\pi/3}\left(\frac{2}{\beta}\right)^{2/3}$$

$$\times \sum_{n=0}^\infty (-1)^n \left(\frac{2}{\beta}\right)^{2n/3} [\bar{A}_n(\xi)\,\mathrm{Ai}(-\xi) - e^{-i\pi/3}\bar{B}_n(\xi)\,\mathrm{Ai}'(-\xi)], \qquad \text{(C.17d)}$$

where $z \equiv (2/\beta)^{1/3}(\lambda - \beta)$ and $\xi \equiv e^{-i\pi/3}z$. The corresponding expansions for $H_\lambda^{(2)}$ and $H_\lambda^{(2)'}$ are obtained by changing the sign of i everywhere, and the first few coefficients are as follows:

$$A_0(\xi) = 1, \qquad B_0(\xi) = 0,$$

$$A_1(\xi) = e^{i\pi/3}\frac{\xi}{15}, \qquad B_1(\xi) = -e^{-i\pi/3}\frac{\xi^2}{60},$$

$$A_2(\xi) = e^{-i\pi/3}\left(\frac{\xi^5}{7200} - \frac{13\xi^2}{1260}\right); \qquad B_2(\xi) = -\frac{\xi^3}{420} + \frac{1}{140};$$

$$\bar{A}_0(\xi) = 0, \qquad \bar{B}_0(\xi) = 1,$$

$$\bar{A}_1(\xi) = -\frac{\xi^3}{60} - \frac{1}{10}, \qquad \bar{B}_1(\xi) = -e^{-i\pi/3}\frac{\xi}{15},$$

$$\bar{A}_2(\xi) = -e^{i\pi/3}\left(\frac{\xi^4}{3360} + \frac{\xi}{60}\right); \qquad \bar{B}_2(\xi) = e^{-i\pi/3}\left(\frac{\xi^5}{7200} + \frac{19\xi^2}{2520}\right). \quad (C.18)$$

Uniform asymptotic expansions

The high-frequency description of the impenetrable sphere in Chapter 2 relies heavily on the uniform expansions of the cylinder functions for real λ provided by Olver (1974). We first define a parameter ζ and a function $\psi(\zeta)$ as follows:

$$\tfrac{2}{3}\zeta^{3/2} \equiv \ln\left(\frac{1 + \sqrt{1-z^2}}{z}\right) - \sqrt{1-z^2},$$

$$\Psi(\zeta) \equiv \frac{5}{16\zeta^2} + \frac{\zeta z^2(z^2+4)}{4(z^2-1)^3}, \qquad (C.19)$$

where z appears in the argument of the following cylinder functions. As $\lambda \to \infty$, uniformly in z and with $|\arg z| < \pi$, Olver obtains

$$J_\lambda(\lambda z) \simeq \frac{1}{\lambda^{1/3}}\left(\frac{4\zeta}{1-z^2}\right)^{1/4}\left(\mathrm{Ai}(\lambda^{2/3}\zeta)\sum_{s=0}^{n}\frac{A_s(\zeta)}{\lambda^{2s}} + \frac{\mathrm{Ai}'(\lambda^{2/3}\zeta)}{\lambda^{4/3}}\sum_{s=0}^{n-1}\frac{B_s(\zeta)}{\lambda^{2s}}\right), \qquad (C.20)$$

$$H_\lambda^{(1,2)}(\lambda z) \simeq \frac{2e^{\mp i\pi/3}}{\lambda^{1/3}}\left(\frac{4\zeta}{1-z^2}\right)^{1/4}\left(\mathrm{Ai}(e^{\pm 2i\pi/3}\lambda^{2/3}\zeta)\sum_{s=0}^{n}\frac{A_s(\zeta)}{\lambda^{2s}}\right.$$
$$\left. + e^{\pm 2i\pi/3}\frac{\mathrm{Ai}'(e^{\pm 2i\pi/3}\lambda^{2/3}\zeta)}{\lambda^{4/3}}\sum_{s=0}^{n-1}\frac{B_s(\zeta)}{\lambda^{2s}}\right). \qquad (C.21)$$

The coefficients are determined from the normalization $A_0(\zeta) = 1$ and a set of integral recurrence relations involving Ψ. The first two terms are

$$A_0(\zeta) = 1,$$

$$B_0(\zeta) = \begin{cases} -\dfrac{5}{48}\zeta^{-2} + \left(\dfrac{5}{24}(1 - z^2)^{-3/2} - \dfrac{1}{8}(1 - z^2)^{-1/2}\right)\zeta^{-1/2}, & z \leq 1, \\[2mm] -\dfrac{5}{48}\zeta^{-2} - \left(\dfrac{5}{24}(z^2 - 1)^{-3/2} + \dfrac{1}{8}(z^2 - 1)^{-1/2}\right)(-\zeta)^{-1/2}, & z \geq 1. \end{cases}$$

$$\text{(C.22)}$$

For applications it is useful to rewrite these by setting $z = \beta/\lambda$ and introducing hyperbolic or trigonometric functions according to the size of z:

$$\begin{aligned} z^{-1} = \lambda/\beta = \cosh\varphi, \qquad \tfrac{2}{3}\zeta^{3/2} &= \varphi - \tanh\varphi, \quad z \leq 1; \\ z^{-1} = \lambda/\beta = \cos\psi, \qquad \tfrac{2}{3}(-\zeta)^{3/2} &= \tan\psi - \psi, \quad z \geq 1 \end{aligned}$$

$$\text{(C.23)}$$

When $|\zeta\lambda^{1/3}| \gg 1$ we can employ the asymptotic expansions of the Airy functions in Appendix B, in which case (C.21) reduces to the Debye expansions with the appropriate substitutions for z.

Appendix D
Spherical angular functions

All properties of angular functions in scattering theory begin with the Legendre polynomials $P_\ell(z)$, $\ell = 0, 1, 2, \ldots$, which are solutions of Legendre's differential equation

$$(1 - z^2)\frac{d^2w}{dz^2} - 2z\frac{dw}{dz} + \ell(\ell+1)w = 0. \tag{D.1}$$

As long as ℓ is a nonnegative integer the explicit polynomial form can be written

$$P_\ell(z) = \sum_{r=0}^{m} \frac{(2\ell - 2r)!}{2^\ell r!(l-r)!(\ell-2r)!} z^{\ell-2r}$$
$$= (-1)^\ell P_\ell(-z), \tag{D.2}$$

where the integer m is $\ell/2$ or $(\ell-1)/2$. Rodrigues' formula provides an alternative definition:

$$P_\ell(z) \equiv \frac{1}{2^\ell \ell!} \frac{d^\ell}{dz^\ell}(z^2 - 1)^\ell, \tag{D.3}$$

and most often one takes z to be a real number between -1 and 1: $z = x = \cos\theta$. (We employ this notation when z is to be restricted to real values.) Another definition is provided by Schläfli's integral,

$$P_\ell(z) = \frac{1}{2\pi} \int_C \frac{(t^2 - 1)^\ell}{2^\ell (t - z)^{\ell+1}} \, dt, \tag{D.4}$$

where C encircles z once in the counter-clockwise direction.

Explicit forms

$$P_0(z) = 1, \qquad\qquad P_1(z) = z,$$
$$P_2(z) = \tfrac{1}{2}(3z^2 - 1), \qquad\qquad P_3(z) = \tfrac{1}{2}(5z^3 - 3z),$$
$$P_4(z) = \tfrac{1}{8}(35z^4 - 30z^2 + 3), \quad P_5(z) = \tfrac{1}{8}(63z^5 - 70z^3 + 15z). \tag{D.5}$$

Special values

$$P_\ell(-x) = (-1)^\ell P_\ell(x), \tag{D.6a}$$

$$P_{-\ell-1}(z) = P_\ell(z), \tag{D.6b}$$

$$P_\ell(\pm 1) = (\pm 1)^\ell, \tag{D.6c}$$

$$P_\ell(0) = \frac{\cos\left(\dfrac{\ell\pi}{2}\right)}{\sqrt{\pi}} \frac{\Gamma\left(\dfrac{\ell+1}{2}\right)}{\Gamma\left(\dfrac{\ell}{2}+1\right)}. \tag{D.6d}$$

Recurrence relations

$$P'_{\ell+1}(z) - P'_{\ell-1}(z) = (2\ell + 1)P_\ell(z), \tag{D.7a}$$

$$(\ell + 1)P_{\ell+1}(z) - (2\ell + 1)zP_\ell(z) + \ell P_{\ell-1}(z) = 0, \tag{D.7b}$$

$$P'_{\ell+1}(z) - zP'_\ell(z) = (\ell + 1)P_\ell(z), \tag{D.7c}$$

$$(z^2 - 1)P'_\ell(z) + \ell P_{\ell-1}(z) = \ell z P_\ell(z), \tag{D.7d}$$

$$P'_\ell(z) = \frac{\ell(\ell + 1)}{2\ell + 1}\frac{1}{1 - z^2}[P_{\ell+1}(z) - P_{\ell-1}(z)], \tag{D.7e}$$

where primes represent differentiation with respect to the argument.

Orthogonality

$$\int_{-1}^{+1} P_{\ell'}(x)P_\ell(x)\,dx = \frac{2}{2\ell + 1}\delta_{\ell'\ell}. \tag{D.8}$$

Completeness

$$\sum_{\ell=0}^{\infty}(\ell + \tfrac{1}{2})P_\ell(x')P_\ell(x) = \delta(x - x'). \tag{D.9}$$

Asymptotics

For $\ell\sin\theta \gg 1$, which excludes $\theta \simeq 0$ and π, Laplace's formula (HMF) is

$$P_\ell(\cos\theta) \approx \left(\frac{2}{\pi(\ell + \tfrac{1}{2})\sin\theta}\right)^{1/2}\cos\left((\ell + \tfrac{1}{2})\theta - \frac{\pi}{4}\right) + O(\ell^{-3/2}). \tag{D.10}$$

This also follows from Eq. (D.40b) below.

Second solution

When z is *not* a real number between -1 and 1, Eq. (D.1) possesses a second linearly independent solution

$$Q_\ell(z) = \frac{1}{2} \int_{-1}^{+1} \frac{P_\ell(t)}{z - t} \, dt = (-1)^{\ell+1} Q_\ell(-z), \qquad \ell = 0, 1, 2, \ldots. \qquad \text{(D.11)}$$

The functions $Q_\ell(z)$ also satisfy the recurrence relations (D.7) and are regular in the entire complex plane cut along $(-1, 1)$.

D.1 Associated Legendre polynomials

The generalizations

$$P_\ell^m(x) = (1 - x^2)^{m/2} \frac{d^m}{dx^m} P_\ell(x), \qquad\qquad\qquad \text{(D.12a)}$$

$$Q_\ell^m(x) = (1 - x^2)^{m/2} \frac{d^m}{dx^m} Q_\ell(x), \qquad m \le \ell, \qquad \text{(D.12b)}$$

are solutions of the differential equation (D.1) when a term $-m^2(1 - z^2)^{-1}w$ is added to the left-hand side, and z is real: $-1 \le x \le 1$. One should note that there is an arbitrariness in phase associated with these definitions and some authors insert a factor of $(-1)^m$ on the right-hand sides of (D.12). We shall adopt the convention of Condon and Shortley (1963), as is customary in most physics discussions, and insert the phase factor into the definition of the spherical harmonics below – either way the same phase factor will arise in the latter. In any event, one can extend m to the negative integers,

$$P_\ell^{-m}(x) = (-1)^m \frac{(\ell - m)!}{(\ell + m)!} P_\ell^m(x). \qquad\qquad \text{(D.13)}$$

Note that $P_\ell^0(x) = P_\ell(x)$.

Explicit forms

$$P_1^1(x) = (1 - x^2)^{1/2},$$
$$P_2^1(x) = 3x(1 - x^2)^{1/2},$$
$$P_2^2(x) = 3(1 - x^2). \qquad\qquad\qquad\qquad \text{(D.14)}$$

Orthogonality

$$\int_{-1}^{+1} P_{\ell'}^m(x) P_\ell^m(x) \, dx = \frac{2}{2\ell + 1} \frac{(\ell + m)!}{(\ell - m)!} \delta_{\ell'\ell}. \qquad \text{(D.15)}$$

Recursion relations

$$(2\ell + 1)xP_\ell^m(x) = (\ell + m)P_{\ell-1}^m(x) + (\ell - m + 1)P_{\ell+1}^m(x), \qquad \text{(D.16a)}$$

$$(1 - x^2)^{1/2}P_\ell^{m'}(x) = \tfrac{1}{2}(\ell + m)(\ell - m + 1)P_\ell^{m-1}(x) - \tfrac{1}{2}P_\ell^{m+1}(x), \quad \text{(D.16b)}$$

$$(x^2 - 1)P_\ell^{m'}(x) = \ell xP_\ell^m(x) - (\ell + m)P_{\ell-1}^m(x)$$

$$= -(\ell + 1)xP_\ell^m(x) + (\ell - m + 1)P_{\ell+1}^m(x). \qquad \text{(D.16c)}$$

Asymptotics

$$P_\ell^m(x) \approx \frac{\Gamma(\ell + m + 1)}{\Gamma(\ell + \frac{3}{2})}\left(\frac{2}{\pi \sin\theta}\right)^{1/2}\cos\!\left((\ell + \tfrac{1}{2})\theta - \frac{\pi}{4} + \frac{m\pi}{2}\right) + O(\ell^{-1}), \quad \text{(D.17a)}$$

$$Q_\ell^m(x) \approx \frac{\Gamma(\ell + m + 1)}{\Gamma(\ell + \frac{3}{2})}\left(\frac{\pi}{2 \sin\theta}\right)^{1/2}\cos\!\left((\ell + \tfrac{1}{2})\theta + \frac{\pi}{4} + \frac{m\pi}{2}\right) + O(\ell^{-1}), \quad \text{(D.17b)}$$

as $\ell \to \infty$ with x and m fixed, and for θ not too close to the forward or backward directions.

D.2 Spherical harmonics

On the unit sphere, with $x = \cos\theta$, $0 \le \theta \le \pi$, and azimuthal angle ϕ, we define the spherical harmonics as

$$Y_\ell^m(\theta, \phi) \equiv \left(\frac{2\ell + 1}{4\pi}\frac{(\ell - m)!}{(\ell + m)!}\right)^{1/2}(-1)^m e^{im\phi}P_\ell^m(\cos\theta), \qquad \text{(D.18)}$$

with $-\ell \le m \le \ell$. One also has the extension

$$Y_\ell^{-m}(\theta, \phi) = (-1)^m Y_\ell^{m*}(\theta, \phi). \qquad \text{(D.19)}$$

Explicit forms

$$Y_0^0 = \frac{1}{\sqrt{4\pi}},$$

$$Y_1^0 = \sqrt{\frac{3}{4\pi}}\,\cos\theta,$$

$$Y_1^{\pm 1} = \mp\sqrt{\frac{3}{8\pi}}\,e^{\pm i\phi}\sin\theta. \qquad \text{(D.20)}$$

Orthogonality

$$\int_0^{2\pi} d\phi \int_0^{\pi} Y_\ell^{m*}(\theta, \phi) Y_{\ell'}^{m'}(\theta, \phi) \sin\theta \, d\theta = \delta_{\ell'\ell}\delta_{m'm}. \qquad \text{(D.21)}$$

The addition theorem

Let x and x' be vectors from a common origin with angles (θ, ϕ) and (θ', ϕ'), respectively, and let γ be the angle between them. Then,

$$P_\ell(\cos\gamma) = \frac{4\pi}{2\ell + 1} \sum_{m=-\ell}^{\ell} Y_\ell^{m*}(\theta', \phi') Y_\ell^m(\theta, \phi). \qquad \text{(D.22)}$$

Eigenfunction representation

Define a differential operator $L \equiv -ir \times \nabla$ and note that, in spherical coordinates,

$$L^2 = L_x^2 + L_y^2 + L_z^2$$
$$= -\left[\frac{1}{\sin\theta} \frac{\partial}{\partial\theta} \left(\sin\theta \frac{\partial}{\partial\theta} \right) + \frac{1}{\sin^2\theta} \frac{\partial^2}{\partial\phi^2} \right], \qquad \text{(D.23)}$$

$$L_x = i\left(\sin\phi \frac{\partial}{\partial\theta} + \cos\phi \cot\theta \frac{\partial}{\partial\phi} \right), \qquad \text{(D.24a)}$$

$$L_y = -i\left(\cos\phi \frac{\partial}{\partial\theta} - \sin\phi \cot\theta \frac{\partial}{\partial\phi} \right), \qquad \text{(D.24b)}$$

$$L_z = -i\frac{\partial}{\partial\phi}. \qquad \text{(D.24c)}$$

The eigenvalues and eigenfunctions of these operators are

$$L^2 Y_\ell^m(\theta, \phi) = \ell(\ell + 1) Y_\ell^m(\theta, \phi), \qquad \text{(D.25a)}$$
$$L_z Y_\ell^m(\theta, \phi) = m Y_\ell^m(\theta, \phi). \qquad \text{(D.25b)}$$

It's also useful to note that the Laplacian operator in the present notation is just

$$\nabla^2 = \frac{1}{r} \frac{\partial^2}{\partial r^2} (r \ ) - \frac{L^2}{r^2} (\ ), \qquad \text{(D.26)}$$

where the appropriate function is to be inserted within the parentheses.

Vector spherical harmonics

Solutions to the vector wave equation in spherical coordinates are often conveniently expressed in terms of vector spherical harmonics:

$$\boldsymbol{X}_{\ell m} \equiv [\ell(\ell + 1)]^{-1/2} \boldsymbol{L} Y_{\ell}^m(\theta, \phi), \qquad (D.27)$$

which is taken to be zero for $\ell = 0$. Both $\boldsymbol{L}$ and Y_{ℓ}^m have been defined in the preceeding section.

Integral relations

With the solid-angle element defined as $d\Omega \equiv \sin\theta\, d\theta\, d\phi$,

$$\int \boldsymbol{X}_{\ell' m'}^* \cdot \boldsymbol{X}_{\ell m}\, d\Omega = \delta_{\ell\ell'}\delta_{mm'}, \qquad (D.28a)$$

$$\int \boldsymbol{X}_{\ell' m'}^* \cdot (\boldsymbol{r} \times \boldsymbol{X}_{\ell m})\, d\Omega = 0, \qquad (D.28b)$$

$$\int (\hat{\boldsymbol{r}} \times \boldsymbol{X}_{\ell 1}^*) \cdot (\hat{\boldsymbol{r}} \times \boldsymbol{X}_{\ell' 1})\, d\Omega = \delta_{\ell\ell'}, \qquad (D.28c)$$

$$\int [f_\ell(kr)\boldsymbol{X}_{\ell' m'}]^* \cdot [g_\ell(kr)\boldsymbol{X}_{\ell m}]\, d\Omega = f_\ell^* g_\ell \delta_{\ell\ell'}\delta_{mm'}, \qquad (D.28d)$$

$$\int [f_\ell(kr)\boldsymbol{X}_{\ell' m'}]^* \cdot [\nabla \times g_\ell(kr)\boldsymbol{X}_{\ell m}]\, d\Omega = 0, \qquad (D.28e)$$

$$\frac{1}{k^2}\int [\nabla \times f_\ell(kr)\boldsymbol{X}_{\ell' m'}]^* \cdot [\nabla \times g_\ell(kr)\boldsymbol{X}_{\ell m}]\, d\Omega =$$

$$\delta_{\ell\ell'}\delta_{mm'}\left[f_\ell^* g_\ell + \frac{1}{k^2 r^2}\frac{\partial}{\partial r}\left(r f_\ell^* \frac{\partial}{\partial r}(r g_\ell) \right) \right], \qquad (D.28f)$$

where f_ℓ and g_ℓ are any linear combinations of spherical Bessel functions that are solutions of the radial part of the scalar Helmholtz, or wave equation, in spherical coordinates.

A sum rule

$$\sum_{m=-\ell}^{\ell} |\boldsymbol{X}_{\ell m}(\theta, \phi)|^2 = \frac{2\ell + 1}{4\pi}, \qquad (D.29)$$

and Table D.1 exhibits several of these multipole angular intensities.

Legendre functions

Equation (D.1) also possesses solutions $P_\lambda(z)$ when λ is not an integer, and is possibly complex. These are defined by the integral representation (D.4)

Table D.1. $|X_{\ell m}(\theta, \phi)|^2$

ℓ	0	± 1	± 2
		m	
1 (dipole)	$\dfrac{3}{8\pi}\sin^2\theta$	$\dfrac{3}{16\pi}(1+\cos^2\theta)$	—
2 (quadrupole)	$\dfrac{15}{8\pi}\sin^2\theta\cos^2\theta$	$\dfrac{5}{16\pi}(1-3\cos^2\theta+4\cos^4\theta)$	$\dfrac{5}{16\pi}(1-\cos^4\theta)$

provided that it is a single-valued representation. To assure this requires a branch cut from -1 to $-\infty$ along the real axis, and then the contour is a loop from a point on this axis to the right of $t = 1$, encircling $t = 1$ and z in the counter-clockwise direction to avoid the cut. Then the *Legendre function* $P_\lambda(z)$ is regular throughout the cut z-plane and is an entire function of λ.

Although (D.2) is no longer valid, since $P_\lambda(z)$ is no longer a polynomial, the recurrence relations (D.6) remain valid. One now has an explicit representation in terms of hypergeometric functions (see HMF for a definition of these):

$$P_\lambda(z) = {}_2F_1\left(\lambda+1, -\lambda; 1; \frac{1-z}{2}\right), \tag{D.30}$$

from which it follows that

$$P_\lambda(z) = P_{-\lambda-1}(z). \tag{D.31}$$

Both these equations are also valid for $\lambda = \ell$, an integer.

Draw an ellipse in the complex t-plane with foci ± 1; let z be a point outside it and let A be the end of the major axis on the right at $t = 1$. Take the contour in Schläfli's integral (D.4) to be C', starting at A and describing a figure eight about $t = \pm 1$, starting above $t = 1$. Then a second solution of Legendre's equation for nonintegal λ is

$$Q_\lambda(z) = \frac{1}{4i\sin(\pi\lambda)}\int_{C'}\frac{(t^2-1)^\lambda}{2^\lambda(z-t)^{\lambda+1}}\,dt, \tag{D.32}$$

which is regular throughout the z-plane cut from $+1$ to $-\infty$. If λ is an integer ℓ, then the cut extends only from $+1$ to -1. This is the Legendre function of the second kind, and, for $|z| > 1$, $|\arg z| < \pi$, and $\mathrm{Re}(\lambda+1) > 0$,

$$Q_\lambda(z) = \frac{\pi^{1/2}}{2^{\lambda+1}}\frac{\Gamma(\lambda+1)}{\Gamma(\lambda+3/2)}z^{-\lambda-1}\,{}_2F_1\left(\frac{\lambda+1}{2}, \frac{\lambda+2}{2}; \lambda+\frac{3}{2}; \frac{1}{z^2}\right). \tag{D.33}$$

This function also satisfies the recurrence relations (D.6).

Interrelations

For $\text{Im}(z) \gtrless 0$,

$$P_\lambda(-z) = e^{\mp i\pi\lambda} P_\lambda(z) - \frac{2}{\pi} \sin(\pi\lambda)\, Q_\lambda(z),$$

$$Q_\lambda(-z) = -e^{\pm i\pi\lambda} Q_\lambda(z). \tag{D.34}$$

$$W\{P_\lambda(z), Q_\lambda(z)\} = (1 - z^2)^{-1}. \tag{D.35}$$

Integral representations

Several representations are given in HMF, and we provide one other here (Watson 1995):

$$P_\lambda(\cos\theta) = \frac{1}{\Gamma(\lambda+1)} \int_0^\infty e^{-t\cos\theta} J_0(t\sin\theta) t^\lambda \, dt, \tag{D.36}$$

for $0 < \theta < \pi/2$ and $\text{Re}(\lambda+1) > 0$. From this one readily obtains the useful inequality

$$\left| P_{i\lambda-1/2}(\cos\theta) \right| \le \frac{e^{\pi\lambda/2}}{\sqrt{\cos\theta}}, \qquad 0 \le \theta < \pi/2, \quad \lambda \ge 0. \tag{D.37}$$

Auxiliary functions

Following Nussenzveig (1965), it is sometimes useful to consider auxiliary functions

$$Q_\nu^{(1,2)}(\cos\theta) \equiv \frac{1}{2}\left(P_\nu(\cos\theta) \pm \frac{2i}{\pi} Q_\nu(\cos\theta) \right), \tag{D.38}$$

with the following interrelations:

$$P_\nu(\cos\theta) = Q_\nu^{(1)}(\cos\theta) + Q_\nu^{(2)}(\cos\theta), \tag{D.39a}$$

$$Q_{\lambda-1/2}^{(1)}(-\cos\theta) = ie^{-i\pi\lambda} Q_{\lambda-1/2}^{(2)}(\cos\theta), \tag{D.39b}$$

$$Q_{\lambda-1/2}^{(2)}(-\cos\theta) = -ie^{i\pi\lambda} Q_{\lambda-1/2}^{(1)}(\cos\theta), \tag{D.39c}$$

$$P_{\lambda-1/2}(-\cos\theta) = ie^{-i\pi\lambda} P_{\lambda-1/2}(\cos\theta) - 2i\cos(\pi\lambda)\, Q_{\lambda-1/2}^{(1)}(\cos\theta)$$

$$= -ie^{i\pi\lambda} P_{\lambda-1/2}(\cos\theta) + 2i\cos(\pi\lambda)\, Q_{\lambda-1/2}^{(2)}(\cos\theta), \tag{D.39d}$$

where here especially one can have $\lambda = \ell + \frac{1}{2}$. Both $Q_{\lambda-1/2}^{(1,2)}$ have poles when λ is a negative half-integer (in the λ-plane), but these cancel out in $P_{\lambda-1/2}$, Eq. (D.39a).

Asymptotic expansions

Robin (1958) has provided the following asymptotic expansions for $|\lambda| \gg 1$, and, when $\epsilon \leq \theta \leq \pi - \epsilon$, $|\lambda|\epsilon \gg 1$:

$$Q^{(1,2)}_{\lambda-1/2}(\cos\theta) \approx \frac{e^{\mp i(\lambda\theta-\pi/4)}}{(2\pi\lambda\sin\theta)^{1/2}}\left(1 \pm \frac{i\cot\theta}{8\lambda} + O(\lambda^{-2})\right), \tag{D.40a}$$

$$P_{\lambda-1/2}(\cos\theta) \approx \left(\frac{2}{\pi\lambda\sin\theta}\right)^{1/2}\left[\cos\left(\lambda\theta - \frac{\pi}{4}\right) + \frac{\cot\theta}{8\lambda}\sin\left(\lambda\theta - \frac{\pi}{4}\right) + O(\lambda^{-2})\right]. \tag{D.40b}$$

For $|\lambda| \gg 1$ in the forward direction, $0 \leq \theta \leq \epsilon$, $|\lambda|\epsilon \lesssim 1$, and $u \equiv 2\lambda\sin(\theta/2)$,

$$P_{\lambda-1/2}(\cos\theta) \approx J_0(u) + \sin^2(\theta/2)\left(\frac{u}{6}J_3(u) - J_2(u) + \frac{J_1(u)}{2u}\right) + O[\sin^4(\theta/2)]. \tag{D.41}$$

Finally, Szegö (1934) has developed a uniform expansion in θ when $|\lambda| \gg 1$:

$$P_{\lambda-1/2}(\cos\theta) \approx \mathscr{P}(\lambda,\theta)$$

$$= \left(\frac{\theta}{\sin\theta}\right)^{1/2}\left(J_0(\lambda\theta) + \tfrac{1}{8}(\theta\cot\theta - 1)\frac{J_1(\lambda\theta)}{\lambda\theta} + O(\lambda^{-2})\right), \tag{D.42}$$

which reduces to (D.40b) when $|\lambda|\theta \gg 1$, yet remains valid as $\theta \to 0$.

D.3 Mie angular scattering functions

In Mie scattering theory the angular distribution of radiation is described by the functions

$$\pi_\ell(x) \equiv \frac{1}{(1-x^2)^{1/2}}P^1_\ell(x) = \frac{dP_\ell(x)}{dx}, \tag{D.43a}$$

$$\tau_\ell(x) \equiv \frac{dP^1_\ell(x)}{d\theta} = -(1-x^2)^{1/2}\frac{dP^1_\ell(x)}{dx}, \tag{D.43b}$$

with $x = \cos\theta$ and $\ell = 0, 1, 2, \dots$.

Recurrence relations

$$\pi_\ell(x) = \frac{2\ell-1}{\ell-1}x\pi_{\ell-1}(x) - \frac{\ell}{\ell-1}\pi_{\ell-2}(x), \qquad \ell > 1, \tag{D.44a}$$

$$\tau_\ell(x) = \ell x\pi_\ell(x) - (\ell+1)\pi_{\ell-1}(x)$$

$$= \ell\pi_{\ell+1}(x) - (\ell+1)x\pi_\ell(x). \tag{D.44b}$$

Special values

$$\pi_\ell(-x) = (-1)^{\ell-1}\pi_\ell(x), \qquad \tau_\ell(-x) = (-1)^\ell \tau_\ell(x). \tag{D.45a}$$

$$\pi_0(x) = \tau_0(x) = 0, \qquad \pi_1(x) = 1, \qquad \tau_1(x) = x. \tag{D.45b}$$

$$\pi_\ell(1) = \tau_\ell(1) = \tfrac{1}{2}\ell(\ell+1). \tag{D.45c}$$

$$\pi_\ell(0) = \ell P_{\ell-1}(0), \qquad \tau_\ell(0) = \ell(\ell+1)P_\ell(0), \tag{D.45d}$$

and $P_\ell(0)$ is given by Eq. (D.6d).

Integrals

From the definitions (D.43) we see that integration of either π_ℓ or τ_ℓ is rather straightforward. In applications, however, integrals of products of these functions, usually over the interval $(-1, 1)$, are encountered. Thus, it is useful to consider the general integral

$$I_n \equiv \int_{-1}^{+1} x^n \left[\pi_\ell(x)\pi_{\ell'}(x) + \tau_\ell(x)\tau_{\ell'}(x)\right] dx. \tag{D.46}$$

The trick is to reduce the product of τ functions to a product of π functions by means of the recurrence relations, for the latter products are simply products of Legendre polynomials, or the associated versions, and we can then make use of the appropriate orthogonality properties. Proceeding in this manner we obtain the first two such integrals:

$$I_0 = \frac{2}{2\ell+1}\ell^2(\ell+1)^2\delta_{\ell\ell'}, \tag{D.47a}$$

$$I_1 = \frac{2\ell(\ell^2-1)^2}{(2\ell+1)(2\ell-1)}\delta_{\ell',\ell-1} + 2\frac{(\ell+2)^2(\ell+1)\ell^2}{(2\ell+1)(2\ell+3)}\delta_{\ell',\ell+1}. \tag{D.47b}$$

In like manner, we define

$$K_n \equiv \int_{-1}^{+1} x^n \left[\pi_\ell(x)\tau_{\ell'}(x) + \pi_{\ell'}(x)\tau_\ell(x)\right] dx. \tag{D.48}$$

Again converting the $\pi\tau$ products into products of πs, we find for the first two integrals

$$K_0 = 0, \tag{D.49a}$$

$$K_1 = 2\frac{\ell(\ell+1)}{2\ell+1}\delta_{\ell\ell'}. \tag{D.49b}$$

Some generalizations of these integrals for $|m| \neq 1$ (broken spherical symmetry) are given by Kim and Lee (1983).

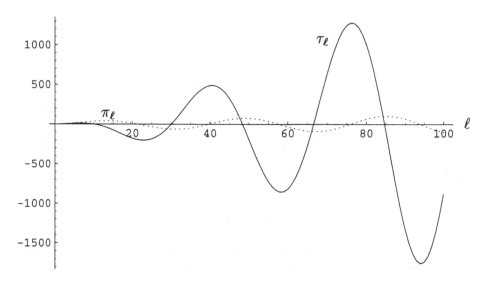

Fig. D.1. The comparative behavior of the angular functions π_ℓ (dashed line) and τ_ℓ (solid line) as functions of ℓ for $\theta = 10°$.

Asymptotics ($\ell \gg 1$)

For $\theta \ll 1$ and $\ell \gg 1$, with $v \equiv (\lambda + \tfrac{1}{2})\theta$, we obtain from Eqs. (D.63) and (D.64) below the expressions

$$\pi_\ell(\cos\theta) \approx \tfrac{1}{2}\ell(\ell+1)[J_0(v) + J_2(v)], \tag{D.50a}$$

$$\tau_\ell(\cos\theta) \approx \tfrac{1}{2}\ell(\ell+1)[J_0(v) - J_2(v)]. \tag{D.50b}$$

Similarly, for $\ell \gg 1$ and away from $\theta = 0$ and π,

$$\pi_\ell(\cos\theta) \approx \left(\frac{2\ell}{\pi\sin^3\theta}\right)^{1/2} \sin\left((\ell+\tfrac{1}{2})\theta - \frac{\pi}{4}\right) + O(\ell^{-1/2}), \tag{D.51a}$$

$$\tau_\ell(\cos\theta) \approx \left(\frac{2\ell^3}{\pi\sin\theta}\right)^{1/2} \cos\left((\ell+\tfrac{1}{2})\theta - \frac{\pi}{4}\right) + O(\ell^{-1/2}). \tag{D.51b}$$

These are derived from earlier results for $P_\ell^1(\cos\theta)$.

Equations (D.51) are often cited as proof that the function τ_ℓ dominates π_ℓ for large ℓ. Even though this becomes more and more generally correct as ℓ increases, it is by no means an absolute truth. Both functions are quasiperiodic in ℓ, and as a consequence there can often be values of ℓ for which this dominance does not hold, as suggested in Fig. D.1.

D.4 Angular functions of complex index

With $x = \cos\theta$, and for $\ell \neq 0$, it is convenient to define new functions

$$p_\ell(x) \equiv [P_{\ell-1}(x) - P_{\ell+1}(x)](1 - x^2)^{-1} = \frac{2\ell + 1}{\ell(\ell + 1)}\pi_\ell(x), \quad \text{(D.52a)}$$

$$t_\ell(x) \equiv -xp_\ell(x) + (2\ell + 1)P_\ell(x) = \frac{2\ell + 1}{\ell(\ell + 1)}\tau_\ell(x). \quad \text{(D.52b)}$$

For integral $\ell = 1, 2, 3, \ldots$ the above Mie results give us

$$p_\ell(-x) = (-1)^{\ell+1}p_\ell(x), \qquad t_\ell(-x) = (-1)^\ell t_\ell(x), \quad \text{(D.53)}$$

along with the special values

$$t_\ell(1) = p_\ell(1) = \ell + \tfrac{1}{2}, \quad \text{(D.54a)}$$

$$t_\ell(-1) = -p_\ell(-1) = (-1)^\ell(\ell + \tfrac{1}{2}). \quad \text{(D.54b)}$$

The bookkeeping convenience aside, however, these definitions have a deeper significance for the electromagnetic scattering problem, beginning with the observation that they exclude the value $\ell = 0$ that is also absent from the Mie solutions. This is what permits a direct definition in terms of Legendre polynomials, which in turn makes it easier to go over to arbitrary values for the index. When ℓ is not an integer we can now make the extension to the backward direction in terms of two additional auxiliary functions:

$$p_\lambda^{(j)}(x) \equiv \left[Q_{\lambda-1}^{(j)}(x) - Q_{\lambda+1}^{(j)}(x)\right](1 - x^2)^{-1}, \quad \text{(D.55a)}$$

$$t_\lambda^{(j)}(x) \equiv -xp_\lambda^{(j)}(x) + (2\lambda + 1)Q_\lambda^{(j)}(x). \quad \text{(D.55b)}$$

Identities

$$t_{\lambda-1/2}^{(1)}(-\cos\theta) = ie^{-i\pi\lambda}t_{\lambda-1/2}^{(2)}(\cos\theta), \quad \text{(D.56a)}$$

$$p_{\lambda-1/2}^{(1)}(-\cos\theta) = -ie^{-i\pi\lambda}p_{\lambda-1/2}^{(2)}(\cos\theta); \quad \text{(D.56b)}$$

$$t_{\lambda-1/2}(-\cos\theta) = -(-1)^j ie^{-i\pi\lambda}t_{\lambda-1/2}(\cos\theta)$$
$$+ (-1)^j 2i\cos(\pi\lambda)\, t_{\lambda-1/2}^{(j)}(\cos\theta), \quad \text{(D.57a)}$$

$$p_{\lambda-1/2}(-\cos\theta) = (-1)^j ie^{-i\pi\lambda}p_{\lambda-1/2}(\cos\theta)$$
$$- (-1)^j 2i\cos(\pi\lambda)\, p_{\lambda-1/2}^{(j)}(\cos\theta), \quad \text{(D.57b)}$$

$$t^{(1)}_{\lambda-1/2}(\cos\theta) - e^{-2i\pi\lambda}t^{(1)}_{-\lambda-1/2}(\cos\theta) = -e^{-i\pi\lambda}\tan(\pi\lambda)\,t_{\lambda-1/2}(-\cos\theta),$$

$$\text{(D.58a)}$$

$$p^{(1)}_{\lambda-1/2}(\cos\theta) - e^{-2i\pi\lambda}p^{(1)}_{-\lambda-1/2}(\cos\theta) = e^{-i\pi\lambda}\tan(\pi\lambda)\,p_{\lambda-1/2}(-\cos\theta).$$

$$\text{(D.58b)}$$

Asymptotic expansions

These are obtained by employing the corresponding expressions for the Legendre functions. For $\epsilon < \theta < \pi - \epsilon$ and $|\lambda|\epsilon \gg 1$, $|\lambda| \gg 1$, we find

$$t^{(j)}_{\lambda-1/2}(\cos\theta) \approx \left(\frac{2\lambda}{\pi\sin\theta}\right)^{1/2} e^{i(-1)^j(\lambda\theta-\pi/4)}$$

$$\times\left(1 + (-1)^j\frac{7i\cot\theta}{8}\frac{1}{\lambda} + O(\lambda^{-2})\right), \qquad \text{(D.59a)}$$

$$p^{(j)}_{\lambda-1/2}(\cos\theta) \approx -\frac{i(-1)^j}{\sin\theta}\left(\frac{2}{\pi\lambda\sin\theta}\right)^{1/2} e^{i(-1)^j(\lambda\theta-\pi/4)}$$

$$\times\left(1 + (-1)^j\frac{3i\cot\theta}{8}\frac{1}{\lambda} + O(\lambda^{-2})\right). \qquad \text{(D.59b)}$$

In the forward direction ($0 \leq \theta \leq \epsilon$, $|\lambda|\epsilon \leq 1$ and $|\lambda| \gg 1$) both $p_{\lambda-1/2}$ and $t_{\lambda-1/2}$ approach λ as $\theta \to 0$. Hence, we need only an expression for $p_{\lambda-1/2}(\cos\theta)$, since $t_{\lambda-1/2}(\cos\theta)$ can be obtained from

$$t_{\lambda-1/2}(\cos\theta) = -\cos\theta\, p_{\lambda-1/2}(\cos\theta) + 2\lambda P_{\lambda-1/2}(\cos\theta). \qquad \text{(D.60)}$$

With $u \equiv 2\lambda\sin(\theta/2)$, then,

$$p_{\lambda-1/2}(\cos\theta) \approx \frac{4\sin(\theta/2)}{\sin^2\theta}\{J_1(u) + O[\sin^2(\theta/2)]\}$$

$$\approx \frac{2}{\theta}J_1(u). \qquad \text{(D.61)}$$

One can differentiate Szegö's uniform expansion (D.42) to obtain the uniform asymptotic expansion

$$p_{\lambda-1/2}(\cos\theta) \approx \left(\frac{\theta}{\sin\theta}\right)^{1/2}\frac{2}{\theta}\left(J_1(\lambda\theta) + \frac{\theta}{\lambda}\frac{J_1(\lambda\theta)/\lambda\theta - J_2(\lambda\theta)}{24}\right), \qquad \text{(D.62)}$$

which remains valid as $\theta \to 0$. Hence, uniformly in θ for $|\lambda| \gg 1$,

$$p_{\lambda-1/2}(\cos\theta) \approx \left(\frac{\theta}{\sin\theta}\right)^{1/2}\frac{2J_1(\lambda\theta)}{\theta}, \qquad \text{(D.63a)}$$

$$t_{\lambda-1/2}(\cos\theta) \approx \left(\frac{\theta}{\sin\theta}\right)^{1/2}2\lambda J_1'(\lambda\theta). \qquad \text{(D.63b)}$$

A more symmetric form can be obtained from the Bessel function recurrence relations

$$2\frac{J_1(\lambda\theta)}{\lambda\theta} = J_0(\lambda\theta) + J_2(\lambda\theta),$$

$$2J_1'(\lambda\theta) = J_0(\lambda\theta) - J_2(\lambda\theta). \tag{D.64}$$

Appendix E
Approximation of integrals

In discussions of the type undertaken in this monograph it is common to deal with functions represented by integrals that are resistant to exact evaluation, yet whose integrands contain one or more parameters that approach specific values in the problem of interest. In such cases it is often possible to find *asymptotic* representations of the function in a series of terms rapidly decreasing in value as $z \to z_0$, say. Even if such series do not converge, they can provide representations of the function for those parameter values to any desired degree of accuracy. With sufficient attention to detail, such expansions can often be differentiated and integrated term by term.

Many asymptotic developments pursued here arise after continuation into the complex plane, in which case additional difficulties emerge because we must insist that asymptotic relations be unique and independent of the path of approach of z to z_0. As an example, consider the behavior of the function $\exp(-z^{-2})$ as $z \to 0$: it vanishes as z approaches zero along the real axis, but becomes unbounded if the origin is approached along the imaginary axis. Although the origin is here an essential singularity of the function, the ambiguity arises quite generally. Asymptotic relations in the complex plane must therfore be stated relative to a specific sector, or wedge, as is done throughout these mathematical appendices. The boundary lines of the wedge are called Stokes lines, and the change in asymptotic behavior as one crosses the edge of the wedge is known as the *Stokes phenomenon*. We emphasize that this behavior is *not* an intrinsic property of a function itself, but rather a property of the functions used to approximate it. The Stokes phenomenon reflects, and is always caused by, the presence of exponential functions in the asymptotic expansion – the arguments of these functions shift the behavior back and forth between real and imaginary parts as one moves about the complex plane. As one might expect, branch cuts often play a role in generating Stokes lines.

With these *caveats* in mind we proceed to investigate a number of exponential-type integrals arising in scattering studies, with one further warning. We shall not provide detailed proofs of the ensuing theorems, but provide adequate reference to more-thorough mathematical treatments. Generally, we outline the derivations and then provide correct statements of the results. Excellent discussions of asymptotic analysis are given by Bleistein and Handelsman (1975), Bender and Orszag (1978), and Morse and Feshbach (1953).

E.1 Laplace integrals

Consider first the Laplace transform

$$L(\lambda) = \int_0^\infty e^{-\lambda t} g(t)\, dt, \quad \lambda > 0, \tag{E.1}$$

for real variables and functions. If $g(t)$ possesses derivatives of all orders it can be expanded uniformly in a Taylor series and integrated term by term to obtain an asymptotic expansion as $\lambda \to \infty$. Quite often, though, $g(t)$ does not behave this nicely and the method fails. However, the basic idea can still be exploited owing to two facts: for large λ the exponential cutoff implies that the only significant contributions come from a neighborhood of the origin; and the Laplace transform of t^n is proportional to λ^{-n-1}. This leads us to

Watson's lemma. Let $g(t)$ satisfy the appropriate conditions for convergence of $L(\lambda)$ and be absolutely integrable on every subinterval of $(0, \infty)$. Suppose that we can find an expansion

$$g(t) \sim \sum_{m=0}^\infty c_m t^{a_m}, \quad t \to 0^+, \tag{E.2}$$

such that $\operatorname{Re} a_m$ is monotonically increasing to infinity as $m \to \infty$, and $a_0 > -1$. Then, if $g(t)$ is also bounded for all finite t,

$$L(\lambda) \xrightarrow[\lambda \to \infty]{} \sum_{m=0}^\infty \frac{c_m \Gamma(a_m + 1)}{\lambda^{a_m+1}}. \tag{E.3}$$

The proof consists simply of splitting the region of integration at some $R > 0$ and examining each of the two resulting integrals as $R \to \infty$, given that $g(t)$ is bounded.

Possibly more interesting is the general Laplace integral, of which (E.1) is a special case, namely

$$L(\lambda) = \int_a^b e^{-\lambda f(t)} g(t)\, dt, \tag{E.4}$$

where all quantities are real and both f and g are considered sufficiently smooth for the validity of any of the following operations. If $f(t)$ is monotonic either Watson's lemma or successive integration by parts can yield an asymptotic expansion for large λ. More commonly, however, $f'(t)$ will have zeros on (a, b). Suppose for the moment that $f(t)$ has a single algebraic minimum at $t = t_0$, $a < t_0 < b$, and that this is the only point on the interval where $f'(t) = 0$. We refer to t_0 as a *critical point* of $f(t)$, and we presume that $f''(t_0) > 0$. Then, rewrite Eq. (E.4) as

$$L(\lambda) = e^{-\lambda f(t_0)} \int_a^b g(t) e^{-\lambda[f(t)-f(t_0)]} \, dt, \tag{E.5}$$

noting that the integrand is now sharply peaked at $t = t_0$ as $\lambda \to \infty$, and essentially negligible elsewhere. The dominant contribution thus comes from a neighborhood of the critical point, and as $\lambda \to \infty$ we expect that $L(\lambda) \approx g(t_0) \exp[-\lambda f(t_0)]$.

To quantify this intuitive expectation we split the interval of integration at $t = t_0$ so that

$$
\begin{aligned}
L(\lambda) &= \int_a^{t_0} e^{-\lambda f(t)} g(t) \, dt + \int_{t_0}^b e^{-\lambda f(t)} g(t) \, dt \\
&\equiv L_a(\lambda) + L_b(\lambda),
\end{aligned}
\tag{E.6}
$$

and $f(t)$ is monotonic on each subinterval. We presume that both $f(t)$ and $g(t)$ can be expanded about t_0, recalling that $f'(t_0) = 0$, and consider only L_b in detail; the treatment is essentially the same for both.

Define a new variable

$$\tau \equiv f(t) - f(t_0), \tag{E.7}$$

which is one-to-one except for a sign ambiguity at $\tau = 0$; this will be important. A change of variables then yields

$$L_b(\lambda) = e^{-\lambda f(t_0)} \int_0^{f(b)-f(t_0)} G(\tau) e^{-\lambda \tau} \, d\tau, \tag{E.8}$$

where

$$G(\tau) \equiv \left. \frac{g(t)}{f'(t)} \right|_{t=f^{-1}[f(t_0)+\tau]}. \tag{E.9}$$

The ambiguity in sign is resolved here by noting that τ increases as t increases from t_0 to b. We can now apply Watson's lemma by extending the upper limit to infinity as $\lambda \to \infty$, after obtaining an expansion for $G(\tau)$, which is mildly tedious. From the expansions of $f(t)$ and $g(t)$ about t_0 (e.g., Eqs. (E.26) below) one can generate by iteration an expansion of $t - t_0$ in powers of τ, which then leads by substitution to corresponding expansions for $f'(t)$ and $g(t)$. Further substitution into (E.9) produces

$$G(\tau) \simeq \frac{g(t_0)}{\sqrt{2f''(t_0)}}\tau^{-1/2} + \left(\frac{g'(t_0)}{f''(t_0)} - \frac{f'''(t_0)g(t_0)}{3[f''(t_0)]^2}\right)\tau^0 + O(\tau^{1/2}), \qquad (E.10)$$

as $\tau \to 0^+$. Finally, application of Watson's lemma in (E.8) yields the asymptotic expansion

$$L_b(\lambda) \xrightarrow[\lambda \to \infty]{} \left(\frac{\pi}{2\lambda f''(t_0)}\right)^{1/2} g(t_0)e^{-\lambda f(t_0)}$$

$$+ \left(\frac{g'(t_0)}{f''(t_0)} - \frac{f'''(t_0)g(t_0)}{3[f''(t_0)]^2}\right)\frac{e^{-\lambda f(t_0)}}{\lambda} + O\left(\frac{e^{-\lambda f(t_0)}}{\lambda^{3/2}}\right). \qquad (E.11)$$

The integral $L_a(\lambda)$ is treated similarly, the only change in $G(\tau)$ being that there is a minus sign in the first term on the right-hand side of (E.10), and the ambiguity in sign here is resolved by noting that τ decreases as t increases, which is reflected in the negative square root. Application of Watson's lemma then produces an expression identical to that of (E.11) but for a minus sign in front of the second term on the right-hand side. The obvious cancellation now provides the expansion of $L(\lambda)$ from (E.6):

$$L(\lambda) \xrightarrow[\lambda \to \infty]{} \frac{2\pi}{\lambda f''(t_0)}g(t_0)e^{-\lambda f(t_0)} + O\left(\frac{e^{-\lambda f(t_0)}}{\lambda^{3/2}}\right). \qquad (E.12)$$

We have belabored this derivation of *Laplace's method* because there are several reasons to understand it in some detail. The technique will provide a fundamental guide for more complicated calculations below; correct treatment of the sign ambiguity was essential for obtaining cancellation of the second terms in the series, and thus for assessing the error correctly; and extensions to more than one critical point and to the case in which critical points occur at the endpoints of the integration interval now follow trivially. Higher-order terms are readily found through a more extensive expansion of $G(\tau)$. As an example of the method's value, the reader is invited to apply it to the integral definition of the Γ-function, which immediately gives Stirling's formula.

E.2 Fourier integrals

The ideas of the preceding section are readily extended to Fourier integrals of the form

$$I(\lambda) = \int_a^b e^{i\lambda t} g(t)\,dt, \qquad (E.13)$$

where everything is real, and the interval can be extended to infinity in either direction; the function $g(t)$ is presumed sufficiently differentiable to justify any of the following operations. In general, the only efficient method for obtaining the asymptotic behavior of integrals of this type is through repeated integration by parts. There is, however, a theorem that often proves quite useful, called the Riemann–Lebesgue lemma.

The Riemann–Lebesgue lemma. If $g(t)$ is absolutely integrable on $(-\infty, \infty)$,

$$\int_{-\infty}^{\infty} |g(t)| \, dt < \infty, \tag{E.14a}$$

then, for real λ,

$$\lim_{\lambda \to \infty} \int_{-\infty}^{\infty} g(t) e^{i\lambda t} \, dt = 0. \tag{E.14b}$$

An associated theorem. If the function $g(t)$ in (E.13) is of bounded variation – i.e., if the sum of the total changes in g in every finite subinterval is finite – then

$$\lim_{\lambda \to \infty} \int_{a}^{b} g(t) e^{i\lambda t} \, dt = O(\lambda^{-1}). \tag{E.15}$$

The proofs of these statements are a little bit messy, so we refer to those given by Cushing (1975).

More interesting for our purposes is the generalization of (E.13) to the form

$$I(\lambda) = \int_{a}^{b} e^{i\lambda f(t)} g(t) \, dt, \tag{E.16}$$

where all quantities are real and $f(t)$ is suitably differentiable. If $f'(t)$ vanishes nowhere on the interval $[a, b]$, as is the case in (E.13), one can obtain the asymptotic behavior as $\lambda \to \infty$ through integration by parts and use of the Riemann–Lebesgue lemma. The result will certainly depend on the endpoints and $I(\lambda)$ will vanish asymptotically as λ^{-1}. Indeed, the integral will always be small for large λ, because, at points where $f'(t)$ does not vanish, the extremely rapid oscillations of the integrand are destructive and lead to an essentially zero result.

Suppose, however, that there exists a point $a < c < b$ such that $f'(c) = 0$. Then, aside from the fact that integration by parts may fail, the destructive interference does *not* occur in a neighborhood of $t = c$. Although this point is just a critical point as defined above, in this context it is called a point of *stationary phase*. In this case the integral must still vanish as $\lambda \to \infty$, by the Riemann–Lebesgue lemma, but it will generally vanish less rapidly than λ^{-1}.

Under the general conditions of the preceding section, and in a manner very similar to Laplace's method, we find that the method of stationary phase gives the leading-order behavior

$$I(\lambda) \xrightarrow[\lambda \to \infty]{} g(c)e^{i\lambda f(c)}\left(\frac{2\pi}{\lambda|f''(c)|}\right)^{1/2} e^{i\pi\mu/4}, \quad \mu \equiv \operatorname{sgn} f''(c). \tag{E.17}$$

Extensions to several stationary points, including the endpoints, are again straightforward, though if $t = c$ is an endpoint one must multiply the right-hand side of (E.17) by $\frac{1}{2}$.

Note that the method of stationary phase gives only the *leading* asymptotic behavior of integrals having stationary points, such as $I(\lambda)$. The error in (E.17) is $O(\lambda^{-1})$, which is algebraically small, in contrast with the exponentially small error in the Laplace method. As a consequence, higher-order corrections are rather complicated, because nonstationary points may contribute. Moreover, it may happen that $f''(c)$ also vanishes and $f'''(c) \neq 0$, in which case $I(\lambda) \sim \lambda^{-1/3}$ as $\lambda \to \infty$, and so on. Thus, as $f(t)$ becomes flatter at $t = c$ the integral vanishes less rapidly as $\lambda \to \infty$.

Despite these shortcomings, the method is quite useful when only the leading-order asymptotic behavior is required. One example is seen in Eq. (2.81), which has precisely the form (E.17). Another is given by the cylindrical Bessel function of the first kind, $J_n(\lambda)$, for fixed n and large λ, which has the integral representation

$$\begin{aligned} J_n(\lambda) &= \frac{1}{\pi} \int_0^\pi \cos(nt - \lambda \sin t)\, dt \\ &= \frac{1}{\pi} \operatorname{Re} \int_0^\pi e^{int} e^{-i\lambda \sin t}\, dt. \end{aligned} \tag{E.18}$$

The reader may wish to show that this verifies exactly the expression (C.1a).

E.3 The method of steepest descents

If higher-order corrections are needed in the asymptotic evaluation of $I(\lambda)$ in (E.16), more powerful tools are secured by continuation into the complex plane. This leads to the method of steepest descents, or the *saddle-point method* (Debye 1909a). Most generally, we consider the integral

$$I(\lambda) = \int_C g(z)e^{\lambda f(z)}\, dz, \tag{E.19}$$

where C is any fixed contour in the complex z-plane, $g(z)$ and $f(z)$ are regular in a domain containing C, and λ is for the moment real. If $f(z)$ is a constant the analysis is trivial, so we exclude that case. With $z = x + iy$ and

$f(z) = u(x, y) + iv(x, y)$, suppose that $z = z_0 = x_0 + iy_0$ is a point where $f'(z)$ vanishes. Usually, though not always, it is safe to presume that $f''(z_0) \neq 0$.

A *direction of descent* from z_0 is one along which $u(x, y)$ decreases from $u(x_0, y_0)$, and a *path of steepest descent* is one on which the tangent is always in the direction in which the rate of descent is maximal, and is therefore described by $-\nabla u$. There will also exist, of course, paths of steepest ascent, described by ∇u. If one visualizes the function $f(z)$ topologically as an undulating surface, then one can descend from z_0 into 'valleys' along paths of steepest descent, and can ascend 'hills' from z_0 along paths of steepest ascent. It is a theorem that curves of steepest descent *and* ascent are those for which

$$v(x_0, y_0) = \operatorname{Im} f(z) = v(x, y). \tag{E.20}$$

That is, $\operatorname{Im} f(z)$ is constant along paths of steepest descent and ascent. This is verified by noting that, near z_0, we can write $\delta f \equiv f(z) - f(z_0) = \delta u + i\, \delta v$, so that $|\delta u| \leq |\delta f|$, with equality if and only if $\delta v = 0$. However, if $\delta v = 0$ then $|\delta u|$ is maximal and will remain so if we let $z \to z_0$ in such a manner as to maintain $\delta v = 0$. Consequently, along such paths (E.20) is valid and describes the paths of steepest ascent or descent (by definition).

We can now describe the general idea of the method. First note that neither $u = \operatorname{Re} f$ nor $v = \operatorname{Im} f$ can be a maximum in the basic domain containing C, which follows from the maximum-modulus theorem (*e.g.*, Knopp (1945)). Since z_0 can be neither a maximum nor a minimum, it must be a *saddle point*. A simple saddle point is illustrated physically by just about any mountain pass, and mathematically in Fig. E.1. The procedure is to locate a saddle point where $f'(z) = 0$, which generalizes the notion of critical and stationary points, and then determine the paths of steepest descent from z_0. We next deform the contour such that C encounters no singularity either of f or of g, and such that C passes through only low valleys except at the point z_0; if C has to sweep across any poles these residues must be added to the eventual value of the integral. Along with these residues, as $\lambda \to \infty$ the vastly dominant contribution to $I(\lambda)$ comes only from a neighborhood of z_0, thereby providing the asymptotic expansion of the integral.

There may, of course, be several saddle points that necessarily contribute to the integral, in which case one merely adds the separate contributions. Higher-order saddle points arise when a succession of higher-order derivatives vanishes at z_0, thereby making the saddle flatter. For example, if $f''(z_0)$ also vanishes we have a saddle point of order 2, called a *monkey saddle*. If $d^n f/dz^n$ is the first nonvanishing derivative at z_0 there are generally n paths

(a) (b)

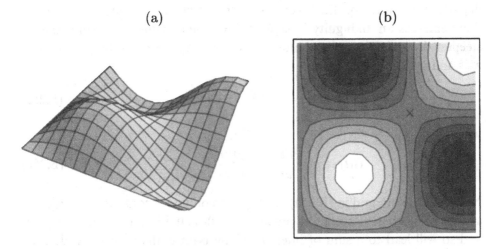

Fig. E.1. (a) An illustration of a saddle point in the complex plane. (b) A contour plot in the vicinity of the saddle point ×, in which lighter shades depict higher altitudes.

of steepest descent. With the notation

$$\left.\frac{d^n f}{dz^n}\right|_{z=z_0} \equiv A e^{i\alpha}, \quad A > 0, \qquad (z - z_0) \equiv \rho e^{i\theta}, \quad \rho > 0, \tag{E.21}$$

the directions of steepest descent are given by

$$\theta = -\frac{\alpha}{n} + (2p + 1)\frac{\pi}{n}, \qquad p = 0, 1, \dots, n - 1. \tag{E.22}$$

The proof follows from the observation that an expansion about $z = z_0$ yields

$$\frac{f(z) - f(z_0)}{\rho^n} \simeq A\frac{e^{i(\alpha+n\theta)}}{n!}[1 + O(\rho)]. \tag{E.23}$$

In the directions of steepest descent the left-hand side of (E.23) must be negative, which implies (E.22). Most often we are interested only in the case $n = 2$, and in that event $\theta = -\alpha/2 + \pi/2,\ -\alpha/2 + 3\pi/2$.

Presume now that a path of steepest descent from a saddle point at $z = z_0$ has been determined, and deform the contour C so that, as $|z| \to \infty$ along the path of steepest descent, $u = \mathrm{Re}\, f \to -\infty$. We can usually do this because the original C need only be maintained in a small neighborhood of z_0. Focusing on the case $f'(z_0) = 0$, $f''(z_0) \neq 0$, we next introduce the transformation

$$f(z) = f(z_0) - t, \tag{E.24}$$

where t is real. In fact, on the path of steepest descent t must be positive as z

departs from z_0. If Eq. (E.24) is used to effect a change of variables in (E.19) there emerges an ambiguity in sign at $t = 0$, depending on the direction of steepest descent. Thus, the contour C is now mapped onto the positive real axis and we can write

$$I(\lambda) = e^{\lambda f(z_0)} \int_0^\infty G(t)\, dt, \tag{E.25a}$$

where

$$G(t) \equiv g(z)\frac{dz}{dt} = -\frac{g(z)}{f'(z)}\bigg|_{z=f^{-1}[f(z_0)-t]}. \tag{E.25b}$$

Note that the integral in (E.25a) corresponds only to a contour from z_0 *down* a path of steepest descent. In practice C will actually cross a saddle point and so will lead to a sum of integrals along both paths of steepest descent, with a phase difference of π.

Equations (E.25) are very similar to (E.8) and (E.9), so that we have reduced the problem to an application of Laplace's method and Watson's lemma. It remains to find an asymptotic expansion of $G(t)$ as $t \to 0^+$ and then employ Watson's lemma. Although interest is usually centered only on the leading term, it will be useful here to include the first correction term as well.

Under the presumed conditions on the functions involved, and with an obvious notation, we expand about $z = z_0$ as follows:

$$g(z) = g_0 + g_0' w + \tfrac{1}{2}g_0'' w^2 + \tfrac{1}{6}g_0''' w^3 + \cdots, \tag{E.26a}$$
$$f(z) = f_0 + \tfrac{1}{2}f_0'' w^2 + \tfrac{1}{6}f_0''' w^3 + \cdots, \tag{E.26b}$$
$$f'(z) = f_0'' w + \tfrac{1}{2}f_0''' w^2 + \tfrac{1}{6}f_0'''' w^3 + \cdots, \tag{E.26c}$$

where $w \equiv (z - z_0) = \rho e^{i\theta}$, $f_0'' = |f_0''| e^{i\alpha}$, $g_0' = [g'(z)]_{z=z_0}$, etc. Combination of these with Eq. (E.25b) and some algebra yields the following expansion in w:

$$-G(t) = \frac{g_0}{w f_0''}\left\{ 1 + \left(\frac{g_0'}{g_0} - \frac{1}{2}\frac{f_0'''}{f_0''}\right)w \right.$$
$$\left. + \left[\frac{1}{2}\frac{g_0''}{g_0} - \frac{1}{2}\frac{f_0'''}{f_0''}\frac{g_0'}{g_0} - \frac{1}{6}\frac{f_0''''}{f_0''} + \frac{1}{4}\left(\frac{f_0'''}{f_0''}\right)^2\right]w^2 + O(w^3)\right\}. \tag{E.27}$$

However, on the path of steepest descent, (E.24) and (E.26b) tell us that

$$-t = \tfrac{1}{2}f_0'' w^2 + \tfrac{1}{6}f_0''' w^3 + \tfrac{1}{24}f_0'''' w^4 + \cdots, \tag{E.28}$$

which can now be iterated to obtain w as a function of t. The specific choice of branch corresponds to the positive square root of t, and we find that

$$w = i\sqrt{2t/f_0''}\left(1 - \frac{i}{3\sqrt{2}}\frac{f_0'''}{(f_0'')^{1/2}}t^{1/2} - \frac{5}{36}\frac{(f_0''')^2}{(f_0'')^3}t + \frac{1}{12}\frac{f_0''''}{(f_0'')^2}t + O(t^{3/2})\right). \quad \text{(E.29a)}$$

Unfortunately, we also need w^{-1} up to $O(t)$:

$$w^{-1} = \frac{1}{i\sqrt{2t/f_0''}} + \frac{1}{6}\frac{f_0'''}{f_0''} + \frac{2^{-1/2}}{12i}\frac{(f_0''')^2}{(f_0'')^{5/2}}t^{1/2} + i\frac{2^{-1/2}}{12}\frac{f_0''''}{(f_0'')^{3/2}}t^{1/2} + O(t). \quad \text{(E.29b)}$$

Substitution into (E.27) produces the desired expansion:

$$G(t) = \frac{g_0 e^{i\pi/2}}{\sqrt{2f_0''}}t^{-1/2} + e^{i\pi}\left(\frac{g_0'}{f_0''} - \frac{1}{3}\frac{g_0 f_0'''}{(f_0'')^2}\right)t^0$$

$$+ \frac{g_0 e^{i3\pi/2}}{\sqrt{2f_0''}}\left(\frac{g_0'}{g_0 f_0''} - \frac{g_0'}{g_0}\frac{f_0'''}{(f_0'')^2} - \frac{1}{4}\frac{f_0''''}{(f_0'')^2} + \frac{5}{12}\frac{(f_0''')^2}{(f_0'')^3}\right)t^{1/2} + O(t), \quad \text{(E.30)}$$

where all functions on the right-hand side are to be evaluated at $z = z_0$.

A lesson learned earlier is that the term in t^0 may well cancel with a similar one in the complete integral across the saddle point, which is why the term in $t^{1/2}$ has been computed to assure the presence of a second-order term. Note also that t has been forced to be real and positive, so that if the integration is taken down the alternative path of steepest descent, the sign of $t^{1/2}$ must be reversed. This then accounts for the ambiguity in sign noted in connection with (E.24). Substitution of (E.30) into (E.25) and use of Watson's lemma now yield the desired asymptotic expansion for the integral up and over the saddle:

$$I(\lambda) \xrightarrow[\lambda\to\infty]{} g(z_0)\left(\frac{2\pi}{\lambda|f''(z_0)|}\right)^{1/2} e^{\lambda f(z_0) + i(\pi - \alpha)/2}$$

$$\times\left[1 - \frac{1}{2}\left(\frac{g_0''}{g_0 f_0''} - \frac{g_0' f_0'''}{g_0(f_0'')^2} - \frac{1}{4}\frac{f_0''''}{(f_0'')^2} + \frac{5}{12}\frac{(f_0''')^2}{(f_0'')^3}\right)\lambda^{-1} + O(\lambda^{-2})\right]. \quad \text{(E.31)}$$

An important first application of this saddle-point technique arose in evaluation of the reflection integral (2.108) for the impenetrable sphere. In the present notation we identify $g(z)$ and $f(z)$ from the unexhibited functions in (2.117):

$$g(z) \simeq e^{i\pi/4}\left(\frac{\beta}{2\pi\sin\theta}\right)^{1/2}\sin z\cos^{1/2}z\left(1 + i\frac{\cot\theta + 2\cot z\left(1 + \frac{5}{3}\cot^2 z\right)}{8\beta\cos z}\right), \quad \text{(E.32a)}$$

$$f(z) = 2i\left[\left(z - \frac{\theta}{2}\right)\cos z - \sin z\right], \quad \text{(E.32b)}$$

where the large parameter is now β (rather than λ), and the saddle point is seen to occur at $z_0 = \theta/2$. Substitution into (E.31) yields the first two terms on the right-hand side of (2.118); the third term requires additional terms of order β^{-2} in (E.32a) and of order λ^{-2} in (E.31), which we do not present here.

We see, then, that for very large λ, the vastly dominant contributions to $I(\lambda)$ come from a neighborhood of the saddle point. The analysis becomes a bit more complicated in the face of certain changes in the scenario, but their treatment is still relatively straightforward. Examples of these complications arise from poles in the neighborhood of the saddle point, and also if $z = z_0$ is also a branch point of $f(z)$, or if the parameter λ is complex, in which case the Stokes phenomenon appears. The reader is referred to the sources mentioned at the beginning of this appendix for these types of generalization.

One complication that must be addressed now, however, is the situation occurring when two saddle points are very close to each other, for it is encountered in the theory of the rainbow. As noted earlier, when two saddle points are separated well enough to be considered independent, their contributions are simply additive. To quantify this notion, rewrite (E.19) as

$$I(\lambda, \epsilon) = \int_C g(z) e^{\lambda f(z, \epsilon)} \, dz, \tag{E.33}$$

where ϵ is an independent parameter describing the separation of two saddle points over which C passes. The *range* of a saddle point is defined as the radius of its complex neighborhood that yields the dominant contribution to the integral, outside which the integrand is effectively negligible. Thus, two saddle points are independent as long as their ranges do not overlap. We now define ϵ as a measure of this overlap, such that for fixed and substantially nonzero ϵ the saddle points are nonoverlapping, coalescing as $\epsilon \to 0$, and thereby producing a second-order saddle point ($f'' = 0$) at $\epsilon = 0$. Near this point the essentially Gaussian character of $I(\lambda)$ becomes of Airy type, with a leading cubic variation in the exponent (e.g., Eqs. (B.3)).

In treating this situation it is necessary to attach some importance to obtaining a smooth match with the previous results for single saddle points; indeed, it is just such a connection that removes the 'fuzziness' from the definition of the range. The required technique has been developed by Chester *et al.* (1957), and this CFU method yields an asymptotic expansion that is uniform in ϵ.

We begin by transforming the exponent function $f(z, \epsilon)$ into an exact cubic through the change of variables

$$f(z, \epsilon) = \tfrac{1}{3} u^3 - \zeta(\epsilon) u + A(\epsilon), \tag{E.34}$$

rather than using (E.24), Then, since $f'(z', \epsilon) = f'(z'', \epsilon) = 0$, the saddle points have the following correspondence:

$$z' \to u = -\zeta^{1/2}, \qquad z'' \to u = \zeta^{1/2}. \tag{E.35}$$

In turn, this implies that ζ and A can be determined from the equations

$$f(z', \epsilon) = \tfrac{2}{3}\zeta^{3/2}(\epsilon) + A(\epsilon),$$
$$f(z'', \epsilon) = -\tfrac{2}{3}\zeta^{3/2}(\epsilon) + A(\epsilon). \tag{E.36}$$

This transformation has one branch that is uniformly regular and one-to-one near $u = 0$, which is chosen so as to preserve the structure of the saddle point. This is, in fact, determined by requiring (E.36) to hold. In place of (E.25b), then, we obtain

$$G(u) = \sum_m [p_m(\epsilon)(u^2 - \zeta)^m + q_m(\epsilon)(u^2 - \zeta)^m], \tag{E.37}$$

where the coefficients p_m and q_m are determined by repeated differentiation of G and evaluation at the saddle points. For example,

$$p_0(\epsilon) = \tfrac{1}{2}[G(\theta_1'', \epsilon) + G(\theta_1', \epsilon)],$$

$$q_0(\epsilon) = \frac{1}{2\sqrt{\zeta}}[G(\theta_1'', \epsilon) - G(\theta_1', \epsilon)],$$

$$p_1(\epsilon) = \frac{1}{4\sqrt{\zeta}}\left(\frac{d}{du}G(\theta_1'', \epsilon) - \frac{d}{du}G(\theta_1', \epsilon)\right),$$

$$q_1(\epsilon) = \frac{1}{4\zeta}\left(\frac{d}{du}G(\theta_1'', \epsilon) + \frac{d}{du}G(\theta_1', \epsilon) - 2q_0(\epsilon)\right),$$

$$p_2(\epsilon) = \frac{1}{16\zeta}\left(\frac{d^2}{du^2}G(\theta_1'', \epsilon) + \frac{d^2}{du^2}G(\theta_1', \epsilon) - 4p_1(\epsilon)\right),$$

$$q_2(\epsilon) = \frac{1}{16\zeta^{3/2}}\left(\frac{d^2}{du^2}G(\theta_1'', \epsilon) - \frac{d^2}{du^2}G(\theta_1', \epsilon) - 12q_1(\epsilon)\sqrt{\zeta}\right). \tag{E.38}$$

Transformation and substitution now yield in place of (E.33)

$$I(\lambda, \epsilon) \simeq 2\pi i e^{\lambda A(\epsilon)} \sum_m [p_m(\epsilon)F_m(\zeta, \lambda, C') + q_m(\epsilon)G_m(\zeta, \lambda, C')], \tag{E.39a}$$

where

$$F_m(\zeta, \lambda, C') = \frac{1}{2\pi i}\int_{C'} (u^2 - \zeta)^m \exp\left[\lambda\left(\tfrac{1}{3}u^3 - \zeta u\right)\right] du, \tag{E.39b}$$

$$G_m(\zeta, \lambda, C') = \frac{1}{2\pi i}\int_{C'} u(u^2 - \zeta)^m \exp\left[\lambda\left(\tfrac{1}{3}u^3 - \zeta u\right)\right] du, \tag{E.39c}$$

and C' is the transformed contour. If C' is chosen to run from $\infty e^{-i\pi/3}$ to $\infty e^{i\pi/3}$ and through the saddle points, then F_m and G_m can be expressed in terms of Airy functions (Chester *et al.* 1957). For example,

$$F_0(\zeta, \lambda, C') = \lambda^{-1/3}\, \mathrm{Ai}(\lambda^{2/3}\zeta), \quad G_0(\zeta, \lambda, C') = -\lambda^{-1/3}\, \mathrm{Ai}'(\lambda^{2/3}\zeta);$$
$$F_1(\zeta, \lambda, C') = 0, \quad G_1(\zeta, \lambda, C') = -\lambda^{-4/3}\, \mathrm{Ai}'(\lambda^{2/3}\zeta), \tag{E.40}$$

and higher-order functions are given by the recurrence relations

$$F_m(\zeta, \lambda, C') = -\frac{2}{\lambda}(m-1)G_{m-2}(\zeta, \lambda, C'),$$

$$G_m(\zeta, \lambda, C') = -\frac{1}{\lambda}\left[(2m-1)F_{m-1}(\zeta, \lambda, C') + 2(2m-1)\zeta F_{m-2}(\zeta, \lambda, C')\right]. \tag{E.41}$$

Thus, $I(\lambda, \epsilon)$ has the asymptotic expansion

$$I(\lambda, \epsilon) = e^{A(\epsilon)}\left[\frac{\mathrm{Ai}(\lambda^{2/3}\zeta)}{\lambda^{1/3}}\left(\sum_{s=0}^{M-1}\frac{a_s(\epsilon)}{\lambda^s} + O(\lambda^{-M})\right)\right.$$

$$\left. + \frac{\mathrm{Ai}'(\lambda^{2/3}\zeta)}{\lambda^{2/3}}\left(\sum_{s=0}^{M-1}\frac{b_s(\epsilon)}{\lambda^s} + O(\lambda^{-M})\right)\right], \tag{E.42}$$

and M is usually rather small. The coefficients a_s and b_s are linear combinations of p_m and q_m and can be obtained from the integrated version of the expansion (E.37). Note from Eqs. (B.10) that for large arguments and $|\arg z| < \pi$, we have $\mathrm{Ai}'(z) \approx -z^{1/2}\,\mathrm{Ai}(z)$, so that the two types of contribution in (E.42) can be comparable. Finally, by employing the asymptotic expansions of Ai and Ai$'$ in Appendix B it can be shown (Ursell 1965) that, under appropriate conditions, the domain of validity of (E.39a) can be extended and matched smoothly with the steepest-descent expansion.

Appendix F

A note on Mie computations

Many of the figures in the text relating to the Mie theory were constructed from computational data generated using the *Mathematica*® system for doing mathematics on a computer. Although this is not a necessary choice of software, it was found to be very convenient and efficient. It is a functional-programming-based language that, although it is not as fast as C or Fortran, is quite user friendly and contains very efficient routines for all the special functions in the preceding appendices. Nevertheless, almost all routines used to compute Mie scattering functions ran fairly rapidly on Pentium II and Pentium III processors. All figures were eventually converted to PostScript® for plotting and were labeled in TeX.

The Mie partial-wave coefficients are given by Eqs. (3.88) and (3.89) in terms of Ricatti–Bessel functions, and it is the latter that encompass most of the computational effort. Although *Mathematica*'s built-in functions were used frequently, we employ iterated recursion relations here because the former are too slow for large indices and arguments. Many years of experience have revealed that their computation is carried out most efficiently if they are recast so that the functions containing the refractive index are replaced by their logarithmic derivatives; ψ'_ℓ/ψ_ℓ. Rayleigh had discovered as early as 1904 that one cannot compute the spherical Bessel functions $j_\ell(z)$ using upward recursion based on Eqs. (A.12) when $\ell \gg z$ (Strutt 1904), and later suggested the use of downward recursion (Strutt 1910). The problem this poses in the present context is with our choice of cutoff in the partial-wave sums for the cross sections – e.g., Eq. (3.99a). Although Fig. A.2 suggests that convergence is rapid when $\ell > z$, Rayleigh's caveat also suggests that we should not compute terms beyond necessity.

Computational difficulties associated with Mie scattering have been studied extensively by Wiscombe (1980). In accord with expected needs in the edge domain, as in Eq. (2.141), a phenomenological cutoff of $L = \beta + 4\beta^{1/3} + 2$

emerges for the calculation of a_ℓ and b_ℓ. It is found that values 1% or so higher are required in locating resonances, and for computation of the internal coefficients c_ℓ, d_ℓ.

Along with this cutoff, Wiscombe also observes that, when there is no absorption, or very little, upward recursion for $\psi_\ell(z)$ is perfectly satisfactory and is the algorithm of choice. (Upward recursion is always stable for the Neumann functions $\xi_\ell(z)$.) This is our computational strategy whenever n is real.

When the refractive index becomes complex, $n = m + i\kappa$, upward recursion for ψ_ℓ becomes unstable near $\kappa \gtrsim 0.01$, and one is well advised to look for trouble before this. In this event downward recursion is absolutely necessary for obtaining reliable values, and we are partial to the algorithm developed by Wang and van de Hulst (1991), which provides reliable computations beyond $\beta = 50\,000$. Beyond this range, and for large batch jobs, one would be well advised to employ the vector algorithms developed by Wiscombe (1979) for vector processors.

Although interest usually lies with $n > 1$, it happens that, for $n < 1$, such as for bubbles in water, downward recursion is also mandatory. The rule of thumb again is to employ downward recursion whenever ℓ substantially exceeds β or $n\beta$. Other than these exceptional cases, one should always prefer upward recursion, because downward recursion is computationally *very* expensive – by orders of magnitude at times.

Other than for the partial-wave coefficients, only determination of high-Q resonances requires extended computational effort. As remarked in the text, we have found it reasonable to use the algorithm of Schiller (1993) as a seed for starting a complex-root finder. The built-in function of *Mathematica* works quite well in many cases.

References

Abramowitz, M. and I. A. Stegun (1964), *Handbook of Mathematical Functions*, National Bureau of Standards, Washington. [Reprinted by Dover, New York, 1972.]

Acker, W. P., D. H. Leach, and R. K. Chang (1989), 'Third-order optical sum-frequency generation in micrometer-sized liquid droplets,' *Opt. Lett.* **14**, 402.

Aden, A. L. and M. Kerker (1951), 'Scattering of electromagnetic waves from two concentric spheres,' *J. Appl. Phys.* **22**, 1242.

Airy, G. B. (1838), 'On the intensity of light in the neighbourhood of a caustic,' *Trans. Camb. Phil. Soc.* **6**, 379.

Aragón, S. R. and M. Elwenspoek (1982), 'Mie scattering from thin spherical bubbles,' *J. Chem. Phys.* **77**, 3406.

Asano, S. and M. Sato (1980), 'Light scattering by randomly oriented spheroidal particles,' *Appl. Opt.* **19**, 962.

Asano, S. and G. Yamamoto (1975), 'Light scattering by a spheroidal particle,' *Appl. Opt.* **14**, 29.

Ashkin, A. (1980), 'Applications of laser radiation pressure,' *Science* **210**, 1081.

Ashkin, A. and J. M. Dziedzic (1977), 'Observation of resonances in the radiation pressure on dielectric spheres,' *Phys. Rev. Lett.* **38**, 1351.

Ballenegger, V. C. and T. A. Weber (1999), 'The Ewald–Oseen extinction theorem and extinction lengths,' *Am. J. Phys.* **67**, 599.

Barber, P. and C. Yeh (1975), 'Scattering of electromagnetic waves by arbirtarily shaped dielectric bodies,' *Appl. Opt.* **14**, 2864.

Barton, J. P., D. R. Alexander, and S. A. Schaub (1988), 'Internal and near-surface electromagnetic fields for a spherical particle irradiated by a focused laser beam,' *J. Appl. Phys.* **64**, 1632.

—— (1989), 'Internal fields of a spherical particle illuminated by a tightly focused laser beam: focal point positioning effects at resonance,' *J. Appl. Phys.* **65**, 2900.

Beck, D. (1962), 'Wide-angle scattering in molecular beams. The rainbow effect,' *J. Chem. Phys.* **37**, 2884.

Bender, C. M. and S. A. Orszag (1978), *Advanced Mathematical Methods for Scientists and Engineers*, McGraw-Hill, New York.

Benincasa, D. S., P. W. Barber, J.-Z. Zhang, W.-F. Hsieh, and R. K. Chang (1987),

'Spatial distribution of the internal and near-field intensities of large cylindrical and spherical scatterers,' *J. Opt. Soc. Am. A* **1**, 822.

Bernstein, R. B. (1966), 'Quantum effects in elastic molecular scattering,' in *Advances in Chemical Physics*, J. Ross (ed.), Vol. 10, Wiley, New York; p. 75.

Berry, M. V. (1966), 'Uniform approximation for potential scattering involving a rainbow,' *Proc. Phys. Soc. London* **89**, 479.

—— (1969), 'Uniform approximations for glory scattering and diffraction peaks,' *J. Phys. B* **2**, 381.

Berry, M. V. and K. E. Mount (1972), 'Semiclassical approximations in wave mechanics,' *Reps. Prog. Phys.* **35**, 315.

Bhandari, R. (1985), 'Scattering coefficients for a multilayered sphere: analytic expressions and algorithms,' *Appl. Opt.* **24**, 1960.

Bickel, W. S. and W. M. Bailey (1985), 'Stokes vectors, Mueller matrices, and polarized scattered light,' *Am. J. Phys.* **53**, 468.

Billet, F. (1868), 'Mémoire sur les dix-neuf premiers arcs-en-ciel de l'eau,' *Ann. Sci. l'Ecole Normale Supérieure* **V**, 67. [He also studied the supernumerary arcs of the first 11 rainbows, in *Comptes Rendus* **56**, 999 (1863); *Ibid.* **58**, 1064 (1864).]

Bleistein, N. and R. A. Handelsman (1975), *Asymptotic Expansion of Integrals*, Holt, Rinehart and Winston, New York.

Bohren, C. F. and D. R. Huffman (1983), *Absorption and Scattering of Light by Small Particles*, Wiley, New York.

Borghese, F., P. Denti, R. Saija, G. Toscano, and O. I. Sindoni (1984), 'Use of group theory for the description of electromagnetic scattering from molecular systems,' *J. Opt. Soc. Am. A* **1**, 183.

Borghese, F., P. Denti, G. Toscano, and O. I. Sindoni (1979), 'Electromagnetic scattering by a cluster of spheres,' *Appl. Opt.* **18**, 116.

Born, M. and E. Wolf (1975), *Principles of Optics*, 5th ed., Pergamon Press, Oxford.

Boyer, C. B. (1959), *The Rainbow: From Myth to Mathematics*, Thomas Yoseloff, New York.

Brewster, D. (1833), *Manuel d'optique*, vol. II, P. Vergnaud (transl.), Paris. [Brewster verified the polarization of the rainbow in 1812, though it was first noticed by Jean-Baptiste Biot.]

Bricard, J. (1940), 'Contribution à l'étude des brouillards naturels,' *Ann. Phys. (Paris)* **14**, 148.

Bridges, J. H. (1964), *The Majus Opus of Roger Bacon*, Minerva-Verlag, Frankfurt am Main. [Reprint of the 1897–1900 edition.]

Brillouin, L. (1949), 'The scattering cross section of spheres for electromagnetic waves,' *J. Appl. Phys.* **20**, 1110.

Brink, D. M. (1985), *Semiclassical Methods for Nucleus–Nucleus Scattering,'* Cambridge University Press, Cambridge.

Bruning, J. H. and Y. T. Lo (1971), 'Multiple scattering of EM waves by spheres part I – multipole expansion and ray optical solutions,' *IEEE Trans. Antennas Propagat.* **19**, 378.

Bryant, H. C. and A. J. Cox (1966), 'Mie theory and the glory,' *J. Opt. Soc. Am.* **56**, 1529.

Bryant, H. C. and N. Jarmie (1968), 'Nuclear glory scattering,' *Ann. Phys. (N.Y.)* **47**, 127.

—— (1974), 'The glory,' *Sci. Am.* **231** (7), 60.

Bucerius, H. (1946), 'Theorie des Regenbogens und der Glorie,' *Optik* **1**, 188.

Buck, U. (1974), 'Inversion of molecular scattering data,' *Rev. Mod. Phys.* **46**, 369.

Buck, U. and H. Pauly (1971), 'Determination of intermolecular potentials by the inversion of molecular beam scattering data. II. High resolution measurements of differential scattering cross sections and the inversion of the data for Na–Hg,' *J. Chem Phys.* **54**, 1929.

Cellini, B. (1571), *Autobiography of Benvenuto Cellini (1500–1571)*, G. Bull (transl.), Penguin, Toronto, 1956.

Chandrasekhar, S. (1950), *Radiative Transfer*, Oxford University Press, Oxford.

Chen, G., R. K. Chang, S. C. Hill, and P. W. Barber (1991), 'Frequency splitting of degenerate spherical cavity modes: stimulated Raman scattering spectrum of deformed dropplets,' *Opt. Lett.* **16**, 1269.

Chen, G., Md. M. Mazumder, R. K. Chang, J. C. Swindal, and W. P. Acker (1996), 'Laser diagnostics for droplet characterization: application of morphology dependent resonances,' *Prog. Energy Combust. Sci.* **22**, 163.

Chen, Y. M. (1964), 'Diffraction by a smooth transparent object,' *J. Math. Phys.* **5**, 820.

Chester, C., B. Friedman, and F. Ursell (1957), 'An extension of the method of steepest descents,' *Proc. Camb. Phil. Soc.* **53**, 599.

Chew, H., D.-S. Wang, and M. Kerker (1979), 'Elastic scattering of evanescent electromagnetic waves,' *Appl. Opt.* **18**, 2679.

Chowdhury, D. Q., P. W. Barber, and S. C. Hill (1992), 'Energy-density distribution inside large nonabsorbing spheres by using Mie theory and geometrical optics,' *Appl. Opt.* **31**, 3518.

Chowdhury, D. Q., S. C. Hill, and P. W. Barber (1991), 'Morphology-dependent resonances in radially inhomogeneous spheres,' *J. Opt. Soc. Am. A* **8**, 1702.

Chýlek, P. (1973), 'Mie scattering into the backward hemisphere,' *J. Opt. Soc. Am.* **63**, 1467.

—— (1976), 'Partial-wave resonances and the ripple structure in the Mie normalized extinction cross section,' *J. Opt. Soc. Am.* **66**, 285.

Chýlek, P., J. T. Kiehl, and M. K. W. Ko (1978a), 'Narrow resonance structure in the Mie scattering characteristics,' *Appl. Opt.* **17**, 285.

—— (1978b), 'Optical levitation and partial-wave resonances,' *Phys. Rev. A* **18**, 2229.

—— (1990), 'Resonance structure of Mie scattering: distance between resonances,' *J. Opt. Soc. Am. A* **7**, 1609.

Chýlek, P. and J. D. Klett (1991), 'Extinction cross sections of nonspherical particles in the anomalous diffraction approximation,' *J. Opt. Soc. Am. A* **8**, 274.

Chýlek, P. and J. Li (1995), 'Light scattering by small particles in an intermediate region,' *Opt. Commun.* **117**, 389.

Chýlek, P., V. Ramaswamy, A. Ashkin, and J. M. Dziedzic (1983), 'Simultaneous determination of refractive index and size of spherical dielectric particles from light scattering data,' *Appl. Opt.* **22**, 2302.

Clebsch, A. (1863), 'Ueber die Reflexion an einer Kugelfläche,' *Z. für Math.* **61**, 195.

Cohen, A. and C. Acquista (1982), 'Light scattering by tilted cylinders: properties of partial wave coefficients,' *J. Opt. Soc. Am.* **72**, 531.

Condon, E. U. and G. H. Shortley (1963), *The Theory of Atomic Spectra*, Cambridge University Press, Cambridge.

Conwell, P. R., P. W. Barber, and C. K. Rushforth (1984), 'Resonant spectra of dielectric spheres,' *J. Opt. Soc. Am. A* **1**, 62.

Crum, L. A. (1994), 'Sonoluminescence,' *Phys. Today* **47** (9), 22.

Cruzan, O. R. (1962), 'Translational addition theorems for spherical vector wave functions,' *Quart. Appl. Math.* **20**, 33.

Cushing, J. T. (1975), *Applied Analytical Mathematics for Physical Scientists*,' Wiley, New York.

Dave, J. V. (1969), 'Scattering of visible light by large water spheres,' *Appl. Opt.* **8**, 155.

Davis, G. E. (1955), 'Scattering of light by an air bubble in water,' *J. Opt. Soc. Am.* **45**, 572.

Davis, L. W. (1979), 'Theory of electromagnetic beams,' *Phys. Rev. A* **19**, 1177.

de Alfaro, V. and T. Regge (1965), *Potential Scattering*, North-Holland, Amsterdam.

Debye, P. (1908), 'Das elektromagnetische Feld um einen vollkommen reflektierenden Zylinder,' *Physik Z.* **9**, 775.

—— (1909a), 'Näherungsformeln für die Zylinderfunktionen für große Werte des Arguments und unbeschränkt veränderliche Werte des Index,' *Math. Ann.* **67**, 535.

—— (1909b), 'Der Lichtdruck auf Kugeln von beliebigem Material,' *Ann. d. Phys.* **30**, 57.

Dehmer, J. L. (1984), 'Shape resonances in molecular fields,' in *Resonances*, D. G. Truhlar (ed.), American Chemical Society, Washington; p. 139.

Delbar, T., G. Grégoire, G. Paic, R. Ceuleneer, F. Michel, R. Vanderpoorten, A. Budzanowski, H. Dabrowski, L. Freindl, K. Grotowski, S. Micek, R. Planeta, A. Strzalkowski, and K. A. Eberhard (1978), 'Elastic and inelastic scattering of alpha particles from ^{40,44}Ca over a broad range of energies and angles,' *Phys. Rev. C* **18**, 1237.

Delvigne, G. A. L. and J. Los (1973), 'Rainbow Stueckelberg oscillations and rotational coupling on the differential cross section of Na + I → Na$^+$ + I$^-$,' *Physica* **67**, 166.

Descartes, R. (1637), *Discours de la méthode*, Paris. [The third appendix, 'Les météores,' is the primary source for his theory of the rainbow. See, also, *Œuvres*, vol. VI, C. Adams and P. Tannery (eds.), Paris, 1897–1913.]

de Zutter, D. (1980), 'Rotating dielectric sphere,' *IEEE Trans. Antennas Propagat.* **28**, 643.

Erdélyi, A., W. Magnus, F. Oberhettinger, and F. G. Tricomi (1953), *Higher Transcendental Functions*, Vols. 1–3, McGraw-Hill, New York.

Erma, V. A. (1968a), 'An exact solution for the scattering of electromagnetic waves from conductors of arbitrary shape. I. Case of cylindrical symmetry,' *Phys. Rev.* **173**, 1243.

—— (1968b), 'An exact solution for the scattering of electromagnetic waves from conductors of arbitrary shape. II. General case,' *Phys. Rev.* **176**, 1544.

—— (1969), 'An exact solution for the scattering of electromagnetic waves from conductors of arbitrary shape. III. Obstacles with arbitrary electromagnetic properties,' *Phys. Rev.* **179**, 1238.

Ewald, P. P. (1916), 'Zur Begründung der Kristalloptik,' *Ann. d. Phys.* **49**, 1. [Based on his Ph. D. thesis, Universität München, 1912.]

Fahlen, T. S. and H. C. Bryant (1966), 'Direct observation of surface waves on water droplets,' *J. Opt. Soc. Am.* **56**, 1635.

—— (1968), 'Optical back scattering from single water droplets,' *J. Opt. Soc. Am.* **58**, 304.

Fiedler-Ferrari, N., H. M. Nussenzverig, and W. J. Wiscombe (1991), 'Theory of near-critical-angle scattering from a curved surface,' *Phys. Rev. A* **43**, 1005.

Fikioris, J. G. and N. K. Uzunoglu (1979), 'Scattering from an eccentrically stratified dielectric sphere,' *J. Opt. Soc. Am.* **69**, 1359.

Fock, V. A. (1965), *Electromagnetic Propagation and Diffraction Problems*, Pergamon Press, Oxford.

Ford, K. W. and J. A. Wheeler (1959a), 'Semiclassical description of scattering,' *Ann. Phys. (N.Y.)* **7**, 259.

—— (1959b), 'Application of semiclassical scattering analysis,' *Ann. Phys. (N.Y.)* **7**, 287.

Franz, W. (1957), *Theorie der Beugung elektromagnetischer Wellen*, Springer-Verlag, Berlin.

Frenzel, H. and H. Schultes (1934), 'Lumineszenz im ultraschallbeschickten Wasser,' *Z. Phys. Chem.* **27B**, 421.

Friedman, W. A., K. W. McVoy, and G. W. T. Shuy (1974), 'Diffraction, refraction, and interference phenomena in heavy-ion transfer reactions,' *Phys. Rev. Lett.* **33**, 308.

Fuller, K. A. (1989), 'Some novel features of morphology dependent resonances of bispheres,' *Appl. Opt.* **28**, 3788.

Fuller, K. A. and G. W. Kattawar (1988a), 'Consummate solution to the problem of classical electromagnetic scattering by an ensemble of spheres I: linear chains,' *Opt. Lett.* **13**, 90.

—— (1988b), 'Consummate solution to the problem of classical electromagnetic scattering by an ensemble of spheres II: clusters of arbitrary configuration,' *Opt. Lett.* **13**, 1063.

Gans, R. (1925), 'Strahlungsdiagramme ultramikroskopischer Teilchen,' *Ann. d. Phys.* **76**, 29.

Glaser, D. A. (1958), 'The bubble chamber,' in *Handbuch der Physik, Band XLV*, Springer-Verlag, Berlin; Fig. 4, p. 331.

Goos, F. and H. Hänchen (1947), 'Ein neuer und fundamentaler Versuch zur Totalreflexion,' *Ann. d. Phys.* **1**, 333.

Gouesbet, G., G. Grehan, and B. Maheu (1985), 'Scattering of a Gaussian beam by a Mie scattering center using a Bromwich formalism,' *J. Optique (Paris)* **16**, 83.

—— (1988), 'Light scattering from a sphere arbitarily located in a Gaussian beam, using a Bromwich formulation,' *J. Opt. Soc. Am. A* **9**, 1427.

Gouesbet, G. and J. A. Lock (1994), 'Rigorous justification of the localized approximation to the beam shape coefficients in generalized Lorenz–Mie theory. II. Off-axis beams,' *J. Opt. Soc. Am. A* **11**, 2516.

Gradshteyn, I. S. and I. M. Ryzhik (1980), *Table of Integrals, Series, and Products*, Academic Press, New York.

Graham, F. L. (ed.) (1975), *The Rainbow Book*, Vintage Books, New York.

Grehan, G., B. Maheu, and G. Gouesbet (1986), 'Scattering of laser beams by Mie scatter centers: numerical results using a localized approximation,' *Appl. Opt.* **25**, 3539.

Guimarães, L. G. (1991), 'Calculo de resonancias do esphalhamento Mie e suas aplicacões,' Ph. D. thesis, CBPF, Rio de Janeiro.

—— (1993), 'Theory of Mie caustics,' *Opt. Commun.* **103**, 339.

Guimarães, L. G. and J. P. R. F. Mendonça (1997), 'Analysis of the resonant scattering of light by cylinders at oblique incidence,' *Appl. Opt.* **36**, 8010.

Guimarães, L. G. and H. M. Nussenzveig (1992), 'Theory of Mie resonances and ripple fluctuations,' *Opt. Commun.* **89**, 363.

—— (1994), 'Uniform approximation to Mie resonances,' *J. Mod. Opt.* **41**, 625.

Gurney, R. W. (1929), 'Nuclear levels and artificial disintegration,' *Nature* **123**, 565.

Gustafsson, T. (1983), 'Shape resonances in the photoionization spectra of free and chemisorbed molecules,' in I. Lindgren, A. Rosén, and S. Svanberg (eds.), *Atomic Physics 8*, Plenum Press, New York; p. 355.

Hahn, D. K. and S. R. Aragón (1994), 'Scattering from anisotropic thick spherical shells,' *J. Chem. Phys.* **101**, 8409.

Hall, W. S. and X. Q. Mao (1995), 'Bounding element method of calculation for coherent electromagnetic scattering from two and three dielectric spheres,' *Eng. Anal. Bound. Elem.* **15**, 313.

Hamid, A.-K., I. R. Ciric, and M. Hamid (1990), 'Electromagnetic scattering by an arbitrary configuration of dielectric spheres,' *Can. J. Phys.* **68**, 1419.

—— (1992), 'Analytic solutions of the scattering by two multilayered dielectric spheres,' *Can. J. Phys.* **70**, 696.

Hertz, H. (1889), 'Die Kräfte elektrischer Schwingungen behandelt nach der Maxwell'schen Theorie,' *Ann. d. Phys.* **36**, 1.

Hightower, R. L. and C. B. Richardson (1988), 'Resonant Mie scattering from a layered sphere,' *Appl. Opt.* **27**, 4850.

Hightower, R. L., C. B. Richardson, H.-B. Lin, J. D. Eversole, and A. J. Campillo (1988), 'Measurement of scattering of light from layered microspheres,' *Opt. Lett.* **13**, 946.

Hill, S. C. and R. E. Benner (1986), 'Morphology-dependent resonances associated with stimulated processes in microspheres,' *J. Opt. Soc. Am. B* **3**, 1509.

Hill, S. C., R. E. Benner, C. K. Rushforth, and D. R. Conwell (1984), 'Structural resonances observed in the fluorescence emission from small spheres on substrates,' *Appl. Opt.* **23**, 1680.

Hill, S. C., D. H. Leach, and R. K. Chang (1993), 'Third-order sum-frequency generation in droplets: model with numerical results for third-harmonic generation,' *J. Opt. Soc. Am. B* **10**, 16.

Hill, S. C., C. K. Rushforth, R. E. Benner, and P. R. Conwell (1985), 'Sizing dielectric spheres and cylinders by aligning measured and computed resonance locations: algorithm for multiple orders,' *Appl. Opt.* **24**, 2380.

Hirschfeld, T., M. J. Block, and W. Mueller (1977), 'Virometer: an optical instrument for visual observation, measurement, and classification of free viruses,' *J. Histochem. Cytochem.* **25**, 719.

Hirschfelder, J. O., C. F. Curtiss, and R. B. Bird (1954), *Molecular Theory of Gases and Liquids*, Wiley, New York.

Hovenac, E. A. and J. A. Lock (1992), 'Assessing the contributions of surface waves and complex rays to far-field Mie scattering by use of the Debye expansion,' *J. Opt. Soc. Am. A* **9**, 781.

Hsieh, W.-F., J.-B. Zheng, and R. K. Chang (1988), 'Time dependence of multiorder stimulated Raman scattering from single droplets,' *Opt. Lett.* **13**, 497.

Huffman, D. R. (1988), 'The applicability of bulk optical constants to small particles,' in P. W. Barber and R. K. Chang (eds), *Optical Effects Associated with Small Particles*, World Scientific, Singapore; p. 279.

Hussein, M. S. and K. W. McVoy (1984), 'Nearside and farside: the optics of heavy ion elastic scattering,' in *Progress in Nuclear and Particle Physics*, D. Wilkinson (ed.), Vol. 12, Pergamon Press, Oxford; p. 103.

Irvine, W. H. (1965), 'Light scattering by spherical particles: radiation pressure, aymmetry factor, and extinction cross section,' J. Opt. Soc. Am. **55**, 16.

Jackson, J. D. (1975), *Classical Electrodynamics*, 2nd ed., Wiley, New York.

Johnson, B. R. (1993), 'Theory of morphology-dependent resonances: shape resonances and width formulas,' *J. Opt. Soc. Am. A* **10**, 343.

Karam, M. A. and A. K. Fung, (1993), 'Electromagnetic energy absorbed within a Mie spere,' *J. Electromag. Waves Appl.* **7**, 1379.

Kattawar, G. W. and C. E. Dean (1983), 'Electromagnetic scattering from two dielectric spheres: comparison between theory and experiment,' *Opt. Lett.* **6**, 543.

Keller, J. B. (1962), 'Geometrical theory of diffraction.' *J. Opt. Soc. Am.* **52**, 116.

Keller, J. B., R. M. Lewis, and B. D. Seckler (1956), 'Asymptotic solution of some diffraction problems,' *Commun. Pure Appl. Math.* **9**, 207.

Kerker, M. (1969), *The Scattering of Light*, Academic Press, New York.

Kerker, M., D. Cooke, W. A. Farone, and R. T. Jacobsen (1966), 'Electromagnetic scattering from an infinite circular cylinder at oblique incidence: I. Radiance functions for $m = 1.46$,' *J. Opt. Soc. Am.* **56**, 487.

Kerker, M., D.-S. Wang, and C. L. Giles (1983), 'Electromagnetic scattering by magnetic spheres,' *J. Opt. Soc. Am.* **73**, 765.

Khare, V. (1975), 'Short-wavelength scattering of electromagnetic waves by a homogeneous dielectric sphere,' Ph. D. thesis, University of Rochester, Rochester, NY.

Khare, V. and H. M. Nussenzveig (1974), 'Theory of the rainbow,' *Phys. Rev. Lett.* **33**, 976.

—— (1977), 'Theory of the Glory,' *Phys. Rev. Lett.* **33**, 976.

Kim, J. S. and S. S. Lee (1983), 'Scattering of laser beams and the optical potential well for a homogeneous sphere,' *J. Opt. Soc. Am.* **73**, 303.

Kleyn, A. W. (1987), 'Rainbow scattering,' *Comments At. Mol. Phys.* **19**, 133.

Knopp, K. (1945), *Theory of Functions, Part I*, Dover, New York.

Kong, P., E. A. Mason, and R. J. Munn (1970), ' "Glorified shadows" in molecular scattering: some optical analogies,' *Am. J. Phys.* **38**, 294.

Krebs, J. (1982a), 'Scattering by slightly nonspherical particles in the high frequency limit. I. Impenetrable particles,' *J. Math. Phys.* **23**, 2494.

—— (1982b), 'Scattering by slightly nonspherical particles in the high frequency limit. II. Transparent particles,' *J. Math. Phys.* **23**, 2502.

Lai, H. M., C. C. Lam, P. T. Leung, and K. Young (1991), 'Effect of perturbations on the widths of narrow morphology-dependent resonances in Mie scattering,' *J. Opt. Soc. Am. B* **8**, 1962.

Lam, C. C., P. T. Leung, and K. Young (1992), 'Explicit asymptotic formulas for the positions, widths, and strengths of resonances in Mie scattering,' *J. Opt. Soc. Am. B* **9**, 1585.

Lamb, H. (1881), 'On the oscillations of a viscous spheroid,' *Proc. Math. Soc. (London)* **13**, 51, 189.

Lambert, R. H. (1978), 'Complete vector spherical harmonic expansion for Maxwell's equations,' *Am. J. Phys.* **46**, 849.

Landau, L. D. and E. M. Lifshitz (1960), *Mechanics*, Pergamon Press and Oxford University Press, Oxford.

Lange, B. and S. R. Aragón (1990), 'Mie scattering from thin anisotropic spherical shells,' *J.Chem. Phys.* **92**, 4643.

Langer, R. E. (1937), 'On the connection formulas and the solutions of the wave equation,' *Phys. Rev.* **51**, 669.

Langley, D. S. and P. L. Marston (1984), 'Critical-angle scattering of laser light from bubbles in water: measurements, models, and application to sizing bubbles,' *Appl. Opt.* **23**, 1044.

—— (1991), 'Forward glory scattering from bubbles,' *Appl. Opt.* **30**, 3452.

Langley, D. S. and M. J. Morrell (1991), 'Rainbow-enhanced forward and backward glory scattering,' *Appl. Opt.* **30**, 3459.

Lee, K. M., P. T. Leung, and K. M. Pang (1998), 'Iterative perturbation scheme for morphology-dependent resonances in dielectric spheres,' *J. Opt. Soc. Am. A* **15**, 1383.

Lentz, W. J., A. A. Atchley, and D. F. Gaitan (1995), 'Mie scattering from a sonoluminescing air bubble in water,' *Appl. Opt.* **34**, 2648.

Leung, P. T. and K. M. Pang (1996), 'Completeness and time-independent perturbation of morphology-dependent resonances in dielectric spheres,' *J. Opt. Soc. Am. B* **13**, 805.

Levine, S. and G. O. Olaofe (1968), 'Scattering of electromagnetic waves by two equal spherical particles,' *J. Colloid Interface Sci.* **27**, 442.

Liou, K.-N. (1980), *Introduction to Atmospheric Radiation*, Academic Press, San Diego, CA.

Liu, C., T. Kaiser, S. Lange, and G. Schweiger (1995), 'Structural resonances in a dielectric sphere illuminated by an evanescent wave,' *Opt. Commun.* **117**, 521.

Lock, J. A. (1987), 'Theory of the observations made of high-order rainbows from a single water droplet,' *Appl. Opt.* **26**, 5291.

—— (1993), 'Contribution of high-order rainbows to the scattering of a Gaussian laser beam by a spherical particle,' *J. Opt. Soc. Am. A* **10**, 693.

—— (1995), 'Interpretation of extinction in Gaussian-beam scattering,' *J. Opt. Soc. Am. A* **12**, 929.

Lock, J. A. and C. L. Adler (1997), 'Debye-series analysis of the first-order rainbow produced in scattering of a diagonally incident plane wave by a circular cylinder,' *J. Opt. Soc. Am. A* **14**, 1316.

Lock, J. A. and G. Gouesbet (1994), 'Rigorous justification of the localized approximation to the beam shape coefficients in generalized Lorenz–Mie theory. I. On-axis beams,' *J. Opt. Soc. Am. A* **11**, 2503.

Lock, J. A., J. T. Hodges, and G. Gouesbet (1995), 'Failure of the optical theorem for Gaussian-beam scattering by a spherical particle,' *J. Opt. Soc. Am. A* **12**, 2708.

Logan, N. A. (1965), 'Survey of some early studies of the scattering of plane waves by a sphere,' *Proc. IEEE* **53**, 773.

Lorenz, L. V. (1890), 'Sur la lumière réfléchie et réfractée par une sphère transparente,' *Vidensk. selsk. Skrifter* **6**, 1.

Mackowski, D. W. (1991), 'Analysis of radiative scattering for multiple sphere configurations,' *Proc. Roy. Soc. (London) A* **433**, 599.

—— (1994), 'Calculation of total cross sections of multiple-sphere clusters,' *J. Opt. Soc. Am. A* **11**, 2851.

Malkus, W. V. R., R. H. Bishop, and R. O. Briggs (1948), 'Analysis and preliminary design of an optical instrument for the measurement of drop size and free-water content of clouds,' NACA Technical Note 1622, Washington.

Marston, P. L. (1979), 'Critical angle scattering by a bubble: physical-optics approximation and observations,' *J. Opt. Soc. Am.* **69**, 1205.

Marston, P. L. and D. L. Kingsbury (1981), 'Scattering by a bubble in water near the critical angle: interference effects,' *J. Opt. Soc. Am.* **71**, 192.

Marston, P. L., K. L. Williams, and T. J. B. Hanson (1983), 'Observation of the acoustic glory: high-frequency backscattering from an elastic sphere,' *J. Acoust. Soc. Am.* **74**, 605.

Martin, R. J. (1993), 'Mie scattering formulae for non-spherical particles,' *J. Mod. Opt.* **40**, 2467.

Mason, E. A., J. J. H. van den Biesen, and C. J. N. van den Meijdenberg (1982), 'Improved calculation of total scattering cross sections in the glory region,' *Physica* **116A**, 133.

Maximon, L. C. (1991), 'On the evaluation of the integral over the product of two spherical Bessel functions,' *J. Math. Phys.* **32**, 642.

Mazumder, Md. M., S. C. Hill, and P. W. Barber (1992), 'Morphology-dependent resonances in inhomogeneous spheres: comparison of the layered T-matrix method and the time-independent perturbation method,' *J. Opt. Soc. Am. A* **9**, 1844.

McVoy, K. W., H. M. Khalil, M. M. Shalaby, and G. R. Satchler (1986), 'Asymmetric deflection functions and the extinction of rainbows: a comparison of alpha-particle scattering from ^{40}Ca and ^{44}Ca,' *Nucl. Phys. A* **455**, 118.

Meeten, G. H. (1997), 'Refraction by spherical particles in the intermediate scattering region,' *Opt. Commun.* **134**, 233.

Metz, H. J. and H.-K. Dettmar (1963), 'Zur Berechnung der Mieschen Streukoeffizienten für reele Brechungsindizes,' *Kolloid-Z. u. z. Polymere* **192**, 107.

Merzbacher, E. (1970), *Quantum Mechanics*, 2nd ed., Wiley, New York.

Mevel, J. (1958), 'Etude de la structure détaillée des courbes de diffusion des ondes électromagnétiques par les sphères diélectriques,' *J. Phys. Rad.* **19**, 630.

Michel, B. (1997), 'Light scattering by inhomogeneous spheres – two new computational methods,' *J. Aerosol Sci.* **28**, S183.

Mie, G. (1908), 'Beiträge zur Optik trüber Medien speziell kolloidaler Metallösungen,' *Ann. d. Phys.* **25**, 377.

Minnaert, M. J. G. (1959), *Light and Colour in the Open Air*, G. Bell and Sons, Ltd, London.

Mishchenko, M. I., L. D. Travis, and D. W. Mackowski (1996), 'T-matrix computations of light scattering by nonspherical particles: a review,' *J. Quant. Spectrosc. Radiat. Transfer* **55**, 535.

Mobbs, S. D. (1979), 'Theory of the rainbow,' *J. Opt. Soc. Am.* **69**, 1089.

Morita, N., T. Tanaka, T. Yamasaki, and Y. Nakanishi (1968), 'Scattering of a beam wave by a spherical object,' *IEEE Trans. Antennas Propagat.* **16**, 724.

Morse, P. M. and H. Feshbach (1953), *Methods of Theoretical Physics*, Vols. 1 and 2, McGraw-Hill, New York.

Nesković, N. (1990), *Rainbows and Catatrophes*, Boris Kidrić Institute of Nuclear Science, Belgrade.

Neubauer, W. G. (1973), 'Observation of acoustic radiation from plane and curved surfaces,' in W. P. Mason and R. N. Thurston (eds.), *Physical Acoustics*, Vol. 10, Academic Press, New York.

Newton, R. G. (1964), *The Complex j-Plane*, Benjamin, New York; Chapter 14.

—— (1976), 'Optical theorem and beyond,' *Am. J. Phys.* **44**, 639.

—— (1982), *Scattering Theory of Waves and Particles*, 2nd ed., McGraw-Hill, New York.

Nicholson, J. W. (1910), 'The scattering of light by a large conducting sphere,' *Proc. Math. Soc. (London)* **9**, 67. [A second paper of the same title appears in *Ibid.* **11**, 277 (1912).]

Nisbet, A. (1955), 'Hertzian electromagnetic potentials and associated gauge transformations,' *Proc. Roy. Soc. (London)* **231A**, 250.

Nussenzveig, H. M. (1965), 'High-frequency scattering by an impenetrable sphere,' *Ann. Phys. (N.Y.)* **34**, 23.

—— (1969a), 'High-frequency scattering by a transparent sphere. I. Direct reflection and transmission,' *J. Math. Phys.* **10**, 82.

—— (1969b), 'High-frequency scattering by a transparent sphere. II. Theory of the rainbow and the glory,' *J. Math. Phys.* **10**, 125.

—— (1972), *Causality and Dispersion Relations*, Academic Press, New York.

—— (1977), 'The theory of the rainbow,' *Sci. Am.* **236** (4), 116.

—— (1988), 'Uniform approximation in scattering by spheres,' *J. Phys. A: Math. Gen.* **21**, 81.

—— (1989), 'Tunneling effects in diffractive scattering and resonances,' *Comments At. Mol. Phys.* **23**, 175.

—— (1992), *Diffraction Effects in Semiclassical Scattering*, Cambridge University Press, Cambridge.

—— (1997), 'Time delay in electromagnetic scattering,' *Phys. Rev. A* **55**, 1012.

Nussenzveig, H. M. and W. J. Wiscombe (1980), 'Forward optical glory,' *Opt. Lett.* **5**, 455.

—— (1991), 'Complex angular momentum approximation to hard-core scattering,' *Phys. Rev. A* **43**, 2093.

Ohanian, H. C. and C. G. Ginsburg (1974), 'Antibound "states" and resonances,' *Am. J. Phys.* **42**, 310.

Olver, F. W. J. (1974), *Asymptotics and Special Functions*, Academic Press, New York.

Oseen, C. W. (1915), 'Über die Wechselwirkung zwischen zwei elektrischen Dipolen und über die Drehung der Polarisationsebene in Kristallen und Flüssigkeiten,' *Ann. d. Phys.* **48**, 1.

Ostrowski, A., W. Tiereth, D. Brandl, Z. Basrak, and H. Voit (1989), 'Evidence for a nuclear forward glory in $^{12}C + ^{12}C$ scattering,' *Phys. Lett. B* **232**, 46.

Owen, J. F., R. K. Chang, and P. W. Barber (1982), 'Morphology-dependent resonances in Raman scattering, fluorescence emission and elastic scattering from microparticles,' *Aerosol Sci. Technol.* **1**, 293.

Pendleton, J. D. (1982), 'Mie scattering into solid angles,' *J. Opt. Soc. Am.* **72**, 1029.

Perntner, J. M. and F. M. Exner (1910), *Meterologische Optik*, W. Braumüller, Vienna.

Pinnick, R. G., A. Biswas, P. Chýlek, R. L. Armstrong, H. Latifi, E. Creegan, V. Srivastava, M. Jarzembski, and G. Fernández (1988), 'Stimulated Raman scattering in micrometer-sized droplets: time-resolved measurements,' *Opt. Lett.* **13**, 494.

Pinnick, R. G., D. E. Carroll, and D. J. Hofman (1976), 'Polarized light scattered from monodisperse randomly oriented nonspherical aerosol particles: measurements,' *Appl. Opt.* **15**, 384.

Pohl, D. W. (1993), *Near-Field Optics*, Kluwer, Dordrecht.

Poincaré, H. (1910), 'Sur la diffraction des ondes hertziennes,' *Rendiconti Circolo Mat. Palermo* **29**, 169.

Probert-Jones, J. R. (1984), 'Resonance component of backscattering by large dielectric spheres,' *J. Opt. Soc. Am. A* **1**, 822.

Pulfrich, C. (1888), 'Ueber eine dem Regenbogen verwandte Erscheinung der Totalreflexion,' *Ann. Phys. Chem. (Leipzig)* **33**, 209.

Purcell, E. M. (1969), 'On the absorption and emission of light by interstellar grains,' *Astrophys. J.* **158**, 433.

Put, L. W. and A. M. J. Paans (1977), 'The form factor of the real part of the α-nucleus potential studied over a wide energy range,' *Nucl. Phys. A* **291**, 93.

Qian, S.-X. and R. K. Chang (1986), 'Multi-order Stokes emission from micrometer-sized droplets,' *Phys. Rev. Lett.* **56**, 926.

Ray, B. (1923), 'The formation of coloured bows and glories,' *Nature* **111**, 183.

Rayleigh, Lord *see* Strutt, J. W.

Reitz, J. R., F. J Milford, and R. W. Christy (1979), *Foundations of Electromagnetic Theory*, 3rd ed., Addison-Wesley, Reading, MA.

Robin, L. (1958), *Fonctions sphériques de Legendre et fonctions sphéroïdales*, Tome II, Gauthier-Villars, Paris.

Roth, J. and M. J. Dignam (1973), 'Scattering and extinction cross sections for a spherical particle coated with an oriented molecular layer,' *J. Opt. Soc. Am. A* **63**, 308.

Rubinow, S. I. (1961), 'Scattering from a penetrable sphere at short wavelengths,' *Ann. Phys. (N.Y.)* **14**, 305.

Ruppin, R. (1975), 'Optical properties of small spheres,' *Phys. Rev. B* **11**, 2871.

Saunders, M. J. (1970), 'Near-field backscattering measurements from a microscopic water droplet,' *J. Opt. Soc. Am.* **60**, 1359.

Schiller, S. (1993), 'Asymptotic expansion of morphological resonance frequencies in Mie scattering,' *Appl. Opt.* **32**, 2181.

Schiller, S. and R. L. Byer (1991), 'High-resolution spectroscopy of whispering gallery modes in large dielectric spheres,' *Opt. Lett.* **16**, 1138.

Schöbe, W. (1954), 'Eine an die Nicholsonformel anschließende asymptotische Entwicklung für Zylinderfunktionen,' *Acta Math.* **92**, 265.

Senior, T. B. A. and R. F. Goodrich (1964), 'Scattering by a sphere,' *Proc. IEE* **111**, 907.

Sharma, S. K. (1992), 'On the validity of the anomalous diffraction approximation,' *J. Mod. Opt.* **39**, 2355.

Shipley, S. T. and J. A. Weinman (1978), 'A numerical study of scattering by large dielectric spheres,' *J. Opt. Soc. Am.* **68**, 130.

Simão, A. G., L. G. Guimarães, and J. P. R. F. de Mendonça (1999), 'Electromagnetic stress on resonant light scattering by a cylinder with an eccentric inclusion,' *Opt. Commun.* **170**, 137.

Snow, J. B., S.-X. Qian, and R. K. Chang (1985), 'Stimulated Raman scattering from individual water and ethanol droplets at morphology-dependent resonances,' *Opt. Lett.* **10**, 37.

Sommerfeld, A. (1949), *Partial Differential Equations in Physics*, Academic Press, New York; p. 282. [This is a translation of his Munich 'Lectures on Theoretical Physics.']

Stein, S. (1961), 'Addition theorems for spherical wave functions,' *Quart. Appl. Math.* **20**, 33.

Stokes, G. C. (1852), 'On the composition and resolution of streams of polarized light from different sources,' *Trans. Camb. Phil. Soc.* **9**, 399.

Stratton, J. A. (1941), *Electromagnetic Theory*, McGraw-Hill, New York.

Streifer, W. and R. D. Kodis (1964), 'On the solution of a transcendental equation arising in the theory of scattering by a dielectric cylinder,' *Quart. Appl. Math.* **21**, 285.

—— (1965), 'On the solution of a transcendental equation arising in the theory of scattering by a dielectric cylinder – part II,' *Quart. Appl. Math.* **23**, 27.

Ström, S. (1975), 'On the integral equations for electromagnetic scattering,' *Am. J. Phys.* **43**, 1060.

Strutt, J. W. (Lord Rayleigh) (1871), 'On the scattering of light by small particles,' *Phil. Mag.* [4] **41**, 447.

—— (1872), 'Investigation of the disturbance produced by a spherical obstacle on the waves of sound,' *Proc. Math. Soc. (London)* **4**, 253.

—— (1881), 'On the electromagnetic theory of light,' *Phil. Mag.* [5] **12**, 81.

—— (1899), 'On the transmission of light through an atmosphere containing small particles in suspension, and on the origin of the blue of the sky,' *Phil. Mag.* [5] **47**, 375.

—— (1904), 'On the acoustic shadow of a sphere,' *Phil. Trans. Roy. Soc.* **203**, 97.

—— (1910), 'The incidence of light upon a transparent sphere of dimensions comparable with the wavelength,' *Proc. Roy. Soc. (London)* **84A**, 25.

—— (1918), 'The dispersal of light by a dielectric cylinder,' *Phil. Mag.* [6] **36**, 215.

Szegö, G. (1934), 'Über einige asymptotische Entwicklungen der Legendreschen Funktionen,' *Proc. London Math. Soc. (2)* **36**, 427.

—— (1959), *Orthogonal Polynomials*, American Mathematical Society, New York.

Takigawa, N. and S. Y. Lee (1977), 'Nuclear glory effect and α–^{40}Ca,' *Nucl. Phys. A* **292**, 173.

Tam, W. G. and R. Corriveau (1978), 'Scattering of electromagnetic beams by spherical objects,' *J. Opt. Soc. Am.* **68**, 763.

Theodoric von Freiburg (1304), 'Über den Regenbogen und die durch Strahlen erzeugten Eindrücke,' in *De iride et radialibus impressionibus*, Toulouse.

Thurn, R. and W. Kiefer (1985), 'Structural resonances observed in the Raman spectra of optically levitated droplets,' *Appl. Opt.* **24**, 1515.

Titchmarsh, E. C. (1948), *Introduction to the Theory of Fourier Integrals*, 2nd ed., Oxford University Press, Oxford.

Toennies, J. P., W. Weiz, and G. Wolf (1976), 'Observation of orbiting resonances in H_2-rare gas scattering,' *J. Chem. Phys.* **64**, 5305.

Toon, O. B. and T. P. Ackerman (1981), 'Algorithms for the calculation of scattering by stratified spheres,' *Appl. Opt.* **20**, 3657.

Tricker, R. A. R. (1970), *Introduction to Meteorological Optics*, American Elsevier, New York.

Trinks, W. (1935), 'Zur Vielfachstreung an kleinen Kugeln,' *Ann. d. Phys.* **22**, 37.

Tsai, W. C. and R. J. Pogorzelski (1975), 'Eigenfunction solution of the scattering of beam radiation fields by spherical objects,' *J. Opt. Soc. Am.* **65**, 1457.

Twersky, V. (1967), 'Multiple scattering of electromagnetic waves by arbitrary configurations,' *J. Math. Phys.* **8**, 589.

Tzeng, H.-M., K. F. Wall, M. B. Long, and R. K. Chang (1984), 'Laser emission from individual droplets at wavelengths corresponding to morphology-dependent resonances,' *Opt. Lett.* **9**, 499.

Ursell, F. (1965), 'Integrals with a large parameter. The continuation of uniform asymptotic expansions,' *Proc. Camb. Phil. Soc.* **61**, 113.

van de Hulst, H.C. (1946), 'Optics of spherical particles,' *Recherches Astron. Obs. d'Utrecht* **11**, (part 1). [Ph. D. thesis, Universiteit Utrecht.]

—— (1947), 'A theory of anti-coronae,' *J. Opt. Soc. Am.* **37**, 16.

—— (1957), *Light Scattering by Small Particles*, Wiley, New York. [Reprinted 1981, Dover, New York.]

—— (1980), *Multiple Light Scattering*, Vols. 1 and 2, Academic Press, New York.

van den Biesen, J. J. H., R. M. Hermans, and C. J. N van den Meijdenberg (1982), 'Experimental total collision cross sections in the glory region for noble gas systems,' *Physica* **115A**, 396.

van der Pol, B. and H. Bremmer (1937), 'The diffraction of electromagnetic waves from an electrical point source round a finitely conducting sphere, with applications to radiotelegraphy and the theory of the rainbow, parts I and II' *Phil. Mag.* [7] **24**, 141 and 825.

Velesco, N., T. Kaiser, and G. Schweiger (1997), 'Computation of the internal fields of a large spherical particle by use of the geometrical-optics approximation,' *Appl. Opt.* **36**, 8724.

Videen, G., D. Ngo, and M. B. Hart (1996), 'Light scattering from a pair of conducting, osculating spheres,' *Opt. Commun.* **125**, 275.

Volz, F. (1961), 'Der Regenbogen,' in *Handbuch der Geophysik*, Band 8, Kapitel 14, Abschnitt 3, F. Linke and F. Möller (eds.), Gebrüder Bornträger, Berlin.

Wait, J. R. (1955), 'Scattering of a plane wave from a circular dielectric cylinder at oblique incidence,' *Can. J. Phys.* **33**, 189.

—— (1963), 'Electromagnetic scattering from a radially inhomogeneous sphere,' *Appl. Sci. Res. B* **10**, 441.

—— (1965), 'The long wavelength limit in scattering from a dielectric cylinder at oblique incidence,' *Can. J. Phys.* **43**, 2212.

Walker, J. D. (1976), 'Multiple rainbows from single drops of water and other liquids,' *Am. J. Phys.* **44**, 421.

—— (1978), 'The amateur scientist,' *Sci. Am.* **239** (4), 179.

—— (1980), 'Mysteries of rainbows, notably their rare supernumerary arcs,' *Sci. Am.* **242** (6), 174.

Wang, R. T., J. M. Greenberg, and D. W. Schuerman (1981), 'Experimental results of dependent light scattering by two spheres,' *Opt. Lett.* **6**, 543.

Wang, R. T. and H. C. van de Hulst (1991), 'Rainbows: Mie computations and the Airy approximation,' *Appl. Opt.* **30**, 106.

Waterman, P. C. (1971), 'Symmetry, unitarity, and geometry in electromagnetic scattering,' *Phys. Rev. D* **3**, 825.

Watson, G. N. (1918), 'The diffraction of electric waves by the Earth,' *Proc. Roy. Soc. (London) A* **95**, 83.

—— (1995), *Theory of Bessel Functions*, 2nd ed., Cambridge University Press, Cambridge.

Weiss, D. S., V. Sandoghdar, J. Hare, V. Lefèvre-Seguin, J.-M. Raimond, and S. Haroche (1995), 'Splitting of high-Q Mie modes induced by light backscattering in silica microspheres,' *Opt. Lett.* **20**, 1835.

Weninger, K. R., B. P. Barber, and S. J. Putterman (1997), 'Pulsed Mie scattering measurements of the collapse of a sonoluminescing bubble,' *Phys. Rev. Lett.* **78**, 1799.

Whittaker, E. T. and G. N. Watson (1963), *A Course of Modern Analysis*, 4th ed., Cambridge University Press, Cambridge.

Williams, K. L. and P. L Marston (1985), 'Backscattering from an elastic sphere: Sommerfeld-Watson transformation and experimental confirmation,' *J. Acoust. Soc. Am.* **78**, 1093.

Wiscombe, W. J. (1979), 'Mie scattering calculations: advances in technique and fast, vector-speed computer codes,' NCAR Technical Note

NCAR/TN–140+STR, National Center for Atmospheric Research, Boulder, CO.

—— (1980), 'Improved Mie scattering algorithms,' *Appl. Opt.* **19**, 1505.

Wiscombe, W. J. and A. Mugnai (1980), 'Exact calculations of scattering from moderately-nonspherical T_N-particles: comparisons with equivalent spheres,' in *Light Scattering by Irregularly Shaped Particles*, D. W. Schuerman, ed., Plenum Press, New York.

Wyatt, P. J. (1962), 'Scattering of electromagnetic plane waves from inhomogeneous spherically symmetric objects,' *Phys. Rev.* **127**, 1837.

Yeh, C. (1964), 'Perturbation approach to the diffraction of electromagnetic waves by arbitrarily shaped dielectric obstacles,' *Phys. Rev.* **135**, A1193.

—— (1965), 'Perturbation in the diffraction of electromagnetic waves by arbitrarily shaped penetrable obstacles,' *J. Math. Phys.* **6**, 2008.

Young, A. T. (1982), 'Rayleigh scattering,' *Phys. Today* **35** (1), 42.

Young, T. (1804), 'Experiments and calculations relative to physical optics,' *Phil. Trans. R. Soc. London* **94**, 12.

Zhang, J.-Z., D. H. Leach, and R. K. Chang (1988), 'Photon lifetime within a droplet: temporal determination of elastic and stimulated Raman scattering,' *Opt. Lett.* **13**, 497.

Zvyagin, A. V. and K. Goto (1998), 'Mie scattering of evanescent waves by a dielectric sphere: comparison of multipole expansion and group-theory methods,' *J. Opt. Soc. Am. A* **15**, 3003.

Name index

Subject index

367